Breakthrough Communities

Breakthrough Communities

Sustainability and Justice in the Next American Metropolis

edited by M. Paloma Pavel

The MIT Press
Cambridge, Massachusetts
London, England

For information about special quantity discounts, please email special_sales@mitpress.mit.edu

This book was set in Sabon on 3B2 by Asco Typesetters, Hong Kong. Printed and bound in the United States of America.

Library of Congress Cataloging-in-Publication Data

Breakthrough communities : sustainability and justice in the next American metropolis / edited by M. Paloma Pavel ; foreword by Carl Anthony.
   p.   cm. — (Urban and industrial environments)
Includes bibliographical references and index.
ISBN 978-0-262-01268-3 (hardcover : alk. paper) — ISBN 978-0-262-51235-0 (pbk. : alk. paper)
1. City planning—United States. 2. Urban renewal—United States.
3. Community development, Urban—United States–Case studies. 4. Sustainable development—United States. 5. Social justice—United States. I. Pavel, M. Paloma, 1951–
HT167.B675   2009
307.1′2160973—dc22   2008039033

10 9 8 7 6 5 4 3 2 1

Contents

Series Foreword

Breakthrough Communities: Sustainability and Justice in the Next American Metropolis is an important and compelling contribution to Sustainable Metropolitan Communities Books, a new imprint at MIT Press. In identifying the language and ideas that underlie the emerging regional equity movement and by pulling together the researchers and the advocates who are helping frame the concepts, the issues, and the organizing involved with this movement, the book serves as a strategic initiative linking research with action. In *Breakthrough Communities*, ideas, language, and stories matter. They complement the research and analysis that situate the environmental, social, and economic changes occurring at regional as well as global levels, and they establish a vivid picture of what those changes represent.

The metropolitan/regional framework is also crucial. The concept of metropolitan or regional equity provides multiple pathways for change. It recognizes environmental problems as rooted in community but linked as well to broader regional boundaries such as watersheds, food sheds, air sheds, economic networks, and transportation corridors. It focuses on the local and global dimensions of social and economic questions that often work themselves out in regional contexts, such as the movement of goods; the location of commercial and industrial facilities in relation to residential communities; the tax, investment, and financial dynamics within a region and between communities; and/or the shifting demographics that reshape and transform regions. The regional equity focus also helps situate the different movements for change, often isolated in their geographic or issue-based silos, as part of a linked process that connects the environment and the economy with daily life experiences.

Sustainable Metropolitan Communities Books are designed to overcome the divide between research and action by providing a narrative as well as an analytical representation of those regional equity issues. By addressing issues of race, ethnicity, class, gender, and connection to place, and also by identifying models for organizing and action and developing a more inclusive and engaging discourse, the books can establish new frameworks for understanding and evaluation and provide pathways for future research and informed action. We welcome you to explore these breakthrough stories and the narrative and analytic discussions to come.

Robert Gottlieb, Occidental College
Series Editor

Foreword

The way we live in cities is at the crux of many environmental, economic, and social challenges. The deteriorating quality of life for poor people who live in the inner city and who are trapped in neighborhoods without the necessary services and resources is connected to the concentration of wealth in affluent urban and suburban neighborhoods, the extraordinary waste in consumer society, and the destruction of the natural resources that surround our cities. How do we find a new way?

In 1987, the World Commission on Environment and Development (WCED) proposed a working framework for sustainable development as "development that meets the needs of the present without jeopardizing the ability of future generations to meet their own needs." Conventional models of economic development had brought devastating consequences to the global environment while failing to meet the challenge of global poverty. As a result, worldwide concern for the environment, economy, and equity emerged as the "three Es" of sustainable development. Despite this historic perspective, the popular conception of sustainable development in the United States remains focused primarily on protecting the biophysical environment—land, air, water, energy, fauna and flora—as resources that exist apart from human society. A dominant ecological narrative that disregards cities has thus obscured the environmental consequences of urban transformation.

Too often we think that the sole forces shaping our cities and suburbs are impersonal market factors or technical expertise, forgetting that the most important aspects of our lives are often the outcome of other social, political, psychological, or spiritual dynamics. As people have worked through the years to escape poverty, to overcome disadvantage, or to establish and defend boundaries of community territory, the results of their efforts have shaped our neighborhoods and enclaves. Stories of these events include forced and voluntary migrations, responses to natural and human disasters, racial discrimination and conflict, heroic efforts to build religious communities, and struggles to resist exploitation.

Today, narratives of people of color and working-class communities and their social movements can help us understand ourselves and others, bring meaning to chaotic events, and evolve a shared moral purpose at a metropolitan regional scale. Once you have a shared story, you can build the will to sustain efforts for change such as re-envisioning abandoned places and designing healthy and effective communities. The personal and community-based narratives of social-justice advocates offer vital sources of sustenance as we move forward to address issues of race and class. Such stories are at the heart of our quest for sustainable communities and build the foundation of this book.

To illustrate the importance of such narratives, I'd like to share a personal story. My interest in the environment and city-building goes back to my third-grade class in an integrated school in West Philadelphia. In 1947, I was one of the few African-American students in a school attended mostly by children and grandchildren of working-class European immigrants. My favorite teacher, Mrs. Aikens, taught us about the constellations in the night sky, the dinosaurs that once roamed the hills and valleys where we lived, and the fossil record in the rocks found along the banks of the Delaware and Schuylkill rivers. We learned to identify the trees in the city by the shape of their leaves and the texture of their bark.

One day, Mrs. Aikens took our class on a trip to see the Better Philadelphia Exhibition on display at a prestigious downtown department store. The expressed purpose of the exhibition—attended by almost 400,000 people in the two months it was open—was to provide a visual image of what the future of Philadelphia would look like based on the new urban design principles that were in vogue following World War II. The exhibition included a huge aerial map of the city, motion pictures illustrating new design concepts, and a full-scale model reproducing an actual street corner in a run-down neighborhood in South Philadelphia. The model was designed to show what was wrong with the city and what could be done about it.

This class trip occurred at that mysterious moment in my childhood when I was considering what I wanted to be when I grew up. My imagination was stirred at seeing streets, houses, trees, and parks in miniature—as if I were larger than life—while also being able to see into the future. The exhibition awakened my aspiration to plan for a future in which my family, my community, and I would feel at home in the world.

In my childhood innocence, I had no way of knowing how much the Better Philadelphia Exhibition's urban renewal plans—either consciously or unconsciously—were bound up with assumptions, attitudes, and strategies to deal with race and class in the city. I learned only much later that the exhibition was in fact the first stage in a public relations campaign designed to dismantle African-American and other working-class neighborhoods in Philadelphia.

At our racially integrated school, we read about the city's founder, William Penn, and his innovations in designing "The City of Brotherly Love." In this narrative, William Penn was presented as a social reformer whose nonviolent ideas and egalitarian commitments gave shape to the early city. We learned about the Vikings, Peter Stuyvesant, and the religious persecution of the Mennonites. Yet, we learned nothing about the Irish, Italian, Polish, Greek, and Russian immigrants from whom the kids in my classroom were immediate descendants. Presumably this information was not in the elementary school teaching syllabus.

Perhaps most problematic for me, we learned nothing about African-Americans. Blacks had been a significant presence in Philadelphia since colonial days. In 1682, when William Penn arrived in the region, he found black people, brought by the Dutch and Scandinavians, already living there. Penn himself had slaves. From the early days, African-Americans owned property, agitated against slavery, and established churches, schools, and businesses. A second wave of African-Americans migrated to Philadelphia between the beginning of the twentieth century and the outbreak of World War II. My own ancestors arrived in the city as a part of these first two waves. In the years following World War II, the influx of the third wave of African-American migrants began to arrive, which would forever change the character of the city's neighborhoods. What effect would this invisible legacy have on my life chances? I learned about none of this in elementary school, or in my later education.

My early experiences in Mrs. Aikens's class instilled in me a rudimentary understanding of the place of human beings in the larger order of the universe and a respect for the nonhuman living community. I was also inspired by William Penn's story and by the Better Philadelphia

Exhibition. Such formative experiences influenced my decision to study architecture and urban planning, through which I believed I could make an impact on the social issues of our time.

The compelling urban development issues in the 1950s, however, had to do with race, class, and the environment. The postwar flight of the white working-class populations to segregated first-ring suburbs was beginning, and white elites were moving to affluent gated communities in the most privileged places throughout the region. The inner core of American cities was being redefined as the ghetto, while at the same time the environmental challenges associated with suburban fragmentation and sprawl were beginning to explode across the landscape.

The missing information in my educational experience left me poorly equipped to deal with these challenges. Why, after twelve generations with African-Americans as a significant part of the population in Philadelphia, had such disparities continued to exist in the quality of neighborhoods and living environments between African-Americans and other populations? What had been the pattern of discrimination that had prevented blacks from exercising greater control over their environments? How had blacks resisted exploitation? Were suburban sprawl and inner city abandonment connected?

Over the course of the nine years it took me to gain a professional degree in architecture and the subsequent twenty years of professional practice and teaching at the College of Environmental Design and the College of Natural Resources at the University of California, Berkeley, I came to the conclusion that neither my education nor conventional models of professional practice had adequately prepared me to address these questions.

An authentic approach to urban sustainability incorporates ecological integrity, beauty, economic viability, and social justice as building blocks for increased livability for all communities. Working with these premises in the late 1980s, a group of colleagues and I formed the Urban Habitat Program—the first U.S. environmental justice organization to focus on metropolitan regional equity. The mission of Urban Habitat was to promote multicultural urban environmental leadership for sustainable communities in the San Francisco Bay Area. The idea was to develop a model of organizing that could be replicated in other regions of the country.

The organization was to be multicultural for a variety of reasons. We wanted to create an example of the kind of society we hoped to develop.

We would have to move beyond the black and white model of racial confrontation toward a more inclusive vision that included other populations. Whites were becoming minorities in many parts of California. Much of the energy of the civil rights movement was passing on to other communities of color: Native Americans, Latinos, Asian-Americans, and Pacific Islanders.

We focused on cities because communities of color had needs and great organizing strengths related to urban issues, and because this focus was missing within the environmental movement. We saw leadership development as being an important ingredient in our program, in part because there was a leadership void in society as a whole. Yet we also felt that organizers in our communities needed new attitudes, skills, and concepts to be effective in the twenty-first century.

At the time we founded Urban Habitat, most established environmental groups were ignoring cities. Instead, the first thing that the Urban Habitat Program did was to treat the city as the whole region, part of the larger ecosystem or watershed. This meant viewing inner-city neighborhoods, the downtown, the suburbs, the surrounding rural areas, the wilderness, the whole metropolitan region as interconnected—not as fragmented parts. Participants already knew a great deal about housing, transportation, workplaces, schools, and churches in our most vulnerable metropolitan neighborhoods. But we also had to learn about land, air, water, biological resources, the patterns of energy consumption and waste, and the economic drivers for the whole region.

For a long time, the discussion in the U.S. environmental movement was dominated by a focus on larger-scale issues that paid no attention to the economy or racial justice. In the late 1980s and early 1990s, a movement emerged in communities of color to address the deteriorating quality of life in inner cities and poor rural communities based on a new concept called environmental justice. Advocates began to argue convincingly that protecting the environment and addressing social and racial justice were intimately connected.

Although mainstream environmental groups were talking about the potential of what might happen if we did not focus on the environment, the environmental justice movement was saying, "The things you are predicting are already happening in our communities." The concerns of mostly white suburban environmentalists (protecting trees and birds) and the concerns of urban and rural people of color environmentalists (toxic pollution, occupational hazards, unemployment, abandoned lots, run-down properties, and lack of decent grocery stores) were linked. A social movement that put these ideas together began to emerge. This book builds on the promising work of this emerging movement.

At Urban Habitat, we quickly discovered that our interest in addressing environmental issues facing communities of color was not unique. All over the country, particularly in the South and Southwest, African-Americans and Latinos were beginning to work on the health challenges of toxic pollution and the disproportionate siting of hazardous waste facilities in communities of color.

The popular conception of environmental justice, however, focuses almost exclusively on important issues of toxic pollution and public health, rather than broader issues of land use and community development. While this work of opposing toxics, which we supported, became well recognized, we were convinced that the solutions would need to be much broader. To address the problems at their roots, we would need to have a say in how the region as a whole would develop.

The elements of the built environment, including houses, streets, roads, parks, schools, retail shops, churches, and office buildings, are among the most important common assets of society as a whole. Thus, any strategy for sustainability must include stewardship of these resources as well as stewardship of the natural environment.

In the 1990s, an important breakthrough in our work at Urban Habitat came when we learned about the work of Myron Orfield, then a state senator in Minnesota. Using geographic information systems (GIS), Myron documented the massive disparities growing up within the nation's suburban areas, and showed how and why it was in the self-interest of suburban constituencies to collaborate on a new vision for metropolitan reform. We also discovered the work of Bruce Katz and his colleagues at the Brookings Institution, who were beginning to map out a new metropolitan agenda that could bring together an astonishing range of constituencies, including urban, suburban, and rural communities. Others who influenced our efforts at Urban Habitat were Manuel Pastor at University of California, Santa Cruz; Amy Dean at Working Partnerships USA; and Angela Glover Blackwell and Joe Brooks at a new and dynamic organization called PolicyLink. These advocates, and many others represented in this book, found new ways to talk with working-class, middle-class, and affluent communities about, and act upon, a shared agenda that could expand opportunities for marginalized populations while improving the quality of life for everyone. We quickly incorporated these insights into our work.

Through its concern with alleviating poverty as well as with investigating environment and development issues, the Ford Foundation learned of our efforts at Urban Habitat. In 2001, Mil Duncan at Ford invited me to join the foundation staff as a program officer to lead its Sustainable Metropolitan Communities Initiative. The long-term goal of this initiative is to reduce the patterns of concentrated poverty in the United States, while influencing the patterns of metropolitan development to conserve natural resources. For the past fifty years, the Ford Foundation has sought to address both urban and rural poverty in the United States and abroad. In the United States, the foundation has focused increasingly on the ways that patterns of residential segregation and concentrated poverty are exacerbated by metropolitan fragmentation and sprawl.

Alarmingly, the land areas occupied by our metropolitan regions are growing faster than the populations that live in those regions. For example, between 1970 and 1990, the population in the Los Angeles metropolitan area grew by 45 percent, while during that same period, the actual land area where people were living grew by 300 percent.[1] What this means is that the land is being consumed six times as fast as the population is growing. A more recent study found that people living in more sprawling regions tend to drive greater distances, breathe more polluted air, face a greater risk of traffic fatalities, and walk and use transit less frequently.[2] Freeways, parking lots, and suburban sprawl gradually spread across the landscape, producing commercial strips that make the quality of life worse and worse for community members, while people spend more hours commuting back and forth.

Organizations participating in the Sustainable Metropolitan Communities Initiative use wide-ranging strategies in their efforts to transform their communities. They are moving beyond the paradigm of organizing within single issues and are coming together across metropolitan regions and lines of race, class, and gender, and coming up with new ways of working together to make our cities healthier, more sustainable, and more livable. They are building institutional capacity for sustainability and regional equity within labor, civil rights, and community development organizations that can play a role in transforming metropolitan regions. Increasingly, social justice and environmental activists are collaborating, acknowledging the importance of regional economic competitiveness as well as regional self-reliance as integral parts of efforts to achieve sustainability.

Specific land use and public-private investment policies and practices for the metropolitan region as a whole provide useful frameworks for organizing efforts. Such frameworks include efforts to make all neighborhoods livable, to make public investment equitable, and to ensure that all communities have access to regional opportunities. Strategies for particular places within metropolitan regions also play a vital role. Examples of such tailored strategies include reinvestment in weak market cities and neighborhoods, wealth-sharing to reduce poverty in hot market cities and neighborhoods, and workforce housing in job-rich suburban communities. Finally, we need to reconceptualize urban-rural linkages and the role of nature in our cities and suburbs.

Breakthrough Communities: Sustainability and Justice in the Next American Metropolis is the story of the remarkable efforts of activists and policy makers across the United States to build a new vision of our metropolitan regions. The communities described are coming together across conventional boundaries of race, class, and jurisdiction to achieve a better quality of life for present-day residents of cities, suburbs, and rural places. Perhaps more important, they are working to build the next American metropolis as an inclusive home for our children and grandchildren, and the community of life—the air, the water, the ecological habitats on which our planet depends.

The editor of this book, Paloma Pavel, is the founder and president of Earth House Center, a social justice and strategic communications collaborative located in Oakland, California. Earth House is dedicated to building healthy, just, and sustainable communities through education, research, and multimedia tools. Paloma's diverse background as both an academic and a social and environmental justice activist makes her a remarkably effective communicator and agent of change. Her graduate studies include both Harvard University and the London School of Economics. She is a consulting psychologist and is recognized internationally for her work in leadership and organizational development and large systems change.

Since 2002, Dr. Pavel has brought a spirit of innovation, creativity, and intellectual inquiry to the development of a national learning-action network in regional equity. Her documentation of this growing movement of regional activists, leaders, and policy researchers includes print, broadcast, and Web media. Her interdisciplinary approach in both re-search and practice focuses on the transformative possibilities of individual leadership development as well as the structural changes required at the level of policy. All of the contributors in this volume could tell their own remarkable stories of collaboration with Paloma Pavel and the Earth House Center team. *Breakthrough Communities* is both a harvest and a feast of this work.

The book reflects a new way to think about cities. It is about people organizing to improve their quality of life and make their neighborhoods more livable for everybody. It includes the voices of people of color, labor activists, and community organizers, and case studies of communities not often included in the debate about sustainability. It recounts stories of people thinking and acting in innovative ways as they face age-old challenges of race, poverty, and the environment, while charting new frontiers of metropolitan sustainability.

Carl Anthony

Notes

1. H. Diamond and P. Noonan, eds. *Land Use in America.* (Washington, D.C.: Island Press, 1996), p. 4.
2. D. Chen, R. Ewing, and R. Pendall, *Measuring Sprawl and Its Impact: The Character & Consequences of Metropolitan Expansion.* (Washington, D.C.: Smart Growth America, 2002).

Acknowledgments

This book is inspired by the breakthrough community leaders and their organizations, whose work is documented here. Many of them provided vision and guidance during the late twentieth and early twenty-first centuries in articulating and birthing the regional equity movement. It is our hope that this book will serve as legacy and testimony to their visionary leadership. Other contributors are new voices writing for the first time. We acknowledge the emerging leaders, some included here and others still unnamed, who will offer new energy for the road that lies ahead in a time of peak oil, climate change, demographic shifts, and globalization. The individual leaders in the book and their organizations can be found in the Resources section of this book, and they have generously offered to be of support to those who contact them. Each of the authors is also a portal to a larger network.

Next I would like to acknowledge my colleague Carl Anthony for his contribution to this project. He has been a partner from the beginning. Almost all of the authors in this volume were his grantees at the Ford Foundation. I had the privilege of meeting them through a grant I received from Ford to provide strategic communication for the Sustainable Metropolitan Communities Initiative. I would also like to thank Susan Berresford, Melvin Oliver, and Mil Duncan, for making resources available for this Initiative. Dessida Snyder, Carl's administrative assistant, handled details and communications essential to project flow. Others at Ford were important sources of inspiration and material support: Sharon Alpert, Connie Buchanan, Jeff Campbell, Mike Conroy, Miguel Garcia, Alan Jenkins, Becky Lentz, Thea Lurie, Suzanne Shea, and Margaret Wilkerson, and consultants Phoebe Eng, Pete Plastrik, Rebecca Riley, and Loren Siegel.

Of particular note is the Conversation on Regional Equity (CORE), facilitated by Manuel Pastor, which occurred in four convenings over a two-year span. Participants included Carl Anthony, Amy Dean, Mil Duncan, Angela Glover Blackwell, Mary Gonzalez, Bart Harvey, Heeten Kalan, Bruce Katz, john powell, Alice Rivlin, and David Rusk. These meetings served to deepen and clarify the movement.

In addition, a second learning-organization effort at Ford, the Regional Equity Development group offered a series of meetings for sharing practical skills and policy analysis to grassroots leaders in various regions including the following organizations: Atlanta Neighborhood Development Partnership, Gamaliel Foundation in Detroit, Los Angeles Alliance for a New Economy, New Jersey Regional Coalition in Camden, and Urban Habitat Program in the San Francisco Bay Area.

Much of the work presented in this volume has been a part of a growing conversation of national and international leaders, organizations, and groups who have participated in three conferences on Advancing Regional Equity. The first two, in Los Angeles in 2002 and in Philadelphia in 2005, were sponsored jointly by PolicyLink and the Funders' Network for Smart Growth and Livable Communities. A third conference, sponsored by PolicyLink, took place in New Orleans in 2008.

Initial research for this volume was prepared for a meeting in Detroit in 2004 of the Funders' Network for Smart Growth and Livable Communities, under the directorship of Ben Starrett. This research catalyzed further investment and work by the philanthropic partners in the Regional Neighborhood Equity Project, which resulted in the monograph *Signs of Promise: Stories of Philanthropic Leadership in Advancing Regional and Neighborhood Equity*. We are grateful in particular for the leadership of Sharon Alpert, Hooper Brooks, Kim Burnett, Anthony Colon, Jesse Leon, Arlene Rodriguez, and Scot Spencer at the Funders' Network.

Clay Morgan, our editor at the MIT Press, and every member of his team have been responsive, creative, and dedicated to bringing the project to completion, offering strategic advice at critical junctures in the process. Assistant editor Mel Goldsipe and copyeditor Suzanne Stradley provided invaluable assistance in getting the manuscript ready for print. Susan Clark did the catalog and jacket copy. Bob Gottlieb and Carl Anthony, my coeditors in the Sustainable Metropolitan Communities book series, have shared a vision of an action-learning series that brings new voices into the conversation and contributes to social and environmental justice. They have been engaged collaborators in bringing this particular volume to fruition.

There are several people who played important roles (offered technical and editorial support) in bringing shape and integrity to this volume. First, I would like to recognize my friend and colleague Jan Thomas for her editorial efforts. Jan was deeply involved with this book through the early drafts and her determination and enthusiasm helped to carry this project forward, both by nurturing relationships with the many contributors and by sensitively refining and focusing their efforts. Alex Artaud worked with us to place the stories in context and bring a flow and consistency to the voices of multiple authors. Joanna Schneier served as a dedicated member of our early team and created an elegant system for coordinating the communications and contributions of more than forty authors. The entire Earth House team has gone the extra mile throughout this process offering research and technical assistance: Rudy Asuncion, Rob Brown, Gloria Cardenas, Dallas Cavanaugh, Shams Karys, Kayla Kirsch, Helene Knox, Azure Kraxberger, Mary Newson, Holly Olsen, Gayle Dawn Price, Daniel Radhakrishna, Dennis Rivers, Elli Nagai Rothe, Mady Stacy, Ernie Tamminga, Lena Wolff, Diana Young, and Marita Zurita. I am profoundly grateful for all their work. The visual messaging of this work has been documented by Camilo Jose Vergara, Rick Butler, James Maxwell, and photomontage artist Keba Armand Konte. Andrea Torrice, director of *The New Metropolis*, has been both a companion on the journey and a significant media collaborator.

I would like to thank my friends and family who have contributed not only their thoughtful reflections and feedback, but also their loving support for my work over time: Frank and Margaret Lester, Mari and Frank Pavel, Frank and Marilyn Pavel, and Frank and Madeline Pavel, Kathryn and Michael Schlesinger, Richard Pavel, Kimberly and Chip Harrison, Annie and Jeff Schultz. My nieces and nephews, Zachary, Amanda, Adam, Bennet, Kristiana, Collin, P. J., Rex, Winston, Kyle, Dylan, and Grace, formed a youth leadership team who tested the Compass for Transformative Leadership and encouraged us to develop a workbook version for emerging leaders. I want to thank my close friends and grassroots community who have provided personal as well as professional support to the development of theory and practice: Catherine Allport, Wim Aspeslagh, Gary David, Anita Doyle, Rani Eversley, Kateri Fleet, Paul Griesgraber, Maureen Kiely, Rosa Lane, Bruce and Chase Mitchell, Penny Rosenwasser, and Pamela Westfal Bochte. Finally, I wish to thank my beloved partner Richard Page who has encouraged and supported this project in countless ways.

Other colleagues and teachers who supported this work are John Adams, Christopher Alexander, Tracy Apple, Angeles Arrien, Nina and Kenny Asubel, Dorothy Austin, Rachel Bagby, Chris Benner, Janine Benyus, Samuel J. Bois, Judith Briggs-Marsh, Rick Butler, Shakti Butler, Peter Calthorpe, Fritjof Capra, Ty Cashman, Ursula Caspary, Lysa Castro, Karen Chappel, Olivia Corson, Sarah Crowell, Carla Dartis, Fusako DeAngelis, K. Lauren De Boer, Drew Dellinger, Diana Eck, China Galland, Michel Gelobter, Susan Griffin, Nancy Halloran, Paul Hawken, Anne Herbert, Elliot Hoffman, Heather Hood, Wendy Johnson, Sam Kaner, Lima Kimura (Web of Life–Japan), Kayla Kirsch, Michael Lerner, Karl Linn, Jon Love, Joanna and Fran Macy, Seth Miller, Holly Minch, Papusa Molina, Faisal Muqqaddam, Louise Music, Tamio and Yoko Nakano, Ricardo Navarro, Naomi Newman, Helena Norberg Hodge, Mother Susan Obrien, Mayumi Oda, Wendy Oser, Grace Paley, Bill Perry, Noel Perry, William Rogers, Belvie Rooks and Dedan Gills, Rumi Sato, Howard Schechter and Barbara Lee, Mary Rowe, Enid Schreibman, Cornelia Schultz, Noriko Senda, Marguerite Spencer, Brother David Steindl Rast, Susan Strong, Lee Swensen and Vijaya Nagarajan, Brian Swimme, Kazuko Takano, Kaz Tanahashi, Lynne Twist, Matthis Wackernagel and Susan Burns, Dick Walker, William J. Williams, Nina Wise, and Sunne Wright McPeak.

We also want to acknowledge other organizations that have contributed to this volume: Active Voice, African American Forum on Race and Regionalism, Alliance to Restore Opportunity to the Gulf Coast Region, Apollo Alliance, Center for Community Innovation, Collective Heritage Institute–Bioneers, Destiny Arts, Ella Baker Center, Forum on Religion and Ecology, Green for All, Hardscrabble Hill, Institute for the Study of Environment and Culture, Pachamama Alliance, Rockwood Leadership Institute, UC Berkeley Center for Community Innovation, UC Davis Center for Regional Change, and Web of Life. We also acknowledge our strategic media partners: *The Next American City*; *Race, Poverty and the Environment* journal; and *Yes!* magazine.

In addition to the Ford Foundation, this book has been made possible through the generous support and conceptual resources of the following organizations: Angeles Arrien Foundation for Cross Cultural Research and Education, Bank of America Foundation, blue moon fund, East Bay Community Foundation, Funders' Network for Smart Growth and

Livable Communities, Kirwan Institute, Mesa Refuge, Mitchell Kapor Foundation, San Francisco Foundation, Surdna Foundation, Tides Foundation, and William Penn Foundation.

The images on the title pages for parts II and IV are photo montages on wood by Keba Konte. The photo on the title page for part I is from Istockphoto/Hidesy and for part III from Istockphoto/Nikada.

We honor and recognize this historic moment of the great turning in which we are participating. It is a privilege to be part of the growing movement for regional equity with leaders and organizations—too many to mention here—who struggle daily for sustainability with justice in regions throughout the United States and globally. We dedicate this work to the beings of the future and to all those who have dedicated their lives to the vision of a more just and sustainable world.

Introduction

M. Paloma Pavel

A new civil rights movement is emerging in communities throughout the United States. This metropolitan regional equity movement presents a vibrant range of vision and voice—a counter to the national, and increasingly global, story of urban sprawl and concentrated poverty. Bold regional organizing and advocacy efforts are linking sustainability and justice through innovative partnerships and policy reforms. New alliances are creating models for the next American metropolis. They demonstrate what is working—and what is possible—through building alliances between inner cities, suburbs, and rural areas within metropolitan regions across the nation. Breakthrough practices and leadership strategies that support healthy communities are transforming policies that affect housing, jobs, land use, and transportation.

Too often, low-income residents and communities of color are saddled with pollution-creating facilities that contaminate air, land, and water. These communities typically lack access to basic infrastructure such as grocery stores, libraries, parks, banks, and vibrant public spaces. Most have no possibility of finding living-wage jobs near their homes and often lack transit options that would make employment elsewhere in the region a viable option. This skew in the distribution of resources and opportunity can be attributed in part to spatial racism, policies that reinforce racially inequitable structures even when individual attitudes of prejudicial behavior may have shifted.

Sustainability and Justice

Drawing from the concept of sustainability and the theory of living systems, metropolitan regions can be considered as whole systems that contain a nested hierarchy of subsystems (Macy 1998). Historically, natural forces such as mountains, harbors, and rivers shaped human settlements. Rivers were often the site of the earliest villages and habitat clusters. Maria Kaika's book *City of Flows* expands this historical reference to underscore the interconnected ecology of natural resources that must flow within regions for human survival, including the flow-through of our water, food, and air (Kaika 2005). These biological principles of living systems deepen our understanding of the interconnection of human settlements that form the metropolitan region.

Today, downtowns, multiple centers of older and newer suburbs, as well as rural areas and the surrounding wilderness are connected to their local, community-based organizations as well as to their larger national and global contexts (Leccese and McCormick 2000). Neighborhoods, whole within themselves, are also part of this larger metropolitan system. To understand metropolitan regions, it is critical to account for their fixed elements such as ecological contexts, buildings, infrastructure, and well-defined patterns of municipal government. It is also necessary to map the dynamic connecting elements such as the flow of people through commuter and migration patterns and the flows of information, capital, energy, resources, and "wastes." Finally, it is important to consider the role of human agency as seen in capital investment, class conflict, and social movements in structuring and restructuring metropolitan regions (Edwards 2005; Feagin 1998; Gottdiener 1985; Hallsmith 2003; Harvey 1983; Susser 2002).

If the quest for sustainability is to be a genuine force for metropolitan transformation, the quest for social equity—and, by extension, the struggle for racial justice—must be integral to the concept. This quest has far-reaching consequences. When taken seriously, it sparks among environmental and racial justice advocates a new public dialogue about the many applications of racial justice and how shared objectives might be realized. Second, it promotes a reexamination of the concept of "smart growth" to ensure that projects receiving wide public acceptance incorporate social equity along with environmental goals. Third, it lays the groundwork for explicit performance standards for equitable development, to be widely shared by the development industry as well as by the general public. Finally, working at the metropolitan scale in the United States, it should create a road map of short- and longer-term strategies, indicators, and policies for how to get to regional equity.

The historic Brundtland Commission Report (1987) provided the first international recognition of the need for sustainable development, defined as "development that meets the needs of the present without compromising the ability of future generations to meet their own needs." This compelling but elusive concept of sustainability inspired extensive research and development of innovative methods for measuring human impacts. One such innovator is Mathis Wackernagel (1996), who provided key leadership in developing the "ecological footprint" model. The footprint is a measurement of the amount of land necessary to support a defined economy at a particular material standard of living. According to Wackernagel, "Modern cities and whole countries survive on ecological goods and services appropriated from natural flows or acquired through commercial trade from all over the world. . . . The Ecological Footprint, therefore, also represents the corresponding population's total 'appropriated carrying capacity.'" In part II of this volume, various case studies demonstrate the application of the ecological footprint and other tools, including geographic information systems (GIS) mapping, to make sustainability visible to communities and compelling as a factor for urban planners and decision makers. Promising efforts in applying measurable social equity indicators that advise governance are also described.

Sustainable communities have also been defined by the "three *e*'s": economically prosperous, environmentally sound, and socially equitable (Wheeler and Beatley 2004). For metropolitan regions to be relevant in the twenty-first century, they need to plan for the future in a way that takes into account all three forces. While the economy was originally the historic driver of urban planning, the second half of the twentieth century saw the rise in "green planning"—creating parks, preserving wetlands, and recognizing the value of open space both to the economy and to the sustainability equation. These ecological conditions that support life have come to be acknowledged and valued as part of the economic competitiveness and social desirability of a region.

Social equity is still too often undervalued or left out of the equation of sustainability. The accepted definition provides a powerful global context for addressing issues of concentrated poverty in the United States. However, environmental organizations in industrialized countries have often misinterpreted the concept of sustainability, ignoring social equity (Portney 2003, 157–175). The Brundtland Report, aptly titled "Our Common Future," explicitly refers to goals of reducing poverty and inequality as central to sustainable development.

To highlight the importance of the equity dimension, social scientist Julian Agyeman coined the term "just sustainability," which he defines as "the need to ensure a better quality of life for all, now and into the future, in a just and equitable manner, while living within the limits of supporting ecosystems" (2005).

Sustainability, as seen through the lens of social equity, also requires healing the land, caring for its vitality, and—in many regions—transforming toxic legacies of someone else's making in order to create a viable economic future (Bullard 2005). The "triple bottom line"—economy, environment, and equity—is not an abstract principle of accounting, nor is it simply a new turn of phrase. Rather, a commitment to the three *e*'s results in policies aligned with conditions that improve the quality of life for all citizens in the future as well as in the present.

Breakthrough Communities

We often think of building new neighborhoods or rebuilding older ones in terms of bricks and mortar, constructing new buildings, planting trees, paving sidewalks, and engaging in other activities to improve the physical appearance of an area; but building a community is first and foremost a social activity based on restoring trust, solidarity, confidence, and faith in the capacity of individuals and groups to implement change. Accomplishing this requires healing the scars of internalized racism, separatism, cynicism, and resignation. It also means restoring awareness of the relationship between human communities and the life-support system of the planet upon which we all depend.

In this book, breakthroughs are described as achievements of community organizing and building solidarity across lines of race, class, and municipal jurisdictions. These achievements are presented through journalistic stories that invite the reader into the lived experience of community advocates and through brief and compelling strategy pieces that summarize issues that need to be addressed and tools needed to bring about change. Wherever possible, the strategy pieces are written by policy analysts who are nationally respected for their research and knowledge and their contributions to the field.

The stories in this book show the ways in which people from many of these communities are working together to reverse the destructive patterns of metropolitan development and to build healthy neighborhoods for all of our communities—urban, suburban, and rural, alike. Like the breakthroughs of scientific discoveries, the innovations described are located at the intersection of many fields (Johansson 2004) because sustainable development is understood as an integration of goals of environment, economic development, and social equity. This frame provides a unifying context for work in fields related to metropolitan land use—ecology, landscape planning, housing, infrastructure, public facilities, workplace planning, parks and open space planning, and growth management. The breakthrough stories in this book also draw from the crosscutting fields of advocacy planning, equitable development, and environmental justice, as defined by the emerging social movement of regional equity.

These stories mark both a continuity with and a radical shift from the worldview prevalent in struggles for racial justice in the 1960s. In that period, Martin Luther King Jr. adapted Gandhi's philosophy of nonviolence to confront the United States' legacy of Southern racism and to move the nation to enact new civil rights legislation. The civil rights movement created a partnership between the federal government and the local urban and rural African-American communities. This phase of the movement was based on the belief that an expanding industrial society could provide opportunities for the African-American population while promoting the welfare of everyone.

Today, most people believe that racial integration is a good idea, but our metropolitan regions—neighborhoods, schools, employment centers, and the prison system—have become resegregated on a vast scale (Bullard 2007a). The federal government has retreated from its historic role of promoting racial justice. We live in a challenging new postindustrial era. Humanity is poised on the brink of unprecedented global transformation, with profound implications for racial and economic justice in the United States (Lane 2006) and beyond. The dynamics of globalization, advancing technology, increased communication, and destruction of the environment are leading to a greater sense of interdependence and vulnerability in local communities.

The Next American Metropolis

Until recently, when people thought of urban areas, they thought of the traditional city, a dense settlement with a single municipal government. But in the second half of the twentieth century, urban settlements have grown more fragmented and decentralized, spreading over a much larger area than the core city (Gottdiener and Hutchinson 1999). Typically, the metropolis in the United States is an urbanized region possessing many jurisdictions and including several geographic rings from the inner city core to the inner suburbs, outer-ring suburbs, and exurbs to agricultural areas at the urban edge. Some analysts expand these rings to include the concept of wilderness, a dwindling but important resource upon which metropolitan regions depend (Katz 2000).

Addressing issues at the metropolitan scale is essential for achieving sustainability. The air shed, like the hydrological cycle, crosses jurisdictional boundaries. Air pollution, caused by stationary as well as mobile sources, does not stop at the city line. Biotic resources become fragmented by the process of suburban sprawl. Patterns of metropolitan development have profound effects on traffic congestion, energy use, and climate change. Metropolitan regions are also the building blocks of the global economy because information, money, people, and goods cross national boundaries (Scott and Storper 2003).

Addressing concentrated poverty in the United States in the twenty-first century requires a shift of geographic imagination and consciousness among advocates of fairness, opportunity, and full participation of disadvantaged populations in the society as a whole (Anthony 2006a). During much of the twentieth century, advocates concerned with race and poverty thought of the city as a compact urban place contained within municipal boundaries. Such a perspective is no longer adequate. Formerly, poverty was isolated in a few African-American inner-city neighborhoods and in rural places like Appalachia. Although poverty persists in many urban and rural neighborhoods, a study of fifteen metro regions by the Institute on Race and Poverty found that by 2000, roughly half of the African-American population and more than 60 percent of Latinos lived in financially stressed suburban areas. Immigrants arriving in the United States in the early twentieth century typically settled in inner-city enclaves. In the twenty-first century, many immigrant populations are bypassing older cities altogether and moving directly to the suburbs, where poverty is now spreading. As David Rusk points out in his influential book *Cities without Suburbs*, "the city is now the region."

To be effective, organizers must come to terms with this new metropolitan landscape. The quest for "regional equity" seeks to implement "just sustainability" at the metropolitan regional scale. The goals of regional equity are to reform those policies and practices that create and sustain social, racial, economic, and environmental inequalities among cities, suburbs, and rural areas, and to integrate marginalized people and places into the region's structures of social and economic opportunity.

Substantial spatial separation—enforced by policy—continues to divide humans across racial and economic lines despite the biological reality that we are all part of an interconnected living system. While "across the highway" has replaced "across the tracks," the myths that foster separation persist, inscribed in the architecture and design of our cities. A metropolitan regional perspective enables us to acknowledge the reality of differentiation and subsystems while also seeing the wholeness of the region as a living system. Linking these interdependent geographic rings, and thereby challenging spatial divisions determined by race and class, has proven to be a powerful regional equity strategy (Orfield 2002). Thus, the quest for regional equity engages in revitalizing inner-city and older suburban neighborhoods and urban markets as assets and key building blocks of a healthy region. It reforms local, regional, and state policies and practices in order to advance social and economic equity within a region. And it links the needs of economically isolated and racially segregated residents with the opportunity structures throughout their region.

Events of the final four decades of the twentieth century undermined the sense of social cohesion among large sections of the American population. Although the civil rights movement challenged the legacy of racism embedded in U.S. history, it also stimulated a national backlash and abandonment of the vision of inclusive communities. Subsequently, we experienced an overemphasis on individualism, reinforced by consumerism, which propped up the illusion that social isolation is sustainable. Conflicting trends of globalization and identity in the opening decade of the twenty-first century are reshaping the everyday lives and relationships of men and women, the elderly, and children in our cities, suburbs, and rural areas (Castells 1997). Persistent poverty is well documented in our inner cities and older suburbs, with associated challenges of joblessness, crime, delinquency, drug trafficking, and changing family structures. New research suggests that these trends are also connected to rising insecurity, loneliness, depression, isolation of the elderly, and stunted development of middle-class children in the nation's newer sprawling suburbs (Morris 2005).

In the first decade of the twenty-first century, social movements are playing an increasingly visible and important role in building and rebuilding a sense of community in America's cities, suburbs, and rural areas. Given the disruptions of the global economy and the technological transformations wrought by the information age, social movements often provide the basis for new forms of identity (Buechler 2000; Castells 2003). Neighborhoods, groups, and communities—building on their ethnic, class, or territorial awareness—come together to fight their common opponents, confronting big-box industries such as Wal-Mart, toxic dumping, and other issues affecting survival and local quality of life. Now, groups that previously forged a shared identity through "saying no" are building new regional power alliances and creating positive and proactive alternatives for the future.

The Lens of Regional Equity

The regional equity movement is creating remarkable new opportunities for community building among an astonishing range of metropolitan social justice actors—environmentalists, labor, blue-collar ethnics, clergy, civil rights advocates, community organizers, immigrant activists, and African-Americans. This burgeoning movement is aggressively challenging institutional racism in the metropolitan landscape and building a new context for multiracial, multiclass, and gender-balanced leadership based on a practical vision that may well prove attractive to established metropolitan elites and decision makers. It is demonstrating a community-building process in which participants respond to an imminent threat, build organizational and leadership capacity, acquire policymaking and litigation tools, and engage in a community-visioning process to develop assets for the region as a whole.

Solutions must take into account the region as a whole because the dynamics that create poverty in our urban cores are regional in scope. Even when extensive resources are directed to lifting a pocket of concentrated poverty, this action alone will not solve the problems. The situation cannot be addressed without taking into account the regional dynamics that result in predictable poverty and other resource inequities (Orfield 2002).

In response to this fragmented geographic and political landscaping, the regional equity movement's multisector coalitions are working to ensure that all communities in the metropolitan region can participate in and benefit from their region's economic growth and activity. Groundbreaking practices and strategies are transforming policies that affect housing, jobs, land use, and transportation. The ultimate goals are to reverse unequal social, racial, economic, and environmental policies and to transform inequitable planning practices in inner cities, suburbs, and rural areas. To build sustainable metropolitan communities, new policies are needed that improve the quality of life in ways that are fair for present and future generations, within the limits of viable ecosystems.

Public policies have reinforced, and in some cases caused, racial segregation and neighborhoods of concentrated poverty in America's cities and suburbs (Jargowsky and Steiner 1997; Massey and Denton 1998). Increasing fragmentation of municipal governments within metropolitan regions has contributed to the development of opportunity-rich areas of residents walled off from the rest of the region (Pastor, Benner, and Rosner 2006). This separation creates vast disparities in housing, schools, tax bases, transportation, and wealth between inner cities and suburbs.

Through the lens of regional equity, the jurisdictional geographic focus of metropolitan planning expands this broader definition of urban to include not only the inner core of a city but also its surrounding outer rings that reach into suburbs and rural areas. During the 1970s and 1980s, as these areas became part of the metropolitan region, many of the older city centers were left behind with few or no resources and could no longer flourish. Job losses and suburban flight left many inner cities abandoned. From this regional perspective of concentric and interdependent rings, it becomes apparent that sprawl, vacant properties, and the lack of affordable housing are all interrelated—as are their solutions (Katz 2000).

Progress in moving toward equity requires a deeper understanding of the disparities that unravel our social fabric. The isolation of those residing in America's hollowed-out urban cores, as well as the social costs of sprawl, are exacerbated by outmoded policies that need to be better understood. Public policies that have resulted in racial segregation and isolation have also been responsible for haphazard growth of low-density development, duplication of public services in the suburbs, destruction of critical habitat, and development of strip malls as well as increased traffic congestion, squandering of energy, and related air and water pollution (Wolch, Pastor, and Dreier 2004).

American cities have seen ebbs and flows of urban rise and demise (Jacobs 1961). Broad demographic forces such as population growth, increased immigration, and domestic migration are changing settlement patterns, lifestyle choices, and consumption trends. Simultaneously, economic forces of globalization, deindustrialization, and technological innovation are restructuring our economy. Together, these complex and interdependent forces are reshaping metropolitan communities.

Compass for Transformative Leadership

The successful organizations documented in this book are employing new patterns of transformative leadership that connect grassroots communities to emerging opportunities in the field of regional equity. These new leaders understand the power of collaboration and demonstrate the capacity to represent their own communities, to change the rules of the game, and to provide leadership in the larger society.

Dominant institutions often have entrenched opposition to change and a substantial interest in maintaining the status quo. Even when such institutions would support change, they often do not know how to go about it. People from more privileged backgrounds may feel helpless, overwhelmed, or unprepared intellectually, emotionally, and politically for confronting the complexities of social and environmental change. Disenfranchised communities often lack role models for bringing about change. They may be torn between goals of their own racial group, those of the larger society, and those of family and community. They may lack confidence, communication skills, or resources for producing effective documents to educate others about their cause. After years of disenfranchisement, some may even feel they do not deserve success.

How can the model of transformative leadership help address these challenges? First, it acknowledges the need for internal transformation within the consciousness of each individual, along with the need to change institutional structures in the larger society. It also encourages learning and critical thinking and nurtures a sense of belonging to a place and a community. Finally, it acknowledges that people from all walks of life have the potential for leadership.

Increasingly, disenfranchised communities are aware of the strengths and limitations of their efforts during the past several decades and are seeking political strategies that can lead to fundamental change. They are open to reexamining social and economic assumptions and exploring new approaches to learning and action. As this book was taking shape, a four-part pattern in the formation of successful grassroots organizations became apparent.

1. The players initially came together in the process of saying no to forces destroying their community.
2. The next stage involved becoming organized and grounded in their unique geography and facing the new dynamics of the twenty-first century.

Figure I.1
Compass for Transformative Leadership.

3. As the group or movement stabilized, members began to explore new horizons of sustainability, metropolitan scale, and patterns of transformative leadership.

4. During the next phase, which I call "getting to yes," groups built creative collaborations that produced long-term benefits for their communities.

Navigating these important leadership challenges prepares participants to develop flexible, long-lasting, and highly effective campaigns and collaborations. The introduction to part II describes the process in more detail. The Compass for Transformative Leadership helps leaders from community organizations identify their own location on a path of learning and action and supports them as they develop projects that help create sustainable and just metropolitan communities.

Plan of the Book

This book is about social justice activists, labor unions, community organizers, environmentalists, and policy advocates joining together to form the emerging regional equity movement. These coalitions are challenging economic and institutional structures that shape spatial injustice in America's metropolitan regions. The stories documented in this anthology are of leaders and organizations achieving breakthroughs by

1. engaging in the discourse and practice of sustainability as an arena for racial and economic justice organizing;
2. linking neighborhood, workplace, and region as a context for social action;
3. developing new models of inclusive leadership for a durable future; and
4. achieving tangible benefits for their primary constituencies while contributing to the long-term health and viability of the regions where they are located.

The anthology is divided into three parts.

Part I provides a historical and conceptual context for the field of regional equity. It describes geopolitical and demographic shifts that have influenced the readiness of various regions to undertake the coalition-building and organizing work needed to win victories in the field. Chapter authors document the struggles of converging movements in transportation justice and environmental justice over several decades. Place-based roots include policy efforts, legislation, and litigation in urban, suburban, and rural contexts. Several of the chapters describe challenges facing this movement, which are philosophical and psychological as well as political. Part I establishes the foundation for the strategies and practices presented in parts II and III.

Part II explores breakthroughs being achieved in metropolitan regions across the United States. The case studies are written by people engaged with issues on the ground. The stories are accompanied by analyses, strategies, and tools being used to achieve specific changes and goals. In part III, visionary voices reflect on how we can best move forward to achieve a more just future, informed by a regional equity perspective.

In 1900, less than 10 percent of the world's population lived in urban areas; today, nearly half do. This trend represents a staggering rise from 1950, when only 30 percent of the global population was urban. By 2030, it is predicted that 61 percent of the world's population will live in urban regions (Davis 2006). As we begin the second decade of the 21st century, our cities and regions face a new set of challenges at a global scale with the threat of global economic collapse, massive extinctions, and climate change. Transformative leadership for sustainable communities will be needed more than ever. Given this rapid expansion, we need a shared understanding and a set of strategies that make it possible to achieve rapid and significant breakthroughs in creating sustainable and socially just metropolitan communities (Satterthwaite 2005).

This book provides a guide to practical, collaborative, and inclusive leadership strategies that can humanize the environmental movement and reverse the dominant narrative of racial segregation and concentrated poverty in our metropolitan regions caused by decades of sprawl and unequal allotment of public benefits. It sheds light on concepts emerging in the regional equity movement around the United States—concepts grounded in the stories and strategies of thinkers and actors on the front lines who are making history as they improve the quality of life in their communities.

I

Roots of the Regional Equity Movement and the Reinterpretation of Metropolitan Space

Introduction to Part I

The meaning of *regionalism* in the field of American planning has changed during the twentieth century. In the early decades of the 1900s, under the influence of Patrick Geddes, Lewis Mumford, Benton MacKaye, and the Regional Plan Association of America, the term referred to a strategy for fitting human settlements into a natural ecological setting. During the Depression, the idea was further influenced by the efforts of the Roosevelt administration to address rural poverty through creation of the Tennessee Valley Authority. It was also impacted by the work of the sociologist Howard Odum and the Southern regionalists who documented the unique folkways of the American South. By the late 1960s, the term came to be applied to metropolitan settings, reflecting the view that issues such as housing, transportation, the environment, and the political governance of large, multijurisdictional urban places should be treated as an integrated system (Sendich 2006).

The movement for regional equity draws from many of these traditions of regionalism, while grafting them on to a root that is outside of conventional representations of regional development, city building, and environmental history—the spatial dimensions of a quest for social and racial justice. Such representations have, until recently, rarely focused on people of color as agents or influences in the shaping of space in the United States. Yet race has impacted the shaping of cities, suburbs, and rural areas of North America since colonial times (Trotter, Lewis, Hunter 2004; Schein 2006).

The quest for regional equity is grounded in the spatial history and cultural geography of the civil rights, labor, and community-organizing movements in the twentieth century. This spatial history includes the economic and social forces that have shaped the geographic environment of African-Americans, other communities of color, working people, and the broader patterns of metropolitan development during this same period.

The formative decades of the civil rights movement occurred from 1876 through the early 1950s, when most blacks lived in the rural South. The great black migrations of the interwar years, federal voting acts, community development initiatives, and urban insurrections of the 1960s represent important strands in this spatial history (Self 2005; McAdam 1999). As Robert Beauregard has noted in his study of the postwar fate of U.S. cities, urban decline was one of the most prominent domestic problems of the 1950s. During this period, the federal government—through discriminatory mortgage programs, urban renewal, federal highway and infrastructure expenditures, tax incentives, the "War on Poverty," and community development programs—played an important role in shaping the cities and surrounding suburbs.

In the late 1960s and early 1970s, a brief and fragile partnership emerged between the federal government and African-American communities to rebuild urban neighborhoods. During this period, despite isolated white hopes for an urban resurgence, race as a negative theme came to unify various fragments of the discourse on cities. The insurrections of the 1960s highlighted the African-American presence in the cities and masked the degree to which whites themselves, through racism, flight, and disinvestment, had been responsible to a significant extent for the decline of the cities (Beauregard 2003). By 1980, the fragile partnership between the African-American community and the federal government had come to an impasse. African-Americans and their progressive allies became effectively isolated from the dominant patterns of metropolitan opportunity and decision making.

In the final decades of the twentieth century, four dynamics set the context for a reinterpretation of metropolitan space as a struggle for racial and economic justice within an emerging regional equity movement. First, the evolution of the environmental movement and public interest in sustainability brought growing awareness of the global threats of pollution, species extinctions, and climate change (Wheeler and Beatley 2004). Second, recognition grew of suburban sprawl and the resulting inequities among suburban communities (Orfield 2002). Third, the impacts of globalization led many policy makers and some business executives to embrace metropolitan planning and collaboration for the sake of enhancing competitiveness (Hershberg 2001). Fourth, changing family structures, as well as shifts in immigration and demographic trends, had important and disproportionate impact on communities of color (Briggs 2005; Orfield 2006).

Environmental devastation and concentrated wealth and poverty are undermining social justice at an unprecedented scale, both domestically and globally. Yet, these forces of destruction also present significant opportunities for change. The movement for metropolitan regional equity is engaging congregations, labor movements, civil rights organizations, and environmental groups in finding ways to meet these challenges. Each of this volume's authors occupies a unique position within the spectrum of this emerging movement. The diversity of their experience and disciplines is reflected in their contributions to this volume; alongside these *divergent perspectives*, regional advocates share a *common analysis* that underlies and motivates their collective action.

This vibrant and growing dialogue between the social and environmental justice movements is a hopeful development. In tandem with these vital strands of activism, an ethos of land stewardship can challenge the race- and class-based structures that threaten the long-term viability of metropolitan regions.

Through education, resistance, and collaborative effort, people are achieving important breakthroughs to improve life in their communities. The chapters in part I explore the roots of the regional equity movement and look at ways that urban, suburban, and rural space affect—and are affected by—the environment, transportation, and land use.

Moving Beyond Apartheid in the Next American Metropolis

The first chapter of section 1 is written by Cynthia M. Duncan, founding director of the Carsey Institute at the University of New Hampshire. Outlining past priorities of community developers, Duncan traces patterns that have led to deep unrest in rural and urban centers and describes the new coalitions that are working to alleviate poverty and restore trust in our basic institutions. She suggests how traditional approaches to community development can be reinforced in vital ways by new rules at the regional level.

Transforming failed infrastructures in urban centers requires strategies that are regional in scale, as well as sensitive to a constituency that is often ignored in the black-white discussion. Angela Glover Blackwell, director of PolicyLink in Oakland, California, and Manuel Pastor, director of the Center for Justice, Tolerance, and Community at the University of California, Santa Cruz, consider how the absence of regional planning locks urban populations out of opportunity, places inner-ring suburbs at risk, and allows sprawl to continue. Evoking the historically rich diversity and changing demographics of the United States, they offer an inclusive perspective that gives communities power to negotiate policy within a regional context.

This approach of working with communities in relationship to one another is also key to john a. powell's contribution. Where Blackwell and Pastor ask what America should look like, powell shines light on the mechanisms and forces that are shaping metropolitan regions today. Distributive equity alone would not suffice to address the imbalances. Reinforcing our common humanity, interdependent models are both relevant and practical. It is powell's belief that transforming our view of "self" and "other" will result only when the existing structures cease to serve any purpose.

1

From Bootstrap Community Development to Regional Equity

Cynthia M. Duncan

Those concerned with poverty have long recognized that the poor often live together in depressed neighborhoods with bad housing, crime and drug problems, few businesses, and few opportunities for good education or jobs that can lead to an escape from poverty. *Community development*, one of the key antipoverty strategies in the United States since the 1960s, works to address these problems by investing in neighborhoods and the families who live there.

For decades, community development corporations have been investing in housing, small business, credit unions, workforce training, and other neighborhood institutions and programs to improve poor communities. While community-level dimensions have figured in the struggles over land and labor rights that have occurred throughout our history, the deliberate work that we call community development emerged as a field only in the 1950s and 1960s. The origins of this work can be traced to Saul Alinsky's community organizing in Chicago's backyards, as well as to the emergence of the civil rights movement in the American South (Brophy and Shabecoff 2001). Both movements organized people on the ground to challenge the ways resources and opportunities were allocated. They sought to include workers and African-Americans in the mainstream, where they would find more opportunities for mobility and for achieving the American Dream of a middle-class life.

The post–World War II period brought prosperity and mobility for many Americans through numerous new jobs, opportunities for education through the GI bill, and programs to help people become homeowners and move to the suburbs. Working-class Americans could build assets and move into the middle class. Yet, large groups were left behind: in the hills of Appalachia, on plantations in the Delta and

American Indian reservations, and in the distressed inner cities that were abandoned by upwardly mobile, white middle-class families in search of a better life.

Many contemporary problems in troubled, racially segregated inner-city neighborhoods, and increasingly in older, inner-ring suburbs, have their roots in the seemingly benign—but effectively racist—housing, school, and transportation policies that emerged after World War II. These policies subsidized suburbanization and resulted in severe inequities in access to good schools, decent housing, and transportation to good jobs. While researchers and policy analysts worried about youth unemployment (Cloward and Ohlin 1960), lack of opportunity for black males (Moynihan 1969) in the cities, and severe chronic rural poverty in Appalachia and the Deep South, the reality of poverty did not fully capture the attention of the mainstream until the 1960s, when American cities experienced widespread unrest.

Persistently Poor Communities

Urban and rural communities across the United States that have endured generations of poverty have a lot in common—whether the Appalachian coalfields or Mississippi Delta plantations; Camden, New Jersey; or East St. Louis. Passed over by public investment, such places are typically isolated from mainstream economic activity. While poor rural areas often are physically remote, poor urban areas are almost as isolated in their own ways. Both types of communities are dumping grounds for environmental toxins. They struggle with substandard housing, abandoned lots, and bad schools. High unemployment, broken school systems, widespread substance abuse, and high crime rates undermine family stability and diminish children's futures.

Development economist Albert Hirschman has said that people who grow up in such places have three choices: "loyalty, exit, or voice (Hirschman 1972)." By "loyalty" he means acceptance of the way things are, and of the power relationships that maintain the status quo. "Exit," of course, means leaving—as those with the resources and the will often manage to do. By "voice" Hirschman means staying, working for change, and advocating for the kind of public and private investment that widens opportunities. Community-based groups across America exercise such a voice every day when they build new housing, develop workforce training, and invest in new business and family-support programs.

Community Development and Regional Equity: Strategies for Combating Poverty

Community development is about building institutions in support of re-silient communities where people can find opportunity, trust one an-other, and build a good life. For many years, community development emphasized development from within—empowerment *within* commu-nities, in effect providing bootstraps by which people could pull them-selves up. While racial injustice was recognized within the field, it was not prominently discussed (O'Connor 2001). Organizing for racial jus-tice and organizing for social justice were more like sister movements than integrated parts of the housing and business development upon which community development work was increasingly focused. Nor was there much emphasis on the environment, although some rural development organizations had projects on natural resource–related de-velopment, and some urban groups found environmental degradation and brownfields an unavoidable part of their landscape of concern.

In the community development field, poverty has typically been viewed in terms of isolation and exclusion, and the concentration of multiple problems in tough neighborhoods with few assets. Community develop-ment corporations have worked hard to create better housing, launch new training programs, establish child-care organizations, and revitalize Main Street. Yet, despite these vital investments, people and neighbor-hoods in America have continued to struggle with persistent, concen-trated poverty and lack of opportunity.

As this collection of stories makes clear, new opportunities for com-munity development are now presenting themselves. Since the 1950s and 1960s, we have learned a great deal about poverty and community, about how people get stuck, and about how places are abandoned and subsequently unravel. Today, many in the field are seeing the wisdom of using a regional strategy when approaching community, building on insights and frameworks such as those presented in this volume.

A *regional equity* approach to development combines *community* efforts to build strong institutions and better infrastructure with *regional* policies to foster equitable public and private investment. The aim is to grow the community in ways that are racially and economically just, as well as environmentally sustainable.

Place matters—you cannot escape poverty or help your children achieve a good life if your community is in shambles, its institutions and services do not work, and people lack a basic trust of one another. Community development is in part about people working in their neighborhoods to turn things around, get rid of crack houses, make schools work better, build decent housing, start local businesses, and train local residents. Developing trust, participation, and local investment are essential to the mission. A regional equity approach combines such community-driven efforts with regional policies that can achieve lasting community change.

Fortunately, community development practitioners do not have to go it alone. New political realities have brought a wide range of community concerns into alignment toward a shared regional equity vision. With environmental and social justice concerns converging, opportunities are emerging for reinvestment and potential new allies to advocate for it. Both public and private leaders are calling for "smart growth." Suburban residents are worried about congestion, sprawl, and the negative ways their neighborhoods are affected by a culture built around the automobile. Business leaders are concerned about their workers' being able to afford reasonably priced housing near their workplaces. Groups in the inner city—frequently teaming up with community development corporations—are organizing for healthy communities and environmental justice through solutions such as mixed-income housing, transit-oriented development, and public transit that links people to jobs and other regional opportunities.

Success means not just overhauling the rules, but creating new structures that allow people to participate in decision making. Those who want to bring about change need to engage not only in local community politics but also in the larger body politic where the "rules of the game" are made. Because all economies are regional, citizen coalitions can play an integral role in making regional decision-making bodies more responsive and effective.

The environmental community typically has more experience working at multiple levels of the political process than do groups working in community development. Unfortunately, in many poor communities, the broken political process is itself part of the problem. This can lead to discouragement and disengagement even among those who seek change.

Conservation, Poverty, and Sustainability

In *Cry, the Beloved Country*, Alan Paton describes a degraded South African landscape that can no longer sustain the people who depend on it:

The great red hills stand desolate, and the earth has torn away like flesh ... Down in the valleys women scratch the soil that is left, and the maize hardly reaches the height of a man. They are the valleys of old men and old women, of mothers and children. The men are away, the young men and the girls are away. The soil cannot keep them any more.

More than fifty years later, *Mountains Beyond Mountains*, Tracy Kidder's account of Paul Farmer's work in Haiti, described the same sort of extreme environmental degradation, with the same consequences of poverty, desperation, and emigration. Clearly, those working for environmental conservation and those working to alleviate poverty must join forces to support sustainable development. This vital symbiosis is as yet embryonic.

While community practitioners and organizers like those whose work is described in this volume are achieving breakthroughs around the country, sustainable development is still a vague, idealistic concept to many Americans, meaning different things to different people. Environmentalists define it as development that does not overexploit natural resources and that does not compromise the ability of future generations to meet their needs. From this perspective, sustainability is about stewardship, and economic factors are not necessarily considered. For others, including many who are working to revitalize poor communities, "sustainable development" refers to projects that can be maintained without subsidy; no reference to environmental protection is necessarily assumed.

More and more people are recognizing that combining environmental stewardship with economic development in poor communities can be the basis for the new coalitions necessary to the success of both. But creating such coalitions is not easy. The mainstream environmental movement's focus on preserving natural resources can threaten those who are dependent on those resources to make a living, even undermining development efforts for improving the lives of the poor. Correspondingly, it may be difficult for community development practitioners, focused on creating jobs for the disadvantaged and helping them to build assets, to pay serious attention to the environment. From their perspective, worrying about the environment may be considered a luxury they cannot afford, unless serious environmental health issues are seen to be at stake.

Conclusion

The social dimensions of equity and justice are inextricably linked to the environmental dimensions, and the integration of these two sets of objectives can be a powerful framework for organizing. In poor rural communities, new community development strategies increasingly encompass working landscapes and waterfronts, while in urban areas, there is increasingly a combined focus on environmental health and green space, on the one hand, and smart growth that includes equitable investment, on the other. There is growing recognition that the transportation, housing, and fiscal policies that created segregation and social isolation are not only unjust but also economically and environmentally unsustainable.

Finally, international concerns about climate change and biodiversity are bringing greater attention—sometimes including resources and funding—to rural communities for the sustainable management of resources to benefit society at large. The result is a new set of criteria for development, a new vision of social and racial equity in place development, and a more profound understanding of the importance of integrating poverty alleviation and environmental stewardship at a global scale.

2

Scaling Up: Regional Equity and the Revitalization of Progressive Politics

Angela Glover Blackwell and Manuel Pastor

The Future of Opportunity

What should America look like? This question, broad as it might seem, is at the heart of thousands—even millions—of individual and collective decisions made every day in the United States and expatriate outposts around the globe. When Congress passes bankruptcy, energy, or health-care reform; when state legislatures allocate funds for education, parks, and transportation; and when local zoning officials approve or deny multi-family, mixed-income housing development proposals, they are shaping the physical, economic, and social fabric of our nation.

The future of our country depends on what we do today to strengthen the fabric of American life. Rising poverty rates, falling or stagnant wages, failing public schools, aging infrastructure, and growing disparities of wealth not only undermine the promise of opportunity but also weaken America's competitive edge.

Across the country, community advocates, policy makers, government officials, business leaders, and researchers have been asking themselves and one another what America should look like. Emerging from these dialogues are new strategies, many of which are regional in scale, that are laying the groundwork for a new broad and inclusive national movement for economic and social equity.

A Vision for a Sustainable America

In our work in this field, we have found that many activists come to regional equity not because of a technical focus on land use or housing, but rather from a profound commitment to civil rights, community empowerment, and social justice. They choose to work at the regional level partly because that is where the challenges are being played out, but also because that is where they see some of the most hopeful prospects for transforming politics. The face-to-face, race-to-race, and place-to-place dialogues that typify many regional collaborations offer hope for a new approach in marked contrast to the corrosive partisan struggles that too frequently characterize national politics. In this light, every word in the term *regional equity movement* is important: the *region* is the level for action, *equity* is the goal, and a *movement* is the way to make change.

Taking movement building seriously means that we need to clarify our underlying values and visions. In our view, a sustainable America must be *competitive, inclusive,* and *democratic.* But we need to move beyond our old misunderstanding of what it means to be "competitive"—beyond the old divide between business and community visions that have the poor competing with the middle class for jobs, and cities competing with suburbs for tax revenue. We need, instead, to pursue our common goal of "competitiveness" in a collaborative community-building context within which individuals, neighborhoods, businesses, and governments work together to build strong communities and regions that offer opportunity for all.

A healthier, more competitive United States must also be *inclusive.* For far too long, African-Americans were excluded from education, employment, and wealth-building opportunities by shortsighted public and private policy decisions, the effects of which remain entrenched and persistent. While some segments of Latino and Asian communities have found their way into prominent places in U.S. society, many have found the doors closed and the paths to incorporation difficult, particularly for new immigrants. While lower-income whites were not systematically excluded by segregation and discriminatory homeownership policies like redlining, many now face similar financial struggles as a result of outsourcing and the shift from a living-wage manufacturing economy to a lower-wage service employment market. Inclusion of all these groups is a moral imperative, but it also makes good business sense: in the new economy, the most important factors fueling competitiveness are the skills and innovative capacities of a nation's and region's workforce.

Finally, getting to a more competitive and inclusive America requires a recommitment to *democratic* decision making. Public and private policies have fostered and sustained inequality. To address these disparities, it is critical to advance a new generation of politics based on the wisdom, voice, and experience of local constituencies. Low-income communities and communities of color are the ones most directly impacted by the nation's challenges and thus are central to the search for solutions. It is vital that they be meaningfully engaged in the policy-making process and that leaders and policy makers focus explicitly on connecting low-income communities and communities of color to opportunity.

Building a Vibrant, Stable Middle Class

The backbone of a competitive, inclusive, and democratic America is a *strong middle class.* Indeed, in an era of border-crossing and technological innovation that allows the very wealthy to be global citizens—with the ability to invest in international real estate, create expatriate networks, and telecommute from anywhere in the world—it is the current, emerging, and aspiring middle classes who have the greatest stake in our nation's future.

Yet, signs indicate America is falling short in maintaining and growing its middle class. For example, a 2007 report found that a majority of African-Americans born to middle-income parents are falling out of the middle class (Isaacs 2007a). Only 31 percent of these children reported family incomes greater than their parents, compared with 68 percent of white children in a similar situation. Worse, nearly half—45 percent—of black children from middle-class families had fallen to the bottom of the economic ladder in one generation, compared with only 16 percent of white children. Another report found that one-third of Americans overall are "downwardly mobile," making less than their parents and failing to surpass their parents' economic standing (Isaacs 2007b).

To build a competitive, inclusive nation and a vibrant, stable middle class, America needs a new progressive movement. Just as global economic and social interaction increasingly transcends national boundaries, the scope of economic activity within the United States has shifted from urban centers to metropolitan regions, crossing and straddling city, county, and town borders. A diverse, growing, multisector movement for regional equity could be the base for advancing a brighter, sustainable future for our nation as a whole.

Regional Equity and Regional Interdependence

Each day, millions of Americans crisscross from city to suburb, suburb to city, and suburb to suburb for work, school, shopping, or entertainment.

Regional equity embraces the new reality of American metropolitan life, acknowledging that cities and suburbs are inextricably intertwined, and the *region* is our economic engine. Most important, regional equity rejects the fear, resentment, and mistrust inherent in a "city versus suburb" mind-set, recognizing that our economic fate is in fact linked across these traditional boundaries.

The regional equity perspective suggests that the region and all of its parts—city, suburb, and rural—are interdependent. Research efforts over the past two decades have suggested that city and suburban economic fates are increasingly interlinked, and that metropolitan regions grow healthier faster when urban-suburban disparity issues are addressed (Pastor et al. 2000). Studies also show that paying attention to the needs of low-income people benefits *all* residents of a metropolitan region (Pastor et al. 2000; Barnes and Ledebur 1998; Voith 1998).

America will not achieve equity goals unless the region thrives, and the region cannot thrive unless it is competitive in the global economy. Economically shackled neighborhoods hinder broader regional success. For regions to realize their full economic potential, we must ensure that *all* communities are communities of opportunity.

The Regional Equity Movement: Linking People to Opportunity

As a progressive effort, the regional equity movement is at its very core about *equity*, about being inclusive. What is transformative in the movement is its authentic, intentional focus on those who are most vulnerable. At the top of the agenda is to shift policy in a direction that is more favorable for disadvantaged communities. For example, a key concept of the movement is that place determines opportunity, that many families remain shut out from the mainstream because they live in places that are isolated from opportunity. Advocates of regional equity are seeking to connect all Americans to communities of opportunity by addressing real, concrete issues—such as creating mixed-income communities that provide affordable housing and access to neighborhood amenities like supermarkets and parks and public spaces that promote healthy living, as well as access to high-quality public schools and sustainable employment. At the local level, the push for inclusionary zoning policy has been one strategy in creating mixed-income and opportunity-rich neighborhoods. New Jersey helped pave the way with the Mount Laurel cases that required wealthier municipalities to make room for affordable housing units. Community groups have also shown interest in advancing more equitable transportation policies to connect people to opportunity. In Los Angeles, for example, the Bus Riders Union has helped to connect residents to broader opportunities through its efforts to shift resources to bus lines and to expand bus service to reach job centers, schools, and hospitals (Center for Justice, Tolerance and Community 2006).

In the past, a traditional weakness of progressive movements has been the temptation to become compartmentalized, too narrowly oriented around single issues, and lacking a broad moral vision. The regional equity movement represents a change in tactics, combining research, community development, and social movement approaches in a broad-based multi-issue effort. To truly build power, regional equity advocates must consciously contextualize their work as a social movement and avoid falling into the trap of working on policy issues only at the expense of building constituencies, organizing, and facilitating regional conversations. Social movements are collective, organized, and sustained efforts to challenge—or transform and recruit as partners—existing authorities, decision makers, power structures, and cultural beliefs and practices. The movements encompass a range of strategies that demand changes and accountability from authorities and institutions in power, while at the same time not *depending* upon only those authorities and institutions for carrying out the movement's goals. Policy change sustains social change over time, engaging both grassroots and leadership.

Communicating Regional Equity: A New Frame of Cooperation and Progress

Regional equity strives to elevate a diverse cross section of interests and constituencies to the "uncommon common ground"—not a lowest common denominator of compromise grudgingly accepted by coalition members with conflicting interests, but rather the *highest* possible point of agreement and collective possibility.

The regional equity movement bridges the divide between communities and the policy process. It transcends traditional boundaries to reach advocates across racial, ethnic, and class lines; to dispel urban, suburban, and rural mistrust; and to unite community residents and leaders from the nonprofit, government, philanthropic, and business sectors. Across the United States, diverse interests are working to create stronger, more equitable cities, suburbs, and regions.

Pitfalls and Challenges of Regional Equity Work

Because the regional equity movement centers on an economic development agenda rather than a purely redistributive agenda, the combination of research, organizing, and action is critical to its success. Finding the balance is crucial: careful research informs the movement's work; organizing fosters long-term civic engagement and builds public will for change; and policy action at the local, state, and federal levels institutionalizes a vision for reform. There is no either/or: a governor and legislature can approve a bill to make changes but, ultimately, legislation and executive action will not produce long-term equity outcomes unless those affected are empowered to participate in the policy process and hold their representatives accountable. This is why being self-aware about the movement-building side of our work—not just the research and the policies—is so important.

Like any social movement, the regional equity struggle is energizing, inspiring, and uniting—but also at times arduous, labor-intensive, and tinged with leadership tensions or organizational politics. With so many equity advocates on the ground fighting for policy change on a daily basis and deeply invested in this work, the regional equity movement must remain conscious of its overarching mission to advance economic and social inclusion, working as a united movement to avoid splintering back into individual issue-based "silos."

Finally, although the black-white paradigm has indelibly shaped American society, regional equity advocates must also be mindful of the nation's rich diversity—Asian-Americans, Latinos, American Indians, indigenous peoples, new and long-standing immigrants, even subgroups within racial or ethnic categories—and keep inclusivity at the forefront of research, organizing, and policy advocacy work (Blackwell, Kwoh, and Pastor 2002).

Hope and Possibility: A Regional Equity Vision for the United States

Regionalism is fresh and exciting in large part because it exists outside of any compartmentalized government structure. Regionalism requires different constituencies to communicate with one another: central cities and suburbs, communities and businesspeople, and political leaders from different jurisdictions all must engage in dialogue to solve common regional problems. In the face of today's corrosive, partisan, and factionalized political climate, the regional equity movement forges a path to a new kind of national politics, incorporating a new style of listening and working together, and featuring series of deep and real conversations in which participants truly begin to understand one another's points of view.

Regionalism and regional equity lift up a new way of viewing and understanding the United States—and indeed the international community—in the face of tremendous change. Rather than approach the necessary seismic social, economic, and demographic shifts with fear—and thus accept inequality as the inevitable consequence of globalization and an omnipotent "market"—the regional equity movement embraces the hope and possibility of meaningful, lasting policy change. Instead of viewing the changes shaping our cities, suburbs, and towns in a context of scarcity, regional equity recognizes that we live in an extraordinarily abundant society and that we must harness that abundance not just for individual wealth but also for the community.

3

Reinterpreting Metropolitan Space as a Strategy for Social Justice

john a. powell

Race and space are deeply intertwined in the United States. This chapter looks at how space is formed and used, how it has been used to create and maintain racial disparities, and how we might be able to reconceptualize space in order to rectify those disparities.

The pervasive pattern of concentrated poverty in our mostly black inner cities, and the concentration of wealth and opportunity in our mostly white suburban areas, can be witnessed across almost every major metropolitan area of this country. This is not by chance, simple personal choice, or bad luck. Instead, it is the predictable consequence of the way we have arranged and managed our physical space. For those willing to see, it is clear that we have racialized space with tremendous disparities. What is not so clear is that this arrangement exacts a heavy toll on us all, as well as on the environment.

People readily understand how important the issue of land ownership and use was during colonization, and how important reclaiming and reimagining the use of that land has been for resistance to colonization. However, here in the contemporary United States, the importance of space in creating systems of domination and exclusion is often seen as irrelevant—confined to the pages of history. Yet, now that we have successfully dismantled most forms of de jure segregation, space and land use policies remain an important way—perhaps the *most* important way—that we racially distribute opportunities and burdens in the United States. I believe that addressing the problem of racialized, concentrated poverty and a racialized distribution of opportunity, as well as enhancing opportunity and life chances for people of all races, are the core civil rights challenges of the twenty-first century.

Structures of Inequity

A strong current of individualism in the United States has led to an enduring myth of meritocracy—the belief that any individual who works hard enough can achieve anything she desires. Even progressives who challenge this assumption have tended to look at social problems in isolation from one another, focusing on individuals.

In contrast, the analyses in this book are part of an emerging structural framework for understanding how racial hierarchy is established through the overlapping and cumulative effects of interrelated institutions. A structural approach can help us to understand how race, space, and opportunity are intertwined, and how our use of space is a primary factor in reproducing disparities along racial lines and depressing the life chances of many of us. A focus on structure should not be seen as an absolute alternative to our understanding of individual agency. Rather, it provides an important adjustment to a restricted ideology that has blinded us to the significance of structures and institutional arrangements.

The problem of *spatial racism* cannot be solved by traditional interventions that focus simply on building up the capacity of impoverished central-city communities. This problem must be understood and addressed within the larger scope of *regional* dynamics. Such an analysis considers historical patterns of intentional segregation, current trends of urban sprawl, the privatization of formerly public space, gentrification, the devolution of federal government power, and increasing municipal fragmentation.

Inner cities and inner-ring suburbs have become synonymous with depressed job growth, inadequate infrastructure investment, and distressed, low-performing schools. These inadequacies come into sharp focus when these areas are compared to outer-ring suburbs, which are rich in opportunity. The skewed distribution of resources and opportunity creates disparities in housing, schools, tax bases, transportation—and, subsequently, wealth.

Through this arrangement of space, we have translated our formerly explicit racist laws into an implicit and pervasive racial and economic segregation. It is now primarily through the use of space that we do our "racing."

How Space Is Racialized

This racialization of space is not unique to the United States. In cities, regions, states, and nations around the world, land has been and continues to be used formally and informally as a mechanism for sorting, ranking, and containing people and opportunity; for restricting interactions between groups; and for limiting social mobility.

The use of land to marginalize and control people of color was a fundamental tenet in the forming of America, starting with the displacement of indigenous people from their homelands. The United States was built upon the practice of using land and space as a racializing mechanism. In this context, the ideal of private property was used to justify the taking of common land from Native Americans in order to construct a system of private property for white males. This system not only trumped the rights of native people; it was also used to establish the enslavement of people of African descent until the Civil War, and to deny white men who did not own property the right to participate in the political process.

The racialized use of space is a deliberate yet ever-changing set of practices that isolates and subordinates marginalized populations. These practices are reflected in the ghettos of the United States, the shantytowns in South Africa, and countless other enclaves of racialized concentrated poverty around the globe. Such practices define and reinforce a de facto racial caste system. Originally, these arrangements were constructed in explicitly racialized ways; today, they are perpetuated through a series of policies that are seemingly "neutral" but lead to consequences that are highly racialized. Our tendency to view present social arrangements as inevitable and permanent allows these policies to become highly resistant to change.

The importance of these seemingly neutral policies has increased in direct proportion to the dismantling of explicit racially hierarchical arrangements. During times of slavery and Jim Crow, the South was one of the least demographically segregated spaces in the country. Segregation was maintained through status, not by geography. Physical separation was not necessary because great psychological and social separation perpetuated racial hierarchy and there was little confusion about a group's place in society. Even today, the South remains much more spatially integrated than other parts of the United States.

With the fall of Jim Crow, a subtler means was needed for reasserting and preserving a racial hierarchy in a way that was not distinguishable as overtly racist. Racializing space serves that purpose. The United States is the best example of how de jure racial hierarchy has been successfully translated into de facto racial hierarchy through the use of space and segregation. Indeed, this process has been so successful in the United States that, following the toppling of apartheid laws in South Africa, white South Africans visited Chicago to observe how racial hierarchy could be maintained even after the apparent dismantling of its legal foundations.

Laying a Foundation for Racialized Space

Much has been written about the racial implications of the construction of suburban space in the United States. We offer a brief overview here, if only to make sure we are standing squarely in the realm of the conscious and the constructed, not the neutral or the natural.

The use of space to create and maintain racial homogeneity in our neighborhoods has been central to the construction of whiteness. From the beginning, the suburbs have been designed as exclusionary white space—the product of social interactions, ideology, legal doctrine, and public policy. The success of this pursuit of spatial isolation would not have been possible without the support of powerful public and private institutional practices and policies. These policies and practices were not accidental, and they have not been redressed to remedy the great suffering they have caused. Hence, we live today with the effects of these earlier arrangements. Thus far, the role the federal government has played, and continues to play, in distributing opportunities and burdens along racial lines has not been widely acknowledged.

In 1933, the federal government established the Home Owners' Loan Corporation, making homeownership more attainable than ever—as long as you were white. *Redlining*, a practice both normalized and nationalized by the federal government, channeled mortgages to outlying white neighborhoods, while racially diverse central-city neighborhoods were considered too risky for investment.

By 1934, the Federal Housing Administration was subsidizing the first round of white flight to the countryside and laying the foundation for sprawl by supporting home ownership in the suburbs and not in the inner cities. The federal government explicitly favored construction in new developments over construction in existing neighborhoods. Further, the government pushed homebuyers to adopt covenants restricting the future sale of those government-subsidized homes to whites only, reinforcing the notion that the suburbs were to be white space *in perpetua*.

The 1940s brought with it policies of "urban renewal." The federal government directed funding to revitalize central-city neighborhoods, but these funds were diverted by local government officials to demolish black neighborhoods and communities. Many of these communities were replaced with high-rise public housing, deepening the isolation of low-income families in pockets of concentrated poverty.

To facilitate the movement of whites to the suburbs, improved transportation and other public infrastructure were needed to support these new communities. By the 1950s, what would come to be called "the largest public works program in the history of the world" was under way (Cashin 2004, 113). The Federal Aid Highway Act of 1956 facilitated the exodus of whites from central cities. As a result of the Highway Act and the policies mentioned earlier, the proportion of the U.S. population living in central cities shifted from 60 percent in 1950 to less than one-third between 1950 and 1970, while the suburban population doubled. It is vital to note that this mass creation of white spaces that could be regulated by whites was occurring at the same time as the U.S. civil rights movement. Government policies encouraged blacks to migrate from farms to urban areas in search of opportunities. Meanwhile, the same government was facilitating the movement of opportunity out of the cities and into the suburbs for a growing white middle class. Importantly, this middle class was itself a result of a number of government programs directed toward whites (Katznelson 2005).

The Highway Act was destructive in multiple ways. Highway construction often required the razing of many vibrant black neighborhoods, and highways were often built along white-black lines to form a potent physical barrier of segregation. The Highway Act also cut funding for public transportation in cities, ensuring that these segregated areas of concentrated poverty would be especially hard to escape. The legacy of this act lives on. Highway spending has surpassed public transit spending by a five-to-one margin over the past fifty years.

Today, spatial racism is evident in the exclusionary land use policies that prevent low-income nonwhites from moving to high-opportunity suburban communities. Suburban housing and land use policies that promote larger lot development have been found to depress the growth of suburban rental housing and limit in-migration of African-American and Latino households (Katznelson 2005). Density restrictions, site restrictions, and land use restrictions make affordable housing difficult to construct. For example, lot size requirements, provisions requiring large setbacks, or a lack of land zoned for multifamily housing add to the cost of housing construction. These racially exclusionary policies, combined with the fragmentation of local school districts in metropolitan areas, work to uphold persistent racial and economic segregation (powell 2002). These policies were enacted by local governments and protected by state and federal courts. These policies and procedures were seen as necessary to protect the new emerging white enclaves.

From the Central Core to the Outer Rings: The Emergence of Sprawl

Even as black people began slowly migrating to the suburbs over the past thirty years, increased urban sprawl usually resulted in a wealthier, whiter area being built just a few more exits down the highway. In many Midwestern cities today, inner-ring suburbs have transitioned to "majority-minority areas," while outer-ring suburbs and exurbs have grown exponentially. This sprawl has been encouraged and supported by multiple policies such as gasoline subsidies, subsidized suburban infrastructure (roads, sewage, water), and tax subsidies for business relocation. Meanwhile, municipal fragmentation, occurring in tandem with sprawl, has resulted in increasingly localized control of increasingly isolated spaces.

Today, sprawl and fragmentation are the two strongest forces perpetuating racialized, concentrated property. Increasing fragmentation of our regions (from 24,500 municipalities and special districts in 1942 to more than 54,000 in 2002) has contributed to the emergence of opportunity-rich areas whose residents are able to create virtual walls and lock themselves off from the rest of the region. Fragmentation is highly correlated with the level of concentrated poverty and concentrated wealth in a region: as fragmentation increases, so does the concentration of poverty. Both of these concentrations are race-coded (powell 2002).

The effects of concentrated poverty are cumulative and mutually reinforcing. It is useful to visualize these effects as a web, with each factor acting as both a cause and a symptom of the other factors. This is an important perspective that brings insight into the isolation of minorities in areas of concentrated poverty and the major self-reinforcing ramifications that minorities experience across many life areas, such as educational attainment, health, income, wealth, access to employment, and incarceration. This view also suggests that if we can create conditions that support the equitable distribution of resources and opportunities, we can reduce racial and economic disparities in almost every life area.

Toward Federated Regionalism

It is certainly the case that both local and regional governance have been used, in different but related ways, to undermine the political power base of poor people of color. Thus, regional government, too, is sometimes seen as implicated in a suburban power grab that does not produce other benefits (Savitch and Vogel 2004, 758–790; powell 2000, 218–246). As we move to address the problems of concentrated poverty, we must also be mindful of the problem of forced dispersal. Holding the proper tension between deconcentration strategies, structured choice, and regionalism is vital. This can be done if our efforts are continually informed by a strong sense of equity.

One strategy for moderating the tension between local and regional policies is federated regionalism. *Federated regionalism* is an attempt at establishing balance between localism and regionalism, while preserving the political and cultural voice of people of color. In this approach, regional and local policy making are integrated. Local authorities retain decision-making power and control over community identity and local governmental responsiveness, while maintaining the ability to tap into the assets and opportunities of the region as a whole. For example, all areas of a region may be able to meet their share of affordable housing needs in different ways. Local school districts may maintain administrative control of school policy, while fiscal resources are shared throughout the region.

In an earlier work on federated regionalism (Savitch and Vogel 2004, 758–790), I noted that the United States has some history of federated solutions. The differences between the House of Representatives and the Senate are prime examples of maintaining a balance between localized interests and larger interests. However, an even more analogous process might be a method of cumulative voting that allows participants either to spread their votes between multiple candidates or choices, or to lump them all in one place for particularly important concerns. This form of regionalism rejects the Hobson's choice[1] often given to communities of color: either lose your political identity and power completely, or lose the ability to make decisions on any sort of regional level and be stuck with increasingly meaningless control.

The regionalism that I propose is not a normative prescription for all racial justice work, but it is a particularly useful lens for understanding how racial hierarchy is perpetuated, and where effective action to alleviate social problems can occur. If we have learned one thing from two centuries of struggle for racial justice in this country, it is that isolated approaches and codified approaches are not likely to work. Structures can shift in ways that preserve white supremacy, even as institutions previously key to that supremacy are successfully attacked. Antimiscegenation laws were once seen as a vital part of maintaining a racial caste system in this country; today, these laws have been successfully dismantled—and yet the racial caste system remains.

We must especially resist positing a universal mode of federated regionalism. Particular strategies should be tailored to the specific problems at hand. For example, in one metropolitan area, it might make sense to institutionalize a veto policy if the voices of people of color are consistently being overwhelmed by a coalition of white suburban voices. In another city, such veto power may serve only to exacerbate the exclusionary operating mode of the outer-ring municipalities.

What we are striving for is to imagine new political processes that allow full democratic participation, so that people have access to the decisions that affect their lives. Differences and commonalities are to be celebrated, and tensions are to be embraced. But these tensions need to play themselves out within an equitable and inclusive framework that allows marginalized voices of all races to be heard.

Exclusionary Space Is a Symbol of the Isolated Self

As we work toward social justice in the twenty-first century, our dreams can be animated not simply by visions of distributive equality, nor even of an equality of opportunity, but more fundamentally, by a transformed view of the self, of relationships, and of the world.

The fiction of an isolated and unitary self underscores our worldview. Indeed, every historical religious tradition has known this and has been dedicated to moving beyond the isolated self, whether through union with God or by processes of inquiry that seek to break the illusion of a stable and separate self.

I do not believe that we can transform our social structures, including our highly racialized spaces, without transforming the self. Nor do I think that we can transform our view of the self without transforming the social systems of domination and exclusion that we have created. I am calling not for a personal and interior project but rather for an expansive and imaginative political and spiritual process. This view will necessarily change how we think of and organize private and common property.

I join political philosopher and theorist Roberto Unger in rejecting any spirituality that is not intertwined with the secular. Unger believes that our spirituality, our moving beyond the limited self, can be worked out only through engagement with others. While this may seem new, it is not. We have the example of various religious luminaries who shared a devotion to service and social justice. Cornel West reminds us of the tradition of prophetic Christianity in this country, of Daniel Berrigan, of Dr. King, of William Sloan Coffin—of those who saw inherent in the message of Jesus a call to transform social structures at a deep level, build a community, and stop accepting, as Dorothy Day noted, this "filthy, rotten system."

This new vision of self—so needed in our structures and institutions—is one of interconnection, of interbeing. We must answer not only the questions "Am I my brother's keeper and am I my sister's keeper?"—but also "Are they indeed my brothers and my sisters?" If so—and I believe it is so—what does this mean for our spirituality and our work for social justice? I believe it means that we need to build this awareness into our institutions and processes. This knowledge needs to animate the way we think about the political work that needs to be done.

In his book *The European Dream*, cultural historian Jeremy Rifkin says, "Freedom is found not in autonomy but in embeddedness. To be free is to have access to many interdependent relationships. The more communities one has access to, the more options one has for living a full and meaningful life. It is inclusivity that brings security—belonging, not belongings" (Rifkin 2004).

The history of land use policies around the world has continually been a history of using space as a tool to separate and dominate. The history of metropolitan dynamics in the United States is one of multiple structures inscribing racial hierarchies into the fabric of our geography. Countering this history will require not only collective action and a bold commitment to the ideals of democracy and justice but also a new way of seeing.

We are at a time of crisis. We are bombarded every day with messages about a ticking ecological clock, about the speeding up of the processes of globalization, about the specter of continued and increased violence around the globe with ever more dangerous weapons. Yet, as scary as times of crisis are, we cannot allow ourselves to be pushed backward into fundamentalism. Times of crisis are times of opportunity, and we have the power to remake the world in tremendous ways.

Future Prospects

As we move forward in the struggle for civil rights, we must understand and confront the way race and poverty are inscribed in space in this country, and the way this space is used to sort and regulate people, opportunity, and freedom. To succeed in creating better regions, more participatory democratic processes, and equal access to opportunity, we must make this work a project of the imagination and a project of the spirit. In so doing, we will need to create a transformative space for those of different races and different levels of income. It is my fervent hope that this anthology sharpens the dialogue, energizes a sense of urgency to actively redress disparities in the spatial distribution of resources and opportunities, and advances regionalism not just as a strategy for building opportunity-rich, sustainable metropolitan communities but also as a strategy that reinforces our common humanity.

Note

1. This refers to a situation in which an apparently free choice that is being offered actually provides no true alternative.

Environment, Transportation, and Land Use in the Quest for Racial Justice

Section 2 discusses three aspects of the quest for racial justice: environmental justice, transportation, and land use. Peggy M. Shepard and Kizzy Charles-Guzmán explore the roots of the U.S. environmental justice movement and how the definition of *the environment* came to be expanded beyond the limited ecological frame. Voices of the *built environment* call on us to calculate these costs to human health and well-being. Formerly excluded populations have spoken from the environment in advocating justice for those who are disproportionately burdened by pollutants and related health concerns.

A lack of transportation opportunities separates people and limits collaborative possibilities with potential mutual benefit. Robert D. Bullard, who directs the Environmental Justice Center at Clark Atlanta University, has written extensively on environmental justice, housing issues, and urban land use. His chapter addresses transportation inequities in the United States and the lack of mobility that has affected generations of people who are marginalized by race and class.

Across U.S. metropolitan regions, a lack of opportunity has brought about pockets of concentrated poverty. Housing policies have steadily shifted the country toward homogeneity, a cumulative process that has encouraged systematic bias and the proliferation of winner/loser communities. Looking beneath the surface, author and Georgetown University Law Center professor Sheryll Cashin examines how this process has brought about a continuation of segregation by race and class within the housing sector.

4

The Roots of Environmental Justice

Peggy M. Shepard and Kizzy Charles-Guzmán

Introduction

Environmental justice is a global movement that challenges the disproportionate burden of pollution and environmental degradation borne by
low-income communities and communities of color. The lack of public
participation by these communities in the decision-making process continues to be a key challenge to the environmental justice movement. For
environmental justice activists and advocates, the urban environment
represents a new perspective on, and definition of, *environment*: the
places where we live, work, play, pray, and learn.

In this chapter, we highlight the history, challenges, and achievements
of the national environmental justice movement. We argue that the environmental justice movement's revision and expansion of environmentalism has developed the foundation for understanding and implementing
initiatives aimed at achieving regional equity.

The Origins and History of Environmental Justice

The environmental justice movement comprises a variety of interests,
stakeholders, and constituents and has engendered far-reaching grassroots struggles that have advanced public policy at the state and federal
levels. It has also spurred the development of an academic field, the application of civil rights laws and other legal tools, and a series of conferences aimed at bringing together impacted communities, activists, and
policy makers.

Though over the years grassroots organizing has been taking place independently in communities across the nation, conferences and gatherings—convened by academics, grassroots organizations, and federal

agencies—have provided important forums for the interaction of key stakeholders in sharing information, developing strategic collaborations and alliances, and articulating specific policy and legislative action plans.

An early instance of such a conference was Working for Environmental and Economic Justice and Jobs, in 1976. This conference was sponsored by the Environmentalists for Full Employment and the United Auto Workers at the latter's Walter and May Reuther Camp, in Michigan. It brought together hundreds of workers, Urban League members, community groups, farmers, and environmentalists in a four-day discussion on the relationships among social, environmental, labor, and health issues. The conference gave rise to a challenge to employers' claims that pollution-control measures inevitably led to economic difficulties (Bryant 2003).

"Environmental racism" was the rallying cry in Warren County, North Carolina, where, in 1982, five hundred people were arrested protesting the dumping of PCB-contaminated soil in an African-American community. Though that initial protest was unsuccessful, two decades of community activism resulted in the government spending $18 million to detoxify 81,500 tons of contaminated soil (Bullard 1993). The struggle gained national attention, resulting in a series of studies and conferences on the disproportionate exposure of people of color and low-income communities to environmental and public health hazards in the United States.

In 1983, the Urban Environment Coalition in New Orleans brought together more than two hundred people of color to a conference entitled Taking Back our Health (Urban Environment Conference, Inc. 1985), and in 1990, the University of Michigan's conference on Race and the Incidence of Environmental Hazards gathered not only key environmental justice scholars but also representatives from the U.S. EPA, the Michigan Department of Environmental Resources, the Michigan State Department of Public Health, and the Governor's Offices. The conference was organized by Professor Bunyan Bryant and led to meetings among seven professors, "the Michigan group," and federal government officials. Its proceedings compiled a wide array of environmental justice case studies from across the country (Mohai and Bryant 1992). Coupled with grassroots advocacy, these efforts helped to move federal agencies to create more equitable environmental policies. After meeting with "the Michigan group," then—U.S. EPA administrator William Reilly formed a task force to address some of the issues raised at the Michigan

Conference—the Work Group on Environmental Equity (Cole and Foster 2001). Reilly also created the Office of Environmental Equity and required all EPA staff to study the allegation that race was the most significant indicator of exposure to environmental hazards (Bryant 2003).

Speaking for Ourselves

In 1991, more than 650 activists gathered in Washington, D.C., in a historic meeting—the First National People of Color Environmental Leadership Summit.[1] This summit was a landmark for the nascent environmental justice movement. Of the attendees, 301 were delegates, people of color activists who drafted and voted on Seventeen Principles of Environmental Justice (Lee 1993, 41–52) that set forth the vision and values of a multiracial, multiethnic, multi-issue national movement. Advocates urged the delegates to go home and build a strong grassroots base of folks who would speak for themselves and enrich the movement from the bottom up, rather than succumbing to the inclination to create a centralized national organization.

Some regional and national organizing had begun to take place, such as the Southwest Network for Environmental and Economic Justice. Based in Albuquerque, New Mexico, this multistate network focused on workers' rights and justice issues along the United States–Mexico border and published an open letter to the major environmental organizations calling on them to account for the narrowness of their agendas and for their not hiring people of color. Some of the leaders of these "Big Ten Green Groups," including the Environmental Defense Fund, National Resources Defense Council, Sierra Club, Audubon Society, World Wildlife Federation, National Wildlife Federation, and Nature Conservancy, among others, attended the summit to give their response. Other groups emerging since that time include the National Black Economic and Environmental Justice Network, based in Detroit; the Northwest Environmental Justice Network, based in Seattle; and the Indigenous Women's Environmental Network, based in Saskatchewan.

Environmental Justice Research and the Role of Academia

A fundamental requirement for advancing the environmental justice perspective is fostering the intersection of academic research and community-based activism. Academic research has helped to support policy aimed at improving the environmental health of poor communities of color.

The juxtaposition of the environmental movement and the civil rights movement can be traced back to the early 1970s, when leading sociologist Nathan Hare coined the term "Black Ecology" in the April 1970 edition of *The Black Scholar* journal (Hare 1970). Roger Davis, who founded the Bronx Environmental Society, Inc., an environmental organization based in New York City, furthered this work.[2] The works of Hare and Davis are some of the earliest attempts to bridge the gap between the natural and human environments. Hare maintained that the environmental movement disregarded the environmental problems faced by urban and metropolitan areas (Hare 1970). Davis presented "black ecology" as a challenge to the white, middle-class character of the environmental movement, attesting that oppressed minorities in urban centers were "endangered species," analogous to species in nature that conservationists sought to protect.[3]

In 1983, Dr. Robert Bullard found that the situating of landfills and incinerators in Houston, Texas, disproportionately affected African-American communities (Bullard 1983). This research led Bullard to a prolific academic career in environmental justice research, advocacy, and activism. Four years later, the United Church of Christ Commission for Racial Justice published "Toxic Wastes and Race in the United States." This 1987 study verified that race was the primary predictor of where a toxic waste facility was sited, and that income was a secondary indicator (Commission for Racial Justice 1987). In the early 1990s, Dr. Bunyan Bryant, founder and chair of the Environmental Justice Initiative at the University of Michigan School of Natural Resources and Environment, and Dr. Paul Mohai coauthored several studies, finding evidence of the disproportionate impact on racial groups in the city of Detroit (Mohai and Bryant 1991, 1–70), and also reviewed fifteen studies published between 1971 and 1992. Their review concluded that nearly every study found inequitable distribution of pollution by income or race, with five out of the eight studies that evaluated the effect of race versus class finding that race was the stronger predictor of the location of toxics and pollution.[4]

Similarly, the research writings of Michael Greenberg,[5] Michael Gelobter,[6] Luke Cole,[7] Lauretta Burke,[8] Robert Bullard,[9] Ivette Perfecto,[10] Marianne Lavelle,[11] Charles Lee,[12] Sheila Foster,[13] and others, strongly support the relationship between race and the exposure to toxic chemicals and polluting facilities, thus providing additional ammunition for the environmental justice movement that spurred consequent public policies at the federal and state levels (Ferris and Hahn-Baker 1995; Clinton 1994, 16–19; Camia 1993, 2257–2260). The academic writings of said prominent authors are the backbone of academic curriculums on topics from environmental studies to environmental justice and ethics, to civil rights law at colleges and universities across the United States. Furthermore, activism and research centers at major Historically Black Colleges and Universities, such as the Environmental Justice Resource Center out of Clark Atlanta University and the Deep South Center for Environmental Justice, formerly out of Xavier University, have provided important technical assistance and support to the environmental justice movement.

Policy Development

Despite the difficulties, organizing and advocacy efforts created the political will to develop new initiatives that have had great impact on public policy regarding environmental protection, enforcement of existing laws, and reprioritizing the national environmental health research agenda. Initial successes focused on the federal government. Dr. Robert Bullard and others served on the Natural Resources Transition Team for Bill Clinton and Al Gore during the Clinton administration. President Clinton issued Executive Order 12898 on Environmental Justice on February 11, 1994. This order mandated all relevant federal agencies to address the disproportionate burden of pollution in communities of color and to include environmental justice as part of their mission (Clinton 1994, 137–144). The order was signed on the same day that the National Institute of Environmental Health Sciences (NIEHS) hosted the Interagency Symposium on Health Research and Needs to Ensure Environmental Justice, attended by 1,100 people, including 400 environmental justice advocates who stressed the importance of community involvement in the scientific research agenda.

The symposium helped catalyze an expanded research agenda at the NIEHS, including new multiyear funding mechanisms such as the Environmental Justice through Communications grant program and community-based participatory research (CBPR) requests for proposals. The goal of the CBPR grants was to connect researchers, academics, clinicians, and community-based advocates to develop and implement research agendas. These new research agendas focused on environmental exposures of community residents, children, and workers, including farmworkers. Most important, it led to the publishing of findings that would be translated into public policy and practice.

As a result of Executive Order 12898, a federal Interagency Working Group on Environmental Justice was established in May 1994 to coordinate policies between and among agencies. The following year, the first interagency public meeting on environmental justice was held at Clark Atlanta University to promote the new federal initiatives and to hear the concerns of affected community residents. In 1995, the Environmental Justice and Transportation: Building Model Partnerships conference was held in Atlanta, the result of a partnership between the U.S. Department of Transportation and the Clark Atlanta University Environmental Justice Resource Center. The conference was part of the DOT's public outreach plan mandated by the executive order.

In 1996, as a result of continuing advocacy, the EPA began to focus on public participation and expanded the Public Participation Rule of the Resource Conservation and Recovery Act (RCRA) to empower communities to become involved in the RCRA permitting process. RCRA provides a framework for national programs to manage hazardous and nonhazardous wastes in an environmentally sound way.[14] In addition, the White House Council on Environmental Quality issued its draft guidance on incorporating environmental justice into the National Environmental Policy Act (NEPA) requirements, and President Clinton issued Executive Order 13007, addressing the protection of native religious and sacred sites.

In 2004, a report from the American Bar Association, "A Fifty-State Survey of Legislation, Policies and Initiatives," reviewed how more than thirty-five states have given force of law—through varied and wide-ranging policies, statutes, or other commitments—as their way to address environmental justice concerns (Cole and Foster 2001). Thus, the environmental justice movement has already established an important legacy, and its work continues to be strengthened by collaborations between activists and researchers.

Advancing the Movement Through the Use of Legal Tools

The case of Kettleman City is one of the earliest examples in which a small, predominantly Latino community in California successfully used the provisions of the California Environmental Quality Act to fight against the toxic waste incinerator proposed by Chemical Waste Management, Inc., the world's largest toxic waste company. This lawsuit was a significant win for the environmental justice movement because it showed that communities could successfully demand the enforcement of existing environmental laws. In the case, the court ruled that the environmental impact report was incomplete in its analysis of the environmental and economic effects of the project. The court also ruled that Kettleman City residents were not meaningfully included in the permitting process—almost 40 percent of the residents were monolingual and the review process did not include Spanish translations (Cole and Foster 2001).

In contrast, the case of Chester, Pennsylvania, was a difficult moment for the environmental justice movement. A small community of people of color within the majority-white county of Delaware, Chester, is the site of a cluster of industrial and waste-processing facilities. The community suffers poor health and mortality rates that are 40 percent higher than in the rest of Delaware County (Cole and Foster 2001). In response to a proposal to site an infectious medical waste sterilization plant in their community next to one of the country's largest incinerators, a proposal they considered a toxic assault, residents of Chester formed the grassroots group Chester Residents Concerned for Quality Living (CRCQL). The group organized protests and meetings but the city's Department of Environmental Protection (DEP) issued a new permit for the construction and operation of another waste treatment facility in Chester. The group then turned to legal action as a tool to combat environmental racism. After a flurry of legal activity and decisions overturning and restoring the permit for the waste treatment facility, the case went to the Pennsylvania Supreme Court, which validated the permit and allowed the facility to reopen (Cole and Foster 2001).

This case was particularly sobering for the environmental justice movement because it highlighted the limitations of relying exclusively on legal action to address environmental injustices—which are often symptoms of larger, structural political problems (Cole and Foster 2001). Legal action, despite bringing public and media attention to grassroots struggles, is an uncertain avenue because corporations, unlike grassroots advocacy groups, can seemingly allocate unlimited financial resources for litigation.

In 1997, CRCQL was able to settle a lawsuit, brought under the Clean Air Act, against the county's sewage treatment plant. In 1998, CRCQL sued the Pennsylvania DEP under Title VI of the Civil Rights Act, alleging that its permitting pattern in the county allowed the proliferation of toxic facilities in an African-American community and constituted racial discrimination. The federal court ruled in favor of the CRCQL, but this victory was short-lived, as the U.S. Supreme Court reversed the ruling (Cole and Foster 2001). Again, the case of Chester showed environmental justice advocacy groups that a Title VI complaint is not always an effective way of dealing with environmental racism.

More recently, the *Hartford Tenants Association v. Rhode Island Department of Environmental Management (DEM)* decision ruled that the "environmental equity" provision of the Industrial Property Remediation and Reuse Act was violated when the DEM approved a petition by the city of Providence to build schools at a former illegal municipal landfill. The court ruled that environmental equity was not considered and that this primarily African-American and Latino community had not been properly notified or meaningfully involved in the processes of investigation and remediation of the contaminated areas (American Bar Association 2005).

Because of the lack of financial resources and political influence in most communities suffering from environmental injustices, the help and support of legal clinics and legal advocates has been invaluable in environmental justice litigation. For example, the National Lawyers Guild and Maurice and Jane Sugar Law Center for Economic and Social Justice (Guild Law Center) in Michigan have been at the forefront of environmental justice litigation in that state since 1994. Aside from their direct advocacy and representation, they have provided technical assistance and training to communities affected by environmental injustices so that they can speak for themselves more effectively. Through the years, the center has been involved in environmental justice cases ranging from the lack of green space in minority communities in New York City (*New York City Environmental Justice Alliance v. Giuliani*) to the siting of an elementary school on a heavily contaminated former industrial site in a predominantly Latino community in Detroit (*Lucero v. Detroit Public Schools*) (National Lawyers Guild and Maurice and Jane Sugar Law Center for Economic and Social Justice 2003). Legal centers such as the Law Guild Center have been instrumental in interpreting and enforcing Title VI regulations, which forbid federally funded programs from carrying out programs or activities that have the effect of discriminating on the basis of race, ethnicity, or national origin. In the *Lucero v. Detroit Public Schools* case, for example, a U.S. federal court judge delivered a breakthrough environmental justice decision by ruling that petitioners had a valid claim under two federal civil rights theories.

Even when litigation has led to successful conclusions, however, their enforcement has met continued challenges. A 1992 study by the *National Law Journal* found that of all environmental lawsuits conducted in the United States within a seven-year period, the EPA imposed penalties that were 46 percent higher in white communities than in communities of color, that the EPA took 20 percent longer to list waste sites in minority communities in the National Priorities List, and that polluters paid 54 percent less in fines when penalized for wrongful activities in minority communities than when penalized for wrongful activities in white communities (Lavelle and Coyle 1992).

Opportunities for Coalition and Movement Building

Despite the challenges already discussed, the environmental justice movement has begun outreach to untraditional allies and is represented in a range of coalitions and government work groups. These include groups that address clean production, environmental health and toxics, brownfields redevelopment, climate justice, energy, children's environmental health, transportation justice, community benefits agreements regarding economic development, smart growth including land use and zoning, and healthy schools.

For more than a decade, the movement has understood the impacts of globalization on residents in this country and abroad and has sought to interact with other activists internationally by sending delegations to United Nations conventions including the 1992 Earth Summit in Rio; the Conference on Racism in Durban, South Africa, in 2001; and the World Summit on Sustainable Development in Johannesburg in 2002.

In all of these initiatives, efforts have been targeted to recruit, mentor, and develop the next generation of grassroots, academic, and policy leaders for environmental justice in order to build metropolitan, regional, and national equity. Environmental equity means that all residents of a given region equally share the environmental burdens and benefits arising from development and technology. Environmental justice opens up an array of possibilities for multisector collaborations, through the leadership of community activists, elected officials, academics, and researchers. It aims at continuing to develop policies that improve equity outcomes for all communities including, but not limited to, policies that create affordable housing, that secure environmental health and safety, that limit public exposure to occupational hazards, that address sustainable transportation needs, and that ensure equal, unrestricted access to environmental amenities.

Scientists can now work in close collaboration with community partners, who in turn are involved in every step of the research process, with the common goal of reducing social and health disparities among groups. A report from the Institute of Medicine, "Environmental Justice: Research, Education and Health Policy Needs," validated the need for a broader research agenda to address these disparities (Committee on Environmental Justice 1999).

Increasingly, government agencies and private foundations have funded community-and-university partnerships to conduct community-based participatory research. Community partners have helped with the formulation of research questions and study design, with the collection of data, with monitoring of ethical concerns, and with the interpretation of the study results. Importantly, in CBPR, the research findings are communicated to the broader community so they can be used to effect needed changes in environmental and health policy. CBPR seeks to build capacity and resources in communities while ensuring that government agencies and academic institutions are better able to understand and incorporate community concerns into their research agendas (Israel, Schurman, and House 1989; Israel, Schurman, and Hugentobler 1992; Hatch et al. 1993; Mittelmark et al. 1993).

Toward Greater Regional Equity

After decades of working to build a strong grassroots base, the environmental justice movement has developed the leaders, tools, expertise, and political support to expand that base, identify shared interests, and coordinate broad-based strategic initiatives that advance equity principles across communities, regions, and metropolitan areas.

Environmental justice advocates have entered into collaborations with businesses, government agencies, civic groups, labor, housing and tenant associations, researchers and academics, community lawyers, environmental groups, and urban planners. They have developed models of collaborative problem solving to advance a toxic-free future, to protect children's environmental health and reduce health disparities, to negotiate community benefits agreements with developers, and to achieve government accountability for environmental protection and enforcement of existing laws. They have also developed partnerships to assess environmental exposure of residents and workers, to promote legislation and state environmental justice policies, and to mobilize residents to develop visions and plans for building parks, healthy schools, and safe communities.

We know that the concerns of the environmental justice movement are shared by a host of potential allies and are defined by the realities of urban environments that are always in the throes of conflict over the priorities of economics, environment, and equity (Campbell 1996). Issues addressed by this movement include land use, transportation, air and water quality, affordable housing, habitat protection, and social equity. These are aspects of sustainable development that are arguably addressed best at the regional scale. Yet, it is often difficult at that scale to create the political will and the capacity to bring about change, because of the scale, size, and complexity of working and strategizing at the regional level and the challenge of understanding how local decisions are linked to regional decision making (Campbell 1996).

To move toward building equity on a metropolitan scale and to create a context for metropolitan planning initiatives, there must be an awareness of economic interdependence (Bollens 2003). Environmental justice movement advocates are already experienced in political organizing and coalition building and are developing expertise in community-driven planning and visioning processes. So far, they have had fewer opportunities to work with regional planning institutions and agencies. In New York, we are considering how to establish sustainability-related goals linked to economic incentives by developing scorecards and indicators of environmental progress and sustainability at the metropolitan and regional levels. There will be ongoing challenges in fostering public involvement, consensus building within communities and with other allies, and the training of residents and allies to understand the interconnections between the local, regional, and global contexts.

Despite the challenges and obstacles, the environmental justice movement has infused a gritty spirit into a new model of environmentalism, a framework that redefines the dominant environmental protection and conservation paradigm to one that is accountable to its grassroots base, replicable, multiethnic, multiracial, and interdisciplinary. The movement has principles, vision, values, and a constituency that continues to challenge the current politics of planning and pollution, and surmounts the inability of the scientific community to obtain and analyze research data that can reveal environmental injustices and lead to comprehensive, precautionary policies and practices that are protective of public health and the environment. The ultimate goal of environmental justice will be attained when we are able to move beyond a paradigm of struggle against entrenched powers to one of broad-based mutuality that extends across all the old boundaries of geography, economics, race, color, and class, and builds robust regions and communities that speak for themselves.

Notes

1. Robert Bullard, "Environmental Justice in the 21st Century." Available online at http://www.ejrc.cau.edu/ejinthe21century.htm.

2. The International Academy of Ikologiks and Advanced Studies, Inc. Available online at http://www.ikologiks.org/directory/aboutus.html.

3. Ibid.

4. See P. Mohai and B. Bryant, "Environmental Injustice: Weighing Race and Class as Factors in the Distribution of Environmental hazards." *University of Colorado Law Review* 63 (1992): 921–923; and B. Bryant and P. Mohai, eds. *Race and the Incidence of Environmental Hazards: A Time for Discourse* (San Francisco: Westview Press, 1992).

5. See M. Greenberg and R. Anderson, "Hazardous Waste Sites: The Credibility Gap" (New Brunswick, N.J.: Rutgers University Center for Urban Policy Research, 1984); and M. Greenberg, "Proving Environmental Inequity in Siting Locally Undesirable Land Uses." *Risk-Issues Health and Safety* 4 (1993): 235.

6. See M. Gelobter, "Toward a Model of Environmental Discrimination." In *Race and the Incidence of Environmental Hazards: A Time for Discourse*, eds. B. Bryant and P. Mohai (San Francisco: Westview Press, 1992).

7. See L. Cole and S. Bowyer, "Pesticides and the Poor in California," *Race, Poverty and the Environment* 2 (1991): 1; and L. Cole, "Empowerment as the Means to Environmental Protection: The Need for Environmental Poverty Law," *Ecology Law Quarterly* 19 (1992): 619.

8. See L. Burke, "Race and Environmental Equity: A Geographic Analysis of Los Angeles," *Geographic Information Systems* 4 (1993a): 44; and L. Burke, "Environmental Equity in Los Angeles," *Technical Report* (Santa Barbara: National Center for Geographic Information and Analysis, University of California, 1993b), pp. 93–96.

9. See the following six sources by R. Bullard: "Environmental Blackmail in Minority Communities," in *Race and the Incidence of Environmental Hazards*, eds. B. Bryant and P. Mohai (San Francisco: Westview Press, 1992); *Dumping in Dixie: Race, Class, and Environmental Quality* (Boulder: Westview Press,

1994); "Grassroots Flowering: The Environmental Justice Movement Comes of Age," *Amicus Journal* (Spring 1994): 32–37; "The Legacy of American Apartheid and Environmental Racism," *St. John's Journal of Legal Commentary* 9 (1994): 445–474; and "Overcoming Racism in Environmental Decision-making," *Environment* 36 (1994): 10–20, 39–44; and "Residential Segregation and Urban Quality of Life," in *Environmental Justice: Issues, Policies and Solutions*, ed. B. Bryant (Washington, D.C.: Island Press, 1995).

10. See I. Perfecto "Hazardous Waste and Pesticides: An International Tragedy" in *Environmental Racism: Issues and Dilemmas*, eds. B. Bryant and P. Mohai (1991); and I. Perfecto and B. Velasquez, "Farm Workers: Among the Least Protected," *EPA Journal* (March/April 1992).

11. See M. Lavelle and M. Coyle, "Unequal Protection: The Racial Divide in Environmental Law," *National Law Journal* 15 (1992): S1–S12.

12. See C. Lee, "Environmental Justice: Building a Unified Vision of Health and the Environment," *Environmental Health Perspectives* 2 (1992): 141–144.

13. See the following three sources: L. Cole and S. Foster, *From the Ground Up: Environmental Racism and the Rise of the Environmental Justice Movement* (New York: New York University Press, 2001); S. Foster, "Impact Assessment," in *The Law of Environmental Justice*, ed. M. Gerrard (Chicago: American Bar Association, 1999), pp. 256–278; and S. Foster, "Environmental Justice in an Era of Devolved Collaboration," in *Justice and Natural Resources*, eds. K. Mutz, G. Bryner, and D. Kenney (Washington, D.C.: Island Press, 2002).

14. Resource Conservation and Recovery Act. Available at http://tis.eh.doe.gov/oepa/law_sum/RCRA.htm.

5

Addressing Urban Transportation Equity in the United States

Robert D. Bullard

In the United States, all communities do not receive the same benefits from transportation advancements and investments.[1] Despite the heroic efforts and the monumental social and economic gains made over the decades, transportation remains a civil rights issue.[2] Transportation touches every aspect of where we live, work, play, and go to school, as well as the physical and natural world. Transportation also plays a pivotal role in shaping human interaction, economic mobility, and sustainability.

Transportation provides access to opportunity and serves as a key component in addressing poverty, unemployment, and equal opportunity goals while ensuring access to education, health care, and other public services. It is basic to many other quality-of-life indicators such as health, education, employment, economic development, access to municipal services, residential mobility, and environmental quality (Bullard and Johnson 1997, 8–9; Lewis 1997, xi–xii).

Transportation equity is consistent with the goals of the larger civil rights movement and the environmental justice movement. Transportation investments, enhancements, and financial resources have provided advantages for some communities, while transportation decision making has made other communities disadvantaged.

Race and class dynamics operate to isolate many low-income and people of color central-city residents from expanding suburban job centers. Transportation dollars have fueled suburban highway construction and job sprawl. Some transportation projects have cut wide paths through low-income and people of color neighborhoods, isolated residents physically from their institutions and businesses, disrupted once-stable communities, displaced thriving businesses, contributed to urban sprawl, subsidized infrastructure decline, created traffic gridlock, and subjected residents to elevated risks from accidents, noise, spills, and explosions from vehicles carrying hazardous chemicals and other dangerous materials.[3] The continued residential segregation of people of color away from suburban job centers (where public transit is inadequate or nonexistent) may signal a new urban crisis and a new form of "residential apartheid."[4]

Old Wars, New Battles

In 1896, the U.S. Supreme Court wrestled with this question of the different treatment accorded blacks and whites. In *Plessy v. Ferguson*, the Supreme Court examined the constitutionality of Louisiana laws that provided for the segregation of railroad-car seating by race (Franklin and Moss 1974, 540–552). The court upheld the "white section" and "colored section" Jim Crow seating law, contending that segregation did not violate any rights guaranteed by the U.S. Constitution.[5]

On December 1, 1955, in Montgomery, Alabama, Rosa Parks ignited the modern civil rights movement. Mrs. Parks refused to give up her bus seat to a white man, in defiance of local Jim Crow laws. Her action sparked new leadership around transportation and civil rights. Mrs. Parks summarized her feelings about resisting Jim Crow in an interview with sociologist Aldon Morris in 1981: "My resistance to being mistreated on the buses and anywhere else was just a regular thing with me and not just that day."

Follow the Dollars

Transportation spending programs do not benefit all populations equally (Bullard and Johnson 1997, 7). By following the transportation dollars in metropolitan regions, one can tell who is "important" and who is not. The lion's share of transportation dollars is spent on roads, while urban transit systems are often left in disrepair. Nationally, 80 percent of all surface transportation funds is earmarked for highways and 20 percent is earmarked for public transportation.[6] Public transit has received roughly $50 billion since the creation of the Urban Mass Transit Administration in the early 1960s, and roadway projects received more than $205 billion between 1956 and the early 1990s (Dittmar and Chen 1995).

Generally, states spend less than 20 percent of federal transportation funding on transit. The current federal funding scheme is biased against metropolitan areas. The federal government allocated the bulk of transportation dollars directly to state DOTs (Puentes and Bailey 2003). Many of the road-building fiefdoms are no friend to urban transit. Just less than 6 percent of all federal highway dollars are suballocated directly to the metropolitan regions. Moreover, thirty states restrict use of the gasoline tax revenue to funding highway programs only (Puentes and Prince 2003). Although local governments within metropolitan areas own and maintain the vast majority of the transportation infrastructure, they receive only about 10 percent of every dollar they generate.[7] Disparate transportation outcomes can be subsumed under three broad categories of inequity: procedural, geographic, and social.

Procedural Inequity

Attention is directed to the process by which transportation decisions may or may not be carried out in a uniform, fair, and consistent manner with involvement of diverse public stakeholders. Do the rules apply equally to everyone?

Geographic Inequity

Transportation decisions may have distributive impacts (positive and negative) that are geographic and spatial, such as rural versus urban versus central city. Some communities are physically located on the "wrong side of the tracks" and often receive substandard transportation services.

Social Inequity

Transportation benefits and burdens are not randomly distributed across population groups. Generally, transportation benefits go to the wealthier and more educated segment of society, while transportation burdens fall disproportionately on people of color and individuals at the lower end of the socioeconomic spectrum. Intergenerational equity issues are also subsumed under this category (Bullard, Johnson, and Torres 2001, 965). The impacts and consequences of some transportation decisions may reach into several generations.

In summary, heavy government investment in road infrastructure may be contributing to an increase in household transportation costs. Lest anyone dismiss transportation as a tangential issue, consider that Americans spend more on transportation than any other household expense except housing. On average, Americans spend nineteen cents out of every dollar earned on transportation expenses (Jakowitsch and Ernst 2003; Floyd 2003, 1C). They spend more on transportation than they do on food, education, and health care. The nation's poorest families spend more than 40 percent of their take-home pay on transportation. This is not a small point since African-American households tend to earn less money than whites. Nationally, African-Americans earn only $649 per $1,000 earned by whites (Thomas 2002, 1–2). This means that the typical black household in the United States earned 35 percent less than the typical white household.

Erasing Transportation Inequities

In general, most transit systems have tended to take their low-income and people of color "captive riders" for granted and concentrated their fare and service policies on attracting middle-class and affluent riders out of their cars. Moreover, transit subsidies have tended to favor investment in suburban transit and expensive new commuter bus and rail lines that disproportionately serve wealthier "discretionary riders." Almost 40 percent of rural counties in this country have little or no public transportation, and "[i]n areas with populations from one million and below, more than half of all transit passengers have incomes of less than $15,000 per year" (Bogren 1990).

In urban areas, African-Americans and Latinos constitute more than 54 percent of transit users (62 percent of bus riders, 35 percent of subway riders, and 29 percent of commuter-rail riders) (Pucher and Renne 2003, 49 and 67). Nationally, only about 5.3 percent of all Americans use public transit to get to work (Garrett and Taylor 1999, 11). African-Americans are almost six times as likely as whites to use transit to get around (Pucher and Renne 2003, 67). Urban transit is especially important to African-Americans since more than 88 percent live in metropolitan areas and more than 53 percent live inside central cities (Cantave and Harrison 2001).

Lack of car ownership and inadequate public transit service in many central cities and metropolitan regions with a high proportion of "captive" transit dependants exacerbate social, economic, and racial isolation, especially for low-income people of color residents who already have limited transportation options. Nationally, only 7 percent of white households do not own a car, compared with 24 percent of African-American households, 17 percent of Latino households, and 13 percent of Asian-American households (Pucher and Renne 2003, 49–77). People of color are fighting to get representation on transportation boards and commissions, and to get their fair share of transit dollars, services, bus shelters and other amenities, handicapped accessible vehicles, and afford-able fares.

Rosa Parks could not sit in the front or back of a Montgomery bus to-day, since that city dismantled its public bus system—which served mostly blacks and poor people (Stolz 2000). The cuts were made at the same time that federal tax dollars boosted the construction of the region's extensive suburban highways. The changes in Montgomery took place amid growing racial geographic segregation and tension be-tween white and black members of the city council. The city described its actions publicly as fiscally necessary, even as Montgomery received large federal transportation subsidies to fund renovation of nontransit improvements.

Suburban Sprawl and Health

In *Sprawl City: Race, Politics, and Planning in Atlanta*, the authors documented that government-subsidized sprawl has substantial social equity, civil rights, and health implications. Sprawl-fueled growth is wid-ening the gap between the haves and the have-nots. Suburban sprawl is fueled by the "iron triangle" of finance, land use planning, and transpor-tation service delivery.[8] Sprawl has clear social and environmental effects.[9] The social effects of suburban sprawl include concentration of urban core poverty, closed opportunity, limited mobility, economic dis-investment, social isolation, and urban/suburban disparities that closely mirror racial inequities (Blackwell 2001, 1273–1277). The environmen-tal effects of suburban sprawl include urban infrastructure decline; increased energy consumption; automobile dependency; threats to public health and the environment including air pollution, flooding, and climate change; and threats to farmland and wildlife habitat (Ehrenhalt 1999, A23).

Many jobs have shifted to the suburbs and communities where public transportation is inadequate or nonexistent. The exodus of low-skilled jobs to the suburbs disproportionately affects central-city residents, particularly people of color, who often face more limited choices of housing location and transportation in growing areas. Between 1990 and 1997, jobs on the fringe of metropolitan areas grew by 19 percent versus 4 percent job growth in core areas (Bullard, Johnson, and Torres 2004). While many new jobs are being created in the suburbs, the majority of job opportunities for low-income workers are still located in central cities.

Suburbs are increasing their share of office space, while central cities see their share declining. The suburban share of the metropolitan office space is 69.5 percent in Detroit, 65.8 percent in Atlanta, 57.7 percent in Washington, D.C., 57.4 percent in Miami, and 55.2 percent in Philadelphia (Lang 2000). Getting to these suburban jobs without a car is next to impossible. It is no accident that Detroit leads in suburban "office sprawl." Detroit is also the most segregated big city in the United States (Farley et al. 1993, 1–2; powell 2003, H12), and the only major metropolitan area without a regional transit system. In Detroit, the Motor City, only 2.4 percent of metropolitan Detroiters use transit to get to work (U.S. Department of Transportation 2004).

Transportation-related sources account for more than 30 percent of the primary smog-forming pollutants emitted nationwide and 28 percent of the fine particulates. Vehicle emissions are the main reason 121 air quality districts in the United States are in noncompliance with the 1970 Clean Air Act's National Ambient Air Quality Standards. More than 140 million Americans—of whom 25 percent are children—live, work, and play in areas where air quality does not meet national standards (American Public Transportation Association 2003). Emissions from cars, trucks, and buses cause 25 to 51 percent of the air pollution in the nation's nonattainment areas. Transportation-related emissions also generate more than a quarter of the greenhouse gases.[10]

Improvements in transportation investments and air quality are of special significance to African-Americans and other people of color who are more likely to live in areas with reduced air quality when compared with whites. National Argonne Laboratory researchers discovered that 57 percent of whites, 65 percent of African-Americans, and 80 percent of Latinos lived in the 437 counties that failed to meet at least one of the EPA ambient air quality standards in 1992 (Wernette and Nieves 1992, 16–17). A 2000 study from the American Lung Association shows that children of color are disproportionately represented in areas with high ozone levels. Additionally, 61.3 percent of black children, 69.2 percent of Hispanic children, and 67.7 percent of Asian-American children live in areas that exceed the 0.08 ppm (parts per million) ozone standard, while only 50.8 percent of white children live in such areas (American Lung Association 2000).

Reduction in motor vehicle emissions can bring about marked health improvements. For example, the Centers for Disease Control and Prevention (CDC) reports that "when the Atlanta Olympic Games in 1996 brought about a reduction in auto use by 22.5 percent, asthma admissions to ERs and hospitals also decreased by 41.6 percent" (Jackson and Kochtitzky 2001, 3). The CDC researchers also concluded that "less driving, better public transport, well designed landscape and residential density will improve air quality more than will additional roadways." Excessive ozone pollution contributed to 86,000 asthma attacks in Baltimore, Maryland; 27,000 in Richmond, Virginia; and 130,000 in Washington, D.C. (Abt Associates for the National Campaign Against Dirty Power October 1999, 31).

Air pollution from vehicle emissions causes significant amounts of illness, hospitalization, and premature death (Fischlowitz-Roberts 2002). A 2002 study in *The Lancet* reports a strong causal link between ozone and asthma (McConnell et al. 2002, 386–391). Ground-level ozone may exacerbate health problems such as asthma, nasal congestions, throat irritation, respiratory tract inflammation, reduced resistance to infection, changes in cell function, loss of lung elasticity, chest pains, lung scarring, formation of lesions within the lungs, and premature aging of lung tissues (Ozkaynak et al. 1996, 2–7).

A 2001 CDC report, "Creating a Healthy Environment: The Impact of the Built Environment on Health" (Fischlowitz-Roberts 2002), points a finger at transportation and sprawl as major health threats. Air pollution claims seventy thousand lives a year, nearly twice the number killed in traffic accidents. Although it is difficult to put a single price tag on the cost of air pollution, estimates range from $10 billion to $200 billion per year (Bollier 1998, 9). Asthma is the number one reason for childhood emergency room visits in most major cities in the country, with the hospitalization rate for African-Americans three to four times the rate for whites (Ozkaynak et al. 1996, 2–7). African-Americans are three times more likely than whites to die from asthma (Centers for Disease Control 2000, 58).

Getting sick is complicated for the nation's uninsured. Blacks and Hispanics are most at risk of being uninsured; blacks and Hispanics now constitute 52.6 percent of the 43 million Americans without health insurance (Mills and Bhandari 2003, 7). Nearly half of working-age Hispanics lacked health insurance for all or part of the year prior to being surveyed, as did almost one-third of African-Americans. In comparison, one-fifth of whites and Asian-Americans ages eighteen to sixty-four lacked coverage for all or part of the year (Collins et al. 2002).

In addition to health and environmental reasons for the United States to move our transportation beyond oil-powered vehicles to more secure and sustainable alternatively fueled ones, there are compelling energy security and economic strength reasons to invest in clean fuels technology. The United States has more than 217 million cars, buses, and trucks that consume 67 percent of the nation's oil. Transportation-related oil consumption in the United States has risen 43 percent since 1975 (Underwood 2001). The United States accounts for almost one-third of the world's vehicles. With just 5 percent of the world's population, Americans consume more than 25 percent of the oil produced worldwide with almost 60 percent of our oil coming from foreign sources (Underwood 2001).[11]

Conclusion

Transportation is a basic ingredient for quality-of-life indicators such as health, education, employment, economic development, and access to municipal services, residential mobility, and environmental quality. Transportation continues to be a civil rights issue. Improvements in transportation investments and air quality are of special importance to low-income persons and people of color who are concentrated in the nation's most polluted urban centers. Transportation investments, enhancements, and financial resources, if used properly, can bring much-needed revitalization to urban areas.

The environmental justice movement has set out clear goals of eliminating unequal enforcement of the nation's environmental, public health, housing, employment, land use, civil rights, and transportation laws. Transportation is a key ingredient in any organization's plan to build economically viable and sustainable communities. State Departments of Transportation and metropolitan planning organizations have a major responsibility to ensure that their programs, policies, and practices do not discriminate against or adversely and disproportionately impact people of color and the poor.

Notes

1. See Preface to *Just Transportation: Dismantling Race and Class Barriers to Mobility*, eds. R. D. Bullard and G. S. Johnson (Gabriola Island, B.C.: New Society Publishers, 1997), pp. xiii–xiv and 7.

2. See Foreword by J. Lewis in *Just Transportation: Dismantling Race and Class Barriers to Mobility*, eds. R. D. Bullard and G. S. Johnson (Gabriola Island, B.C.: New Society Publishers, 1997), pp. xi–xii. See also M. Garrett and B. Taylor. 1999, "Reconsidering Social Equity in Public Transit," *Berkeley Planning Journal* 13 (1999): pp. 6–10. Available at http://www.ced.berkeley.edu/pubs.bpj/backissues13.html or http://www.uctc.net/papers/701.pdf. "The incongruence between transit ridership patterns and subsidy policies has both social and special consequences that can potentially reinforce existing patterns of racial, ethnic, and economic segregation."

3. See D. Kong, "Filipino Americans Work to Preserve Heritage," *Honolulu Star-Bull* (December 26, 2002). Available at http://starbulletin.com/2002/12/26/news/story8.html. "By the 1930s, Stockton was home to the largest Filipino population outside the Philippines. But a cross-town freeway cut through the neighborhood in the early 1970s, and the once-vibrant enclave is now just a shadow of what it was."

4. See R. D. Bullard, "Introduction: Anatomy of Sprawl," in *Sprawl City: Race, Politics, and Planning in Atlanta*, eds. R. D. Bullard, Glenn S. Johnson, and Angel O. Torres (Atlanta: Island Press, 2000), pp. 3–4. "Apartheid-type employment, housing, development, and transportation policies have resulted in limited mobility, reduced neighborhood options, decreased residential choices, and diminished job opportunities for African-American and other people of color who are concentrated in cities."

5. *Plessy v. Ferguson*. 1896. 163 U.S. 537, at 548. "[W]e think the enforced separation of the races, as applied to the internal commerce of the state, neither abridges the privileges or immunities of the colored man, deprives him of his property without due process of law, nor denies him the equal protection of the laws, within the meaning of the fourteenth amendment."

6. See T. W. Sanchez, R. Stolz, and J. S. Ma, "Moving to Equity: Addressing Inequitable Effects of Transportation Policies on Minorities," a joint report of the Civil Rights Project at Harvard University and the Center for Community Change (2003). Available at http://www.civilrightsproject.ucla.edu/research/transportation/trans_paper03.php. See also Surface Transportation Research and Development Needs for the Next Century, Testimony Before the House Committee on Science, Subcommittee on Technology (April 23, 1997). Hank Dittmar, Executive Director, Surface Transportation Policy Project, notes that "highway and vehicle research ... account for more than 80% of available funding."

7. See Highway and Transit Needs: The State and Local Perspective, Testimony Before the House Transportation & Infrastructure Committee, Subcommittee on Highways, Transit & Pipelines (May 7, 2003), p. 3. Statement of

Victor H. Ashe, Mayor of Knoxville, Tennessee, explaining that "metropolitan areas get to make decisions only on about ten cents on every transportation dollar they generate."

8. See W. W. Buzbee, "Urban Sprawl, Federalism, and the Problem of Institutional Complexity," *Fordham Law Review* vol. 68, no. 112 (October 1999): 57–136. "Providing that the main strategy of TEA-21 is to avoid patronage-driven transportation decisions by mandating a more open and participatory planning process as a condition for receipt of federal dollars."

9. Ibid., describing the social and environmental deterioration in the Atlanta area due to sprawl.

10. Testimony Before the Senate Finance Committee, Statement of J. S. Cannon, on behalf of INFORM, Inc. (July 10, 2001).

6

Race, Class, and Real Estate

Sheryll Cashin

Extending an environmental justice agenda requires a way to address the equity issues connected to where we live—an issue that Hurricane Katrina made visible to the world. The scenes from New Orleans in the wake of Katrina were devastating. Collectively as a nation we watched, and wept, and thought, "This is not America." Yet it is. It is the America that normally hides from public view. The ghetto was moved by natural forces, and man-made ineptitude, from the invisibility of a low-lying, marginal area to the public square of the Superdome and cable television. The brutal, unvarnished ugly truth of what it meant to be literally left behind by the nation was revealed.

One silver lining in this heartrending national moment is that it could spark renewed interest in comprehending how race and class inequality became entrenched in the very architecture of cities and metropolitan regions. In short, race and class segregation in our housing markets begets inequality of opportunity. It creates "winner" and "loser" communities within a single metropolitan region. Sometimes these communities of disparate fortune share a jurisdictional boundary or are right across the road from each other.

Housing or geography is at the core of the opportunity structure in America. Where you live largely defines what type of people you are exposed to on a daily basis, and hence, who you have the opportunity to relate to. It defines what schools you go to, what employers you have access to, and whether you are exposed to a host of models for success.

Since 1970, with each passing decade we have made glacially slow improvements in opening up housing markets and decreasing racial segregation. During this period we have also seen a marked increase in class segregation. In the 2000s, it remains the case that the neighborhood where you live is highly likely to reflect both your race and class.

How Housing Segregation by Race and Class Became the "Natural" Order

In my book *The Failures of Integration*, I identify three main factors that contribute to race and class segregation in U.S. housing markets: 1) the pull of personal preferences, 2) the push of discrimination, and 3) a host of public and private institutional policies that value homogeneity over inclusion.[1]

Although in opinion polls the majority of all races say they would prefer an integrated neighborhood, similar majorities also state a preference for living in a neighborhood in which their own race is a majority or plurality. Whites, blacks, Latinos, and Asians typically do not want to be vastly outnumbered by "others." That prospect is inherently threatening, particularly for whites. As a result, residential integration necessarily has been limited.

The most recent national audit of racial discrimination in housing, conducted in 2000, shows that while there has been considerable improvement since the last audit in 1989, whites are still consistently favored over racial minorities.[2] Researchers for the study, commissioned by the Department of Housing and Urban Development, sent minority and white testers out to attempt to rent or purchase homes in twenty-three metropolitan areas with significant black or Latino populations. The test partners had virtually the same income, assets and debt liabilities, and education levels.

Overall, *Latinos experienced more discrimination than blacks*. The group most discriminated against were Latino renters; landlords favored whites over this group one quarter of the time. They favored whites over black renters one fifth of the time. In particular, whites were more likely to receive information about available housing units and had more opportunities to inspect available units. The numbers were only slightly better for blacks and Latinos seeking to buy, rather than rent, a home.

Our separatism exists, but it is not inherently natural. Through a series of public and private institutional choices, we created a separatist social order. It did not have to be this way; separation was not our preordained fate. At the dawn of the twentieth century, economic and racial *integration* was the norm. It was not at all uncommon in American cities to find blacks living in close proximity to other races, or to find blue-collar workers living among the elite. This was especially the case in southern cities.

Four seminal public policy choices made in the twentieth century contributed mightily to the racially and economically divided landscape, the bastion of affluence and of need, now familiar to metropolitan America. First, we adopted a system of local governance premised on a religion of local autonomy that has fueled the proliferation of new, homogeneous communities. Chief among the local powers that are wielded to exclude undesired uses of land and undesired populations is the zoning power.

Second, the federal government, through its Federal Housing Administration (FHA) mortgage insurance program, adopted and propagated the orthodoxy that homogeneity was necessary to ensure stable housing values. The FHA, the largest insurance operation in the world in its heyday, essentially chose to underwrite mortgages only for new single-family homes in predominately white neighborhoods, inventing and propagating the notion of redlining—a legacy we live with to this day—and initially locking out whole races and classes of people from the largest wealth-producing program in our nation's history (Hall 1988, 291–294; Jackson 1985, 203–218).

Third, the interstate highway program—the largest public works program in the history of the world—opened up easy avenues for escape from the city while at the same time destroying vital black, Latino, and white ethnic neighborhoods. Fourth, the federal government, through a number of urban development programs, created the black ghetto. Urban renewal, famously renamed by black folks as "Negro Removal," destroyed mostly black-occupied housing strategically located near the central business district, ostensibly to help cities prepare for a postindustrial economy and to eliminate "blight." The federal government spent about $3 billion to remove almost 400,000 units of affordable, largely black-occupied housing that was strategically located. Those people who were displaced had to move somewhere, which typically meant to public housing or more marginal neighborhoods.

Any one of these policies individually would have altered the metropolitan landscape in a way that advantaged some and greatly disadvantaged others. But these policies were *cumulative.* Coupled with the federal government's tepid resistance to housing discrimination, these policies worked in concert to create a systemic bias in favor of racial and economic segregation rather than inclusion.

Private Actors

Private actors, particularly those in the real estate industry, have contributed mightily to the racial and economic segmentation of our life space. Most critically, the real estate, banking, and insurance industries embraced the federal government's orthodoxy that racial and economic homogeneity were necessary to protect property values. Private developers tend to develop to meet a certain class niche.

However, something even more insidious is going on. Every zip code in America has been racially profiled. Marketing companies create databases that rate each zip code based upon its demographics. In turn, all of the actors that shape real estate markets—land use planners, real estate developers, financial institutions, insurance companies, and retailers—rely on these databases to decide where to invest, develop, and do business. One company, for example, has developed the Claritas PRIZM system of categorization—forty socioeconomic rankings of "zip quality," ranging from ZQ1 (known as "Blue Blood Estates") to ZQ40, "Public Assistance."[3] All of these profiling databases establish a hierarchy of neighborhood types that skew investment decisions heavily in favor of predominately white suburban communities.

The Costs of Our Separate and Unequal Society

The separatism in our society was not inevitable; it has resulted from a series of conscious choices. It is true that individuals acting on their personal prejudices and preferences might have chosen, in the absence of exclusionary public and private policies, to cluster among people of their own race and economic class. Yet, it would not have been possible for millions of individuals acting independently to create the regime of systematic stratification and exclusion that reigns today.

The impact on African-Americans, particularly those who are poor, has been seminal. But in truth, no one is spared the costs of American separatism. People of all races and classes suffer the anxieties of living in a society premised upon there being winners and losers. In a rapidly diversifying America, the concern to move oneself and one's children into "safe" havens and the best schools is felt acutely precisely because society is not fundamentally committed to bringing *all* children and *all* communities along.

In a separatist system that sets up "winner" and "loser" communities, "winner" and "loser" schools, and even "winner" and "loser" tracks within schools, those in the middle of the income spectrum have to work harder and harder to get into or stay in the "winner" column. Those who can afford it pay a steep premium in the form of housing costs for predominately white, affluent neighborhoods, which equate to a gold standard of quality that is difficult for others to attain. Many parents, especially those in diverse metro areas, feel they have no choice but to pay this cost as insurance against the risk of their child being left behind.

For suburbanites, this separation and its attendant sprawl results in more traffic congestion, air pollution, environmental degradation, loss of open space, and deaths from auto accidents, as well as underutilization of many sewers, utilities, and schools in central cities and older suburbs. Many suburbanites spend anywhere from one to three hours each day driving to and from work, not to mention constant driving for every human need in car-dependent, inefficiently designed communities that intentionally segregate not just people but types of land uses. The ultimate cost of this car-dependent existence may be the loss of time with family or friends.

Our separated schools also impose heavy social costs. Seven out of ten black and Latino public school students attend schools that are predominately minority, while eight out of ten white children attend schools that are predominately white (Frakenberg, Lee, and Orfield 2003, 31–34). The growing re-segregation of American schools means that the average experience for white public school students is a middle class one, while the average experience for black and Latino children is one of high or concentrated poverty. High poverty schools tend to have an oppositional culture that denigrates learning.

In addition to the increased financial and social costs that come with educating minority children in a high poverty context, this separatism contributes to an achievement gap on the part of black and Latino students for which society also ultimately pays. Ghetto residents bear the heaviest wages of separatism. They typically endure a violent alternative culture and a social distress that even the most ambitious would find hard to overcome. Ghetto life contributes mightily to racial stereotyping; it rationalizes fear of black people and encourages the continuation of separatist policies.

Finally, race and class separation has contributed to a divisive politics, one where geographic advantage and disadvantage heavily correlate to one's partisan tendencies. Because we tend to live in political jurisdictions that differ by race and class, elected officials often respond to specific niches with seemingly antithetical interests. As a result, it is difficult to form a broad political consensus for anything approaching mutual responsibility or pursuit of a common, greater good.

What to Do About It

The stratospheric costs we are enduring, individually and collectively, as a result of race and class separation reflect our failure to deal with the truly hard questions left over from the civil rights movement. The civil rights movement largely stopped once the barriers of formal Jim Crow segregation were dismantled. While the delegitimation of discrimination was the chief success of the civil rights movement, we never reached any national consensus about whether integration of the races and classes—that is, the sharing of neighborhoods, schools, and life space—was an important objective to be affirmatively pursued.

I believe there is no substitute for taking "the hard path" not yet chosen, explicitly tackling segregation. Our nation's history shows that only when we choose the hard path of attacking frontally issues of race and racial inclusion do we make meaningful progress on achieving the values we profess to uphold. Indirect approaches are no substitute for a frontal attack on what is ailing us as a nation. They will simply delay the inevitable adjustment that is needed—as happened when the infamous case of *Plessy v. Ferguson* installed the "separate but equal" doctrine for sixty years, until the Supreme Court jettisoned it in *Brown v. Board of Education*.

Hope through Coalition Building

I dare to imagine an America that has experienced a transformative integration of the races. By this I do not mean the assimilationist model of the 1970s. Instead, I envision an America where the majority of citizens, especially whites, have developed a true comfort with racial difference, or what I call "cultural dexterity." A culturally dextrous person—former President Bill Clinton comes to mind—can walk into a room and be completely outnumbered by a different race or ethnic group and experience this with a sense of wonder, even celebration, rather than fear.

There is no shortage of sound ideas for bringing about more race and class integration in our neighborhoods by creating more integrated islands. What is missing is an insistent movement to alter our present separatist course in all communities. We lack a broad advocacy base for inclusive and more equitable public policies. Any movement for race and class integration must come from the grass roots. As Myron Orfield has so eloquently stated, national and state political leaders "do not create social movements around race," rather they "mediate energy for change that is created below the surface."[4]

Revolutionary change can be wrought by powerful new multirace, multiclass coalitions that pursue smart new policies. Without doing the difficult, labor-intensive work of building sustainable coalitions that command at least 51 percent representation in any given policy-making arena, no change will be forthcoming. As my hero Frederick Douglass eloquently stated in the context of the abolition movement, "Power concedes nothing without a demand. It never did, it never will."

Any effort that is serious about reducing regional inequity must have racial and economic integration as its central goal. Among the policies that could bring about more race and class integration at the neighborhood level is inclusionary zoning (or fair share affordable housing). Since 1973, for example, Montgomery County, Maryland, has required that 15 percent of all new housing developments with more than thirty-five units be affordable to low- and moderate-income families. In addition, the black ghetto can be broken up through policies that give people trapped in high-poverty communities the assistance they need to find decent housing in middle-class settings. On the school front, I recommend universal choice options. Why should a family's ability to choose a good school for their child be limited to their ability to pay their way into an exclusive neighborhood?

These are just a few of the revolutionary policies we have not considered because of our separated way of living and the defensive parochialism it engenders. Such possibilities are achievable, however, if the majority of people who now suffer under American separatism organize and act to reclaim democratic processes. It is time for people who care to imagine a different, more inclusive order—one that will benefit everyone.

Notes

1. For a detailed treatment of the issues raised in this chapter, see S. Cashin's book, *The Failures of Integration: How Race and Class Are Undermining the American Dream* (Cambridge: Public Affairs, 2004).

2. See M. A. Turner S. L. Ross, G. C. Galster, and J. Yinger, "Discrimination in Metropolitan Housing Markets: National Results from Phase I HDS 2000," Final Report, November 2002. Available at http://www.huduser.org/Publications/pdf/Phase1_Annex_1.pdf.

3. See J. T. Metzger, "Clustered Spaces: Racial Profiling in Real Estate Investment." Paper prepared for the International Seminar on Segregation and the City, Lincoln Institute of Land Policy, Cambridge, Massachusetts, July 26–28, 2001.

4. See M. Orfield, "Comment on S. A. Bollen's 'In Through the Back Door: Social Equity and Regional Governance,'" *Housing Policy Debate* 13 (2003): 659 and 666 (noting that Abraham Lincoln initially opposed the abolition of slavery and Lyndon Johnson initially opposed civil rights, but both leaders were ultimately forced by grassroots movements to pursue the progressive course).

Geographic Context, Sustainability, and Regional Equity

Chapters in the final section of part I discuss particular geographic contexts as they relate to issues of sustainability and regional equity. The chapter begins with an application of the regional equity concept to small town rural communities that struggle with poverty. Although rural America suffers from many of the same imbalances experienced by residents of the urban core, the fate of the rural poor receives much less focus than the situation of those who confront urban poverty. Cynthia M. Duncan and Priscilla Salant examine the vulnerabilities of rural communities and explore the advantages of a "working landscape" approach, blending conservation and antipoverty strategies.

Concentrated poverty persists at the urban core of major U.S. metropolitan regions. Our society's division of race and class became all the more starkly visible when Hurricane Katrina devastated the Gulf region. In their chapter, Amy Liu and Bruce Katz, both of the Metropolitan Policy Program at the Brookings Institution in Washington, D.C., review Katrina's lessons for policy makers. Pointing to the central role of federal and state agencies, the authors argue that civic leadership in support of equitable housing, jobs, and access to opportunity is essential to a region's competitiveness. In addition, researchers have found that reducing disparities within a region brings important benefits to all economic sectors.

As the American economic and political landscape undergoes dramatic changes, notes Amy B. Dean, president of Building Partnerships, USA, the metropolitan region is rapidly growing more important as a center of power and influence. Through its Central Labor Councils, the American labor movement has a great opportunity not only to adapt to this new geography of decision making but also to provide a new role for working people in the development of regional economic strategy. In her contribution to chapter 9, Dean describes how a pilot project of the Central Labor Council in Silicon Valley, California, created a model that gave working families a voice in addressing issues of the region's economic development. With a three-part strategy developed further in cities such as Atlanta, Boston, and Denver, Dean outlines how policy analysis, coalition building, and political action can help working communities.

7

Development and Opportunity in Small Towns and Rural Communities

Cynthia M. Duncan and Priscilla Salant

Persistent Poverty in Rural America

Fifty million Americans live in small towns and rural communities, and 7.5 million of them live in poverty. Forty percent of all rural families have incomes below $35,000—low-income families who struggle to make ends meet. Working poverty is pervasive; two-thirds of the poor are in families with at least one person working. Rural poverty has persisted in some areas for decades. With the notable exceptions of Appalachia, the Ozarks, and Oklahoma, this persistent rural poverty particularly affects Americans of color: African-Americans in the South, Hispanics in the Southwest and along the Texas border, and Native Americans in the West. More than 20 percent of all rural blacks, Hispanics, and Native Americans are poor. Although conditions and opportunities vary by region, the South stands out as a region with high rural poverty, low education, and high joblessness (see figure 7.1).

Many of the same dynamics of race, culture, and power that created our hollowed-out inner cities are at play in poor, left-behind rural places: extreme isolation for the poor, chronic underinvestment, no middle class, a few powerful people allocating resources and opportunities, and a large vulnerable population that is excluded from the mainstream and caught up in an oppressive climate with little access to the personal or institutional resources that could enable them to turn things around. An elite group of families, the "haves," control jobs and assets in the community, while the "have-nots" depend on them for any opportunity. Like our inner cities, the "success stories" are usually those who have gotten out, leaving for greater opportunities and freedom in cities while further handicapping those who remain behind in areas of concentrated poverty.

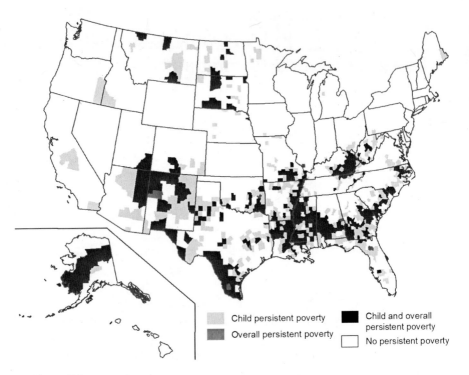

Figure 7.1
Poverty has persisted across rural America for decades, especially among communities of color.

The story of rural America, however, is about more than persistently poor communities. New demographic trends and economic restructuring are reshaping rural America in the twenty-first century. This is creating some rural regions that are indeed growing. Smart policies in these places could generate new opportunities for equitable development. Yet, it remains true that other rural communities continue to struggle with stagnation and decline, as decades of young people's out-migration pile up alongside the ongoing difficulties that follow from historical underinvestment in people and places.

Typology of Rural American Communities

Looking at demographic and socioeconomic trends in the early twenty-first century, we see *three broad types of rural communities* emerge in America:

First are the *rural areas on the outer rim of America's suburbs*, home to about 33 million residents, of whom roughly 14 percent are poor. These exurban communities at the metropolitan fringe are threatened by sprawl as families seek more affordable housing, even at the cost of extremely long commutes. "People drive until they qualify for a mortgage loan," one regional leader working in the rural West told us. The population of exurban communities has increased by about 13 percent since 1990. New residents, some of whom are themselves low-income, put pressure on housing costs that further press long-term residents and dramatically change the character and environment in these communities. With the right state and local policies, these changes could offer opportunities for long-term residents in otherwise lagging economies, while containing sprawl and congestion.

Second are *high-amenity rural communities*—some near metropolitan areas and others remote but accessible—where the population has grown by about 30 percent since 1990. These areas are now home to about 11 million people, of whom roughly 14 percent are poor. Their lakes, seashore, mountains, and other natural assets attract visitors as well as urban residents who relocate to continue working or to retire (see figure 7.2). Some early research indicates that under current policies, wage and employment gains are offset by increases in the cost of living (Hunter, Boardman, and Saint Onge 2005). Large numbers of visitors and amenity migrants change the housing market, exacerbating inequalities and making affordable housing for long-term residents a serious problem. In many cases, the new economy depends heavily on low-wage workers, often Hispanic, for whom good social services are vital but in short supply. Here, as well, policies and practices exist that could turn these trends into opportunities, while failure to do so could create divided communities.

Third are the *remote rural "frontier" communities* that have fewer natural amenities to attract vacationers and retirees. Such remote counties include America's heartland and the long-term poverty counties in Appalachia, the Delta, the Southwest, and the northern plains. These communities are home to about 12 million people. Often such areas have relied on natural resources such as mining, timber, ranching or agriculture, or—since rural industrialization of the 1960s and 1970s—on low-wage manufacturing. Some of these communities are characterized by high poverty rates, others by chronic out-migration over the years. They tend to have weak economies. Many have experienced cyclical downturns as productivity increases and global competition intensifies in

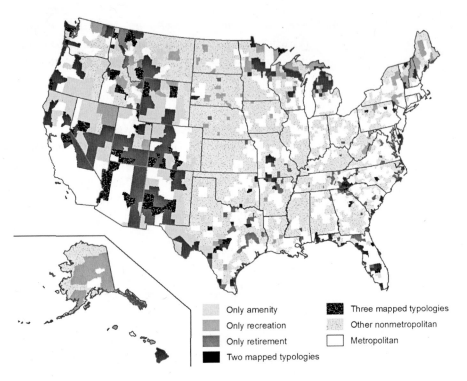

Figure 7.2
Amenity-rich rural counties.

historically important industries. Persistently poor counties are plagued by low education and high unemployment. The poverty rate in this group of communities is about 16 percent.

Patterns of Rural Settlement: Demographic Trends

Demographers William Frey (2002, 349–355) and Kenneth Johnson and John Cromartie (2005) place these rural settlement patterns in the context of larger population redistribution trends in America: urbanization, suburbanization, and interregional migration from the Rust Belt to the Sun Belt. Between 1940 and 1970, the number of people living on U.S. farms declined by 700,000 a year (Johnson 2006). The out-migration trends continue today, especially of those who are younger and better educated. This has been accompanied by growing suburbanization on the outskirts of metropolitan areas, creating sprawl and the exurbaniza-

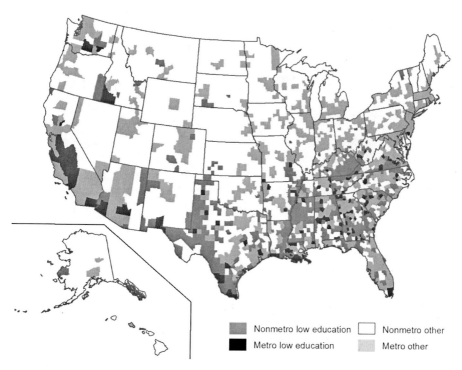

Figure 7.3
Low education disadvantages whole regions in the new economy.

tion that continues to threaten the character of rural communities on the metropolitan fringe.

The West and South's share of the nation's population grew from 44 to 58 percent between 1950 and 2000 as manufacturing jobs spread across the nation, people moved to amenity-rich areas, and, as Johnson and Cromartie put it, "public investment in transportation and communication infrastructure disproportionately benefited the South" (Johnson and Cromartie 2005, 28). Yet, much of the benefit left rural areas behind, going instead to growing metropolitan areas in the regions. Nine out of ten residents in the rural South live in low-education counties, 85 percent live in persistently poor counties, 70 percent live in low-employment counties, and 33 percent live in counties that are losing population (see figure 7.3).[1]

Today, these demographic and economic trends, which have been decades in the making, portend a critical transition for the nation's rural

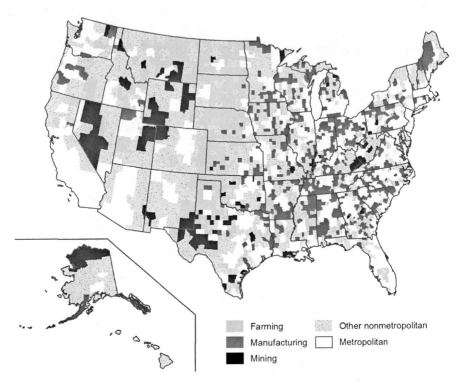

Figure 7.4
Manufacturing-dependent counties, 1998–2000.

areas. Exurban and high-amenity rural areas are experiencing new pressures of population growth and changing socioeconomic profiles, while other areas are coping with unemployment and chronic lack of opportunity as they continue to lose young people—and with them, core community capacity. In many industries on which rural communities have traditionally depended, productivity continues to improve, often bringing increased output accompanied by declining demand for labor. These industries, such as farming, timber, mining, and manufacturing, make up a shrinking share of the total economy and are no longer the economic drivers they used to be (see figure 7.4).

Globalization is increasing competitive pressures on almost every sector of the economy. U.S. firms now compete with countries that have lower land and labor costs, lower environmental standards, and in some cases, public subsidies. Jobs and smaller firms are disappearing. Importantly, the disappearance of these jobs often reflects a transition from more low-skill jobs to a concentration of a few high-skill jobs within industries (Gibbs, Kusmin, and Cromartie 2005).

This pattern is playing out in every region. In rural Appalachia, the coal industry is still producing coal, but productivity increases mean fewer jobs and fewer larger companies. In rural parts of the plains states, California's central valley, and several other areas, agriculture remains an economic engine, but production requires fewer people overall and industrial concentration is unrelenting. The rural Northeast has seen steady, permanent job losses in manufacturing over the past decade, eroding the blue-collar middle-class core in many communities. In the case of pulp and paper, production levels remain, but with fewer, more productive workers. And the rural South is coping with devastating declines in low-skill manufacturing jobs, especially in textiles and apparel, as well as the decline of the tobacco industry. These trends, which will only intensify as global trade increases, are hitting rural areas especially hard because their local economies are more specialized, their labor force is less skilled and more blue collar, and their infrastructure is less developed. Between 1997 and 2003, 1.5 million rural workers permanently lost their jobs due to fundamental changes in industries that have historically been the mainstay of the rural economy. Almost half of these losses occurred in the South. The rate of job loss is increasing as industries employ lower-wage workers outside the United States or increase their productivity by substituting capital for labor. Nearly half of all rural jobs that disappeared were in manufacturing, compared with one-third in the nation as a whole (Glasmeier and Salant 2006).

Rural Development Beyond the Agribusiness Frame

Those concerned with rural development have long lamented the fact that most rural policy at the federal level is agricultural policy benefiting agribusiness—a concern eloquently articulated in Bonnen's 1992 article, "Why Is There No Coherent U.S. Rural Policy?" (1992, 190–200). In recent years, thoughtful analysts have emerged to challenge this status quo and advocate for more appropriate rural development policy.

America was settled as a rural, resource-developing country, and for over a century agriculture has been the defining feature of rural policy. Our modest rural development programs are still within the Department of Agriculture, with the Farm Bill still the big omnibus that affects rural America. Bonnen (1992) shows how these policies, which were designed to preserve the middle class in rural America and prevent a peasantry, instead created and subsidized large agribusiness operations while undermining rural America's alternative community capacity and infrastructure building.

Recent rulings at the World Trade Organization (WTO), such as the decision that the United States cannot subsidize big agriculture's production of cotton, threaten U.S. dominance in agriculture. Meanwhile, the forces of globalization have pulled away many of the good blue-collar manufacturing jobs that fueled rural industrialization, such as food processing.

Policies to Support Working Landscapes and Equitable Development

Thus far, the regional equity perspective has been applied largely within metropolitan areas, where policy analysts, progressive community development practitioners, and organizers see the opportunities to connect jobs and housing and encourage inclusionary zoning. However, parallel applications can be identified for each of the three types of rural communities described earlier. A regional equity approach would emphasize mixed-income housing and small-to-medium-scale working landscape enterprises in exurban and high-amenity counties, while featuring sustainable stewardship of natural resources—coupled with investments in education and training—in remote frontier communities.

The Federal Reserve's Mark Drabenstott (2003) argues for a regional approach that encourages and supports entrepreneurship and small business, invests in research and technology such as pharmaceutical crops and broadband, and reorients land grants to identify and support postagriculture economic engines. David Freshwater and Eric Scorsone (2002) make a similar analysis of the changes shaping rural America and argue for partnerships between small cities and open countryside to attract investment and immigrants and to stimulate development. In their conclusion to *Challenges for Rural America in the 21st Century* (2003), David L. Brown and Louis E. Swanson note that their book's chapter authors emphasize the important role of community collaboration, even as they recognize the global challenges to its effectiveness as a rural development strategy.

In a seminal analytical piece, John Stauber (2001) proposes public policy that accomplishes three things: protects the rural middle class, reduces chronic rural poverty, and protects the natural environment.

His strategies include restructuring rural education institutions to invest in human capital in remote and poor areas, as well as investing in infrastructure that links rural areas to centers and facilitates comparative advantages to attract new immigrants to remote places.

All of these rural analysts stress the need for investment in human capital, for investment in small businesses, and for partnerships that recognize regionalism's capacity to enhance communities' strengths. Increasing education and skills is ever more important in rural areas in the face of globalization, especially in remote areas where underinvestment has occurred for decades.

Working Landscapes as a Conservation Strategy

Among the most promising policy directions for rural communities is a regional equity approach that deliberately integrates conservation and stewardship with community development. Development practitioners are doing innovative experimental work in shifting from a paradigm of promoting rural development through small-scale urbanization and rural industrialization to one of supporting the development of sustainable working landscapes. This is an essential regional equity strategy for rural America (Wells 2002).

Once primarily the focus of extraction and production activities, rural areas are increasingly the aspiration of people seeking amenity-driven recreation, vacation homes, and retirement residences. While natural resources remain a cornerstone for development, the policy challenge is to draw on these resources in ways that promote equitable development opportunities, while at the same time building commitment for investment in human capital in remote places with less promising natural resources.

Urban residents have long looked to rural areas for food, timber, and minerals. Today, more Americans recognize that stewardship of our natural resources is essential to the well-being of everyone, rural and urban alike. More ecologically sound production practices protect the health of both current and future generations. There is growing recognition that rural people can manage natural resources in a way that also provides essential urban benefits—such as clean water, clean air, and a healthy and safe food supply.

Increasingly, environmentalists are recognizing that efforts to protect natural areas and wildlife cannot be undertaken at the expense of working people who have relied on those areas for generations—whether in the northwestern timber and fishing areas, or the northeastern forests and fisheries. Policies need to encourage and support rural residents' sustainable management of the nation's watersheds, forests, farmland, waterways, and marine resources, while acknowledging their need to use these resources, in part to provide goods and services that sustain urban populations and in part to make their rural communities viable economically.

Promising but fledgling examples exist of a working-landscapes approach to rural development. Coastal Enterprises, Inc., in Maine encourages sustainable small farms, working waterfronts, and sustainable forestry. ShoreBank Pacific in Washington also pursues a sustainable fisheries strategy, encouraging small entrepreneurial fisheries ventures that can fill niches in the market. The Northern Forest Center seeks community economic development strategies, often based on culture and heritage assets in addition to the beauty of the woods, in the northern forest stretching from northeastern New York through Vermont, New Hampshire, and Maine. In North Carolina, the Conservation Fund uses mapping, planning, land acquisition, and conservation training of youth to link good stewardship with job development and ecotourism. In the South and Midwest, community forestry efforts are part of the development work undertaken by the Mountain Association for Community Economic Development in Kentucky and Rural Action in Ohio. In the Rocky Mountain West, Wallowa Resources is promoting community, forest, and watershed health. The Sonoran Institute is experimenting with conservation-based development throughout the West. The American Farmland Trust also encourages sustainable small farms, both to preserve farmland and to promote rural development. Others in the West are working to protect ranchland through a strategy that integrates working lands with stewardship.

Conclusion

While it is a promising direction for numerous regions, significant questions remain about a working-landscape strategy's potential to achieve equitable outcomes and jobs and businesses that can sustain a middle class. Many areas where deep poverty plagues rural America do not have the resources to be transformed into working landscapes. Often these places resort to hosting prisons or waste facilities that urban areas are eager to "export." In cases where jobs are scarce and high poverty is severe, drug use and even violence are growing at disturbing rates. For example, reporters say that Harlan County, Kentucky, a longtime coal-producing county, is consumed by drug problems, and the methamphetamine crisis in rural America is well reported, if not well studied and understood. And in Indian Country and some poor Mississippi Delta counties, casinos have been developed, with mostly minimal benefit to long-term residents. These new businesses present complicated scenarios for equitable community development and poverty alleviation.

Although not a solution for all rural areas that struggle with poverty, a working-landscape strategy can shift the landscape of possibilities for many communities. Such an approach can move rural development beyond constantly reminding the rest of America that rural is not just about agriculture anymore, and beyond the dead end of calling for rural America's "fair share" and maintenance in the face of declining industries. Instead of arguing for retention of old, underfunded programs based on an earlier economy when rural industrialization and big agriculture reigned ("we want roads too, we want hospitals, our elderly need services too, our schools need support too"), rural advocates can campaign for policies that build on rural communities' common strengths and comparative advantages.

Natural resources have always been rural America's defining characteristic and greatest asset. In the twenty-first century, natural resources remain vital to rural communities' future. Policies that support working landscapes integrate concerns about rural livelihoods and community resiliency with broader national and global interests in environmental stewardship and health, while also encouraging political collaborations across inner-city, suburban, and rural boundaries.

Linking with those who are concerned about environmental stewardship and health can bring about new alliances for rural development, while nurturing tourism and recreation. Such a strategy can contribute to building an economy that sustains a middle class on the signature asset of all rural areas: the natural resource base that in the twenty-first century has both productive and amenity value. This strategy can move us forward in the essential work of sustaining our natural resource base for the health of current and future generations.

Note

1. See Economic Research Service, "Measuring Rurality," for a description of these county types based on poverty, employment, and out-migration (Economic Research Service, USDA). Available at http://www.ers.usda.gov/Briefing/Rurality/Typology.

8

Katrina Is Everywhere: Lessons from the Gulf Coast

Amy Liu and Bruce Katz

In late August 2005, Hurricane Katrina crashed into the Gulf Coast, tearing through homes, office buildings, bridges, oak trees, and all things in her path. In the New Orleans area, the fury of Katrina's Category 4 impact toppled the levees, immersing whole swaths of urban and suburban neighborhoods along Lake Pontchartrain in deep floodwaters. In all, the hurricane wrought severe damage in the New Orleans area and in towns large and small throughout Louisiana, Mississippi, and parts of Alabama. More than one million families were immediately left homeless, more than one hundred thousand homes and businesses were destroyed, and the death toll exceeded one thousand persons.[1] Within days, Hurricane Katrina, with its 145-mile-per-hour winds, had become the nation's worst natural disaster since tragedy struck Galveston, Texas, in 1900.

For years to come, the chaotic response to Hurricane Katrina will serve as a case study and cautionary tale for government, nonprofit, and private sector leaders about the challenges of preparing for emergencies and disasters. But Hurricane Katrina also exposed a social disaster, one that often escapes attention. In vivid television images, millions of viewers across the United States and the globe watched scenes from New Orleans as African-American families and young children scrambled to evacuate with their few belongings. Black families sat stranded on their rooftops waiting to be rescued. Officials shuttled families, children, and the elderly—nearly all African-Americans—into the city's convention center and football stadium for temporary shelter, only to have conditions worsen there. What went wrong? Why were there so many families with so few resources or so few options to evacuate? And why were these families—the poor, the working class, and people of color—most disproportionately impacted by the storm?

A quick scan of Census data confirmed the social and racial disparities in New Orleans. In 2000, more than two-thirds of the residents in the city of New Orleans were African-American. Of those, nearly one in three lived below the federal poverty line, while less than 5 percent of white New Orleanians were poor. Further, more than one in three African-American households (or 39,340 families) in the city lacked access to a car, compared with only 15 percent of whites, in part explaining the severely uneven prestorm evacuations.[2]

These disparities are not unique to New Orleans. Nationwide, local leaders from community organizers to chambers of commerce struggle to close the gap in wealth and opportunities among families and among communities. Often simple morality motivates efforts to address regional inequities: it is just the right thing to do. Yet, leaders are also beginning to recognize that metro areas will not reach their full *economic* potential if disparities by race, class, and place persist. Reducing regional disparities is essential to achieving economic prosperity.

This chapter discusses how poverty and unequal opportunity are not unique to New Orleans but persist in many regions around the country. It shows how regional disparities—often depicted as the pattern of concentrated poverty in the core coupled with the out-migration of people and jobs to surrounding suburbs—play out in many U.S. metro areas. This regional growth and development pattern undermines opportunities for city residents, exacts fiscal and economic costs to the region, and can hurt the overall regional quality of life.

However, in New Orleans and in other metro areas, these trends are not inevitable. In fact, an array of state and federal policies has helped facilitate these uneven growth patterns and can, in turn, be used as powerful tools to help address regional inequalities and build prosperous, inclusive communities. This chapter closes with a review of promising efforts by federal, state, and local leaders to do just that. Still, more progress must be made and additional structural reforms must be embraced to build truly economically vibrant communities that provide better opportunities for all residents, not just in New Orleans but for metropolitan areas across the nation.

Regional Disparities Persist

The way in which our cities and metro areas grow—in social, economic, and physical terms—shapes the quality of opportunities provided to their residents and businesses. A closer look at patterns in New Orleans before the hurricane struck, and in metro areas across the country, shows that sharp divisions by race, class, and geography still exist in many U.S. metro areas, limiting opportunities for families and hampering the economic potential of communities.

New Orleans Before the Storm

In 2000, the New Orleans area was growing in ways quite similar to many U.S. metro areas. First, pockets of severe poverty were found in the urban core. Among the fifty largest cities, New Orleans had the second-highest concentration of poverty in the country in 2000. Approximately 38 percent of the city's poor residents lived in extremely poor neighborhoods.[3]

As poverty hardened over the years in the heart of the city, middle-class families—including black households—and jobs moved out, especially to the surrounding parishes. For instance, the New Orleans region saw a net out-migration of African-Americans in the 1990s, topping off a four-decade trend (Frey 2004). But the city's "white flight" was more dramatic. According to the census, New Orleans lost more than half of its white residents between 1970 and 2000, while its share of African-Americans grew by 27 percent.

The upshot: the city population shrank by 18 percent between 1970 and 2000. Meanwhile, suburban Jefferson Parish doubled in population to match that of its neighboring central city. And New Orleans flipped from being 45 percent African-American in 1970 to being more than two-thirds black by 2000. Not surprisingly, jobs also suburbanized. In 1970, New Orleans was home to two-thirds of the region's jobs, but by 2000, that share had dropped to less than half, at 42 percent.

This self-reinforcing cycle of poverty concentrations and suburbanization, coupled with a stagnant, predominantly low-wage economy in greater New Orleans, helped create wide social and economic disparities among groups of residents in the region. Most fundamentally, the city had become residentially segregated. By 2000, New Orleans—a racially integrated community in the 1950s—had become a place divided by race, with the typical African-American living in a neighborhood that was 82 percent black.

Social and economic divisions also emerged. For instance, in 2000, the median household income for an African-American in the city was just half that of whites—$21,461 versus $40,390. Only 13 percent of black adults had a college degree or higher, compared to 48 percent of whites. The majority of white households in New Orleans were homeowners, at 56 percent, while only 41 percent of black households owned their own homes. Beyond the color lines, those who lived in extremely poor neighborhoods possessed far lower household incomes, engaged less in the labor force, and were less likely to be college-educated than those living in other neighborhoods.

New Orleans's uneven growth patterns helped explain the disparate impacts of the storm. When Hurricane Katrina hit land in late August 2005, those most affected by the storm were low-income families, the working poor, and African-American families who lived in isolated neighborhoods located in the lowest-lying parts of the city.

Beyond New Orleans: Regional Disparities in U.S. Metropolitan Areas

Such uneven growth patterns are not unique to New Orleans. Regional disparities exist in many metro areas, albeit at varying degrees (see table 8.1).

Further, central cities continue to shoulder the burden of housing the nation's poorest neighborhoods. Of the nearly 8 million people living in concentrated poverty in 2000, approximately six million, or 75 percent, lived in the largest cities.

Decentralization of People and Jobs to the Suburbs

Despite the concentrations of low-income residents within their borders, cities generally did well in the 1990s, adding both residents and jobs, for the first time in decades for some. Yet, suburban growth still dwarfed these overall city strides.

Granted, the 1990s proved to be a good decade for city residential and economic growth. The population of the top one hundred cities grew by 9 percent during the 1990s compared with 6.3 percent during the 1980s. Seventy-four of the largest cities added population during the past decade, compared with 62 in the 1980s. These population gains have continued. The latest population estimates from the Census show that approximately two-thirds of the nation's largest cities added new residents between 2000 and 2004 (Frey 2005). Cities similarly exhibited economic strength in the 1990s. Of the 114 large cities tracked by the

Table 8.1
Top Ten Cities in Concentrated Poverty, 2000

City	Concentrated Poverty Rate*—Total	Concentrated Poverty Rate—Blacks	Extreme-Poverty Neighborhoods**
Fresno, CA	43.5	44.9	22
New Orleans, LA	37.7	42.6	47
Louisville, KY***	36.7	53.2	11
Miami, FL	36.4	67.6	23
Atlanta, GA	35.8	41.0	28
Long Beach, CA	30.7	26.8	17
Cleveland, OH	29.8	35.6	52
Philadelphia, PA	27.9	27.1	54
Milwaukee, WI	27.0	39.3	42
New York, NY	25.9	33.7	248
U.S. Total	10.3	18.6	2,510

* The concentrated poverty rate reflects the proportion of all poor people city-wide who lived in extreme-poverty neighborhoods. ** Extreme-poverty neighborhoods had more than 40 percent of their residents living below the federal poverty threshold in 2000. *** Louisville, KY, defined as of Census 2000, prior to its merger with surrounding Jefferson County, KY. Source: Census 2000.

U.S. Department of Housing and Urban Development, 102 of them experienced at least modest job growth between 1992 and 2001.

For all this overall strength in population and job growth in the nation's cities, the suburbs continued to capture greater shares of economic and residential life. In the largest metro areas, the suburbs grew at twice the population rate of central cities over the decade—18 percent versus 9 percent. Even Sun Belt cities like Phoenix, Dallas, and Houston were eclipsed by their suburbs.

Like people, jobs have also continued to shift outward into the suburbs. Across the largest one hundred metropolitan areas, on average only 22 percent of jobs were located within three miles of the city center; meanwhile, more than 35 percent of jobs were situated more than ten miles away from the central core.

The added complexity is that as more and more families and jobs have moved to the suburbs, American suburbs are no longer homogeneous, muddying the simple city-versus-suburb dichotomy. Suburbs are now home to more people of color and low-income and working families, exhibiting more of the social and economic traits of cities (Frey 2001; Berube and Forman 2001).

Poverty Concentration and Decentralization Sharpened Disparities by Race, Class, and Space

Despite the increasing diversity of the nation's suburbs, many metro areas remain sharply divided by race, ethnicity, and class. The adverse effect of these growth patterns is that many low-income families live severely segregated from opportunity. Families, particularly in the cities, lack opportunities to access good jobs, raise their children in good schools, and build wealth.

While both jobs and people have moved increasingly to the suburbs, this has not translated to improved job accessibility for blacks and Latinos. A spatial mismatch still persists between where blacks and Latinos live and where jobs are located.

Living in neighborhoods isolated from the rest of the community not only affects workers but also affects children. Many studies have shown that schools predominantly composed of low-income students are at a greater risk of failing than more economically integrated schools, as exhibited by low standardized test results and high dropout rates.

Segregated neighborhoods affect a family's ability to accumulate wealth, especially as expressed in the value of one's home. Not surprisingly, there is a wide wealth gap among race and ethnic groups across the country. According to a recent report on wealth disparities by the Pew Hispanic Center, while white households had a median net worth of $88,651 in 2002, black households had a net worth of only $5,988 and Hispanics only $7,932 (Kochhar 2004).

No doubt, the level of minority homeownership has improved over the years, jumping from 42 percent in 1995 to just over 50 percent in 2004.[4] But African-American homeowners own homes that have less market value than homes owned by whites. In 2000, the median home value for black homeowners was $80,600, but the median value was $122,800 for homes owned by whites. The difference between home values is not simply because African-Americans can afford less of a home than whites. Median home values vary even for black and white households in the same income bracket.[5]

In short, uneven growth and development patterns remain too common in our cities and metro areas, including in New Orleans, isolating low-income and minority households from wealth-building opportunities. Leaving these disparities unchecked may unleash serious fiscal and economic consequences for metropolitan areas.

Reducing Regional Disparities Matters to Economic Competitiveness

For many residents of New Orleans and other cities, the neglect and isolation from opportunity is an injustice. They argue that low-income families, working-poor families, and families of color should have access to more fair and decent opportunities.

That is undeniable. Yet, the benefits of addressing disparities go further than helping disadvantaged families. The presence of regional disparities by race, class, and space can undermine the fiscal and economic health of whole regions. Promoting greater regional equity is not just the honorable thing to do; it also makes sound fiscal and economic sense (Sohmer 2005).

A mix of new and long-standing academic evidence makes the case for this emerging understanding of the link between regional equity, economic competitiveness, and fiscal efficiency.

Reducing Disparities Is Fiscally Smart

A host of academic literature finds that alleviating poverty or lifting the prospects of those at the bottom of the income ladder can reduce public costs as well as generate more revenue for localities.

Alleviating Poverty Increases Incomes and Expands the Tax Base

While there are many strategies to reduce extreme poverty, efforts to increase educational attainment rates will have the biggest impacts on income levels, which in turn contribute to a stronger tax base and greater local revenues. In particular, studies have shown that the more you learn, the more you earn. According to a Census report, a person with a college degree will earn an estimated $1 million more over a lifetime than a person with only a high school degree (Cheeseman Day and Newberger 2002). The reason: an educated worker will have greater access to highly skilled, higher-paid jobs than those with less formal education.

Alleviating Poverty Reduces Taxpayer Costs

Similarly, as overall poverty and concentrated poverty is reduced, this reduces the direct and indirect costs that taxpayers have to pay for services associated with poverty (Pack 1998, 1995–2019; Joassart-Marcelli, Musso, and Wolch 2005, 336–356; Gyorko and Summers 1997). Direct costs include the provision of welfare and health care, which are increasingly being passed on to city taxpayers as federal and state payments for these services are diminished (Joassart-Marcelli, Musson, and Wolch 2005). An even larger burden to localities is the indirect costs of poverty—court costs, police, fire, and general administration—that often incur high expenses in high-poverty cities (Pack 1998).

Closing Disparities Increases Consumer Spending in the Local Economy

Closing income and economic disparities among racial and ethnic groups would result in higher incomes for households, translating to more dollars circulating in the economy. One study goes further, finding that reducing income disparities would also result in large gains in minority purchasing power. The report by the Minority Business Development Agency at the U.S. Department of Commerce projects minority purchasing power through the year 2045 under two scenarios—one maintaining current income disparity levels and one where income disparities are eliminated. Currently, with existing disparities, the minority share of the country's total purchasing power is 25.8 percent, but with no disparities it would jump to 30.4 percent. By 2045, the minority share of purchasing power would improve upon the status quo by 13 percentage points if disparities were closed (U.S. Department of Commerce Minority Business Development Agency 2000).

Reducing Spatial Disparities Saves on Costs from Services, Infrastructure, and Disinvestment

Half a century of academic literature has shown that sprawling development patterns impose greater net fiscal costs than more compact ones. Low-density development creates greater demand for more infrastructure such as schools, roads, water, and sewer extensions, while also generating greater demand for police, fire protection, and emergency medical services. However, on the flip side, abandoning older communities also imposes great fiscal costs to taxpayers. Especially in slow-growing metropolitan areas where families tend to "move up and out" to new housing being developed in the suburban and exurban fringe, the net result is abandonment of existing homes in the urban core. With increased housing vacancies and little to no growth in property values, many older communities are stuck with reduced revenues, higher taxes, and overstressed services. The lesson: reducing spatial disparities and growing metropolitan areas in more compact ways with healthier urban cores can save money for taxpayers as well as local governments.

Reducing Disparities Promotes Economic Competitiveness

Beyond helping to improve the fiscal health of localities, research also finds that closing social, economic, and spatial disparities can improve a region's overall economic performance.

Reducing Disparities Strengthens the Future Workforce

Concentrated poverty limits educational opportunities for children, especially as they grow to become a community's future workers. According to sociologist M. Corcoran, if parental income is low, children are more likely to attend fewer years of school, more likely to not graduate from high school, and more likely to not attend college (Corcoran 1995, 237–267). The same study reiterated how the challenges associated with poverty can be passed on through generations. For instance, a black child growing up in poverty is 2.5 times as likely to become a poor adult than a black child growing up in a nonpoor household. Thus, addressing childhood poverty and concentrated poverty will help to improve the educational opportunities of children and ensure that regions can begin to prepare for a more skilled workforce, especially as the nation's baby boomers begin to retire in 2010.

Reducing Disparities Bolsters Suburban Wealth

Reducing poverty in the central city is good for families, for a city's fiscal health, and for generating a region's workforce. It can also pay direct dividends to its surrounding suburbs (Muro and Puentes 2004). Economist Richard Voith has found that, in nearly all metro areas, income gains in central cities correlates with increases in the population, income levels, and housing prices in neighboring jurisdictions. Similarly, Manuel Pastor and his colleagues' review of trends in seventy-four metro areas found that reducing poverty in central cities led to income growth for the whole metropolitan area (Pastor et al. 2000).

Reducing Spatial Disparities Improves Economic Productivity

There is a host of academic literature about the economic value of clustering firms and jobs in a region, rather than producing a low-density spread of employment opportunities. For instance, doubling employment density can lead to a 6 percent increase in labor productivity. Higher-density employment centers have also been linked to increased patent activities, an indicator of innovation. Overall, dense labor markets, efficient transportation systems, and high clustering of jobs bring about positive knowledge spillovers, both within and across industries (Muro and Puentes 2004).

In sum, there are plenty of good fiscal and economic reasons to try to reduce regional disparities.

Reforming Federal and State Policies Can Help in Producing Regional Equity

Addressing the geographic, racial, and income inequalities in a region is possible because the factors that have contributed to these growth patterns are not inevitable. Federal and state policies set the rules for how housing, jobs, and opportunity are distributed in a region. For the most part, these policies have facilitated the decentralization of the economy while concentrating low-income families in cities, ultimately undermining the value of other programs aimed to assist them.

In the broadest and most fundamental sense, the federal government sets the larger context for demographic, economic, and income growth in the United States. Federal action on monetary policy and immigration policy affects the pattern of investment and migration in cities and metro areas. Federal tax policies to support working families, as well as minimum wage laws, help frame the opportunity structure for workers in the United States. All these policies affect regional equity.

States play an even more direct and multipronged role in the distribution of growth and opportunities within metro areas.

First, states set the geography of governance. They decide the number of general-purpose local governments and whether such governments have fixed or movable borders. As David Rusk and others have documented, the higher the number of local governments with "fixed" borders in a region, the higher the economic and racial segregation in a community.

Second, states set the powers of local governance, including the types of responsibilities to be empowered at each level of government—local, county, or even regional entities. For instance, state rules governing local land use planning would affect regional equity, as would the elimination of local exclusionary zoning.

Third, states establish the fiscal playing field for municipalities and school districts. They decide the form of taxes that municipalities can impose and determine whether to equalize tax revenues across jurisdictions or across school districts.

Fourth, states help set the backbone for city and metropolitan growth with their investments in roads, infrastructure, main streets, downtowns, public parks, and green spaces. Their economic development subsidies—particularly those favoring urban reinvestment, infill development, or greenfield development—also do much to shape the locus of opportunities in a region.

Finally, states are a major player when it comes to human capital investments, which are crucial to closing social and economic disparities. They govern the state's systems of K–12 education, higher education, and workforce development, and particularly how those investments meet changing demographics and link to the changing needs of the economy.

In short, federal and state policies set the rules of the development game, dictate the geography for decision making, and, at the state level, are major investors in economic development and human capital. The effect of all these policies is to favor low-tax, high-service suburban areas for investment and opportunity over fiscally stressed urban cores, further exacerbating concentrated poverty there.

Promising Progress Exists but Continued Reforms Are Needed

Since the 1990s, federal and state policy makers have made some gains in addressing the inequitable growth patterns and opportunities found in the nation's cities and metropolitan areas. Furthermore, there is a growing recognition, especially at the state and regional level, that doing so will enhance state and regional economic competitiveness.

Federal transportation policies, beginning in the early 1990s, have devolved greater responsibility for planning and implementation to metropolitan planning organizations and have given states greater flexibility to apply highway funds toward public transit and other modes of transportation. Federal money spent on transit almost doubled during the 1990s. New light-rail systems are being built in metro areas as diverse as Salt Lake City, Denver, Dallas, and Charlotte. The shift to greater regional decision making and investments in broader transportation alternatives will enhance a region's ability to meet the needs of low-income workers and ensure they have access to jobs, no matter where they are located.

With the expansion of Low Income Housing Tax Credits and Section 8 housing vouchers, federal housing policies are doing a better job of increasing the supply of affordable housing throughout a region, while also undoing the damage of concentrated poverty caused by failed public housing policies.

In the mid-1990s, the federal government doubled the annual size of the Earned Income Tax Credit (EITC) to more than $30 billion, making it the nation's largest antipoverty program.

Similar progress can be seen at the state level. In 1998, the State of New Jersey put in place a set of more streamlined building codes to spur the rehabilitation of older buildings in cities. The success of that state provision in promoting urban redevelopment spurred other states to adopt similar provisions. More recently, Governor Ed Rendell in Pennsylvania and former governor Mitt Romney in Massachusetts have embraced "Fix It First" policies in transportation and are applying the same principles to other investments, such as parks and economic development. In response to private and nonprofit interests in boosting the assets of working families, fifteen states and the District of Columbia in 2003 put in place their own versions of the EITC to supplement the federal program.

Moreover, regional business and civic leaders are beginning to understand that closing regional disparities by race, class, and geography will go a long way in bolstering state and regional economic competitiveness.

In 2003, the Human Services Coalition in Miami-Dade County launched a Prosperity Campaign to raise the assets of low-income workers by connecting them to existing federal benefits, such as the EITC and the child-care tax credit. That year, the Prosperity Campaign generated $62 million in revenue for working families and the local economy. Since then, chambers of commerce and financial services institutions have joined the campaign by promoting free tax-preparation services and the availability of existing federal benefits for their workers.

The program was seen as a "win-win for employers, employees, and the local economy, gaining the support of the business community."[6] The success of the Prosperity Campaign in the Miami area fueled the creation of six other prosperity campaigns across the state.

In 2005, the newly formed Itasca Group of forty business leaders in Minneapolis–St. Paul named the addressing of regional disparities a top priority. Members of the business group realized that while the Twin Cities is, by all accounts, a successful economic region, persistent disparities by race, ethnicity, and class may be undermining their region's future competitive position, especially in the quality of its human capital. In sum, these promising reforms signal the growing awareness by policy makers and regional civic leaders that investing in underserved people and communities can and must be dealt with head-on, not only to improve the prospects of families and workers but also to prepare a state and region for a stronger economic future.

Unfortunately, ongoing efforts are in motion to reverse these general accomplishments at a time when more systemic reforms are needed. Since 2000, the federal government has advocated for the elimination of funding to build mixed-income housing and to reduce the size and effectiveness of the affordable housing voucher program. The federal government has also called for reducing the federal match for mass transit, while disinvesting in buses and reducing requirements in clean fuels. Further, federal programs that support working-poor families are under attack, with cuts in funding for Medicaid, food stamps, and child care. At the state level, the vast majority of departments of transportation have yet to carry out the full promise and vision of the 1991 federal transportation law. Also, states still invest the bulk of their economic development dollars in greenfields far from the urban core. The bottom line: it remains essential to protect and advance meaningful progress on regional equity, with deeper policy reforms and even new political coalitions.

Conclusion

While public attention to Hurricane Katrina and rebuilding the Gulf Coast region may wane over time, the daily challenges faced by disadvantaged Americans and communities across the country will remain and cannot be ignored. Federal, state, and local leaders must use the momentum provided by Hurricane Katrina to not just evaluate disaster preparedness but to examine the structural challenges that keep segments of our communities from fully taking part in the nation's prosperity.

The work to create a more equitable society is a long-term fight. As noted here, there are promising strides that have been made at the federal, state, and regional levels to address the racial, class, and spatial disparities found in our nation's metropolitan areas. Yet, New Orleans reminds us that more persistent and fundamental reforms are needed to ensure that low-income and working-poor families have access to better opportunities. By doing so, the gains will not only be enjoyed by these families but will also yield enormous returns to the overall economic health of metropolitan areas.

Notes

1. See "Katrina's Death Toll Climbs Past 1,000" in the Associated Press (September 22, 2005); and FEMA Update: Hurricane Katrina Update (September 13, 2005).

2. Unless otherwise noted, all New Orleans statistics are from the U.S. Census Bureau, 2000 Decennial Census.

3. For a detailed discussion of the demographic trends before Hurricane Katrina, see "New Orleans after the Storm: Lessons from the Past, a Plan for the Future" (Washington: Brookings Institution, 2005).

4. Despite high minority homeownership levels, homeownership rates between whites and minorities remain wide, further explaining the wealth gap. Approximately 51 percent of black households owned their homes in 2004, while an impressive 78 percent of white households did. Joint Center for Housing Studies of Harvard University, "The State of the Nation's Housing 2005" (Cambridge: Harvard University, 2005).

5. For instance, the median home value for white households earning between $35,000 and $79,999 is $135,300. However, black homeowners in the same income bracket have a median house value of $105,000—a $30,300 difference. These calculations were made using the Census's Advanced Query System, which allows the user to create cross tabulations normally not available through SF1 and SF3 reporting and mapping.

6. A Community and an Economy That Work for All: The Florida Prosperity Campaign. Available at www.prosperitycampaign.com/greatermiami.html.

9

Growing Together, or Growing Apart?
Central Labor Councils and Regional Equity

Amy B. Dean

Introduction

The decline in union membership in recent decades has often been cited as evidence that American unions have outgrown their usefulness and are now obsolete. This perception is widely off the mark. Organized labor remains a key player in the growing movement for equity on a metropolitan scale. In fact, Labor is the only worker-centric group in America that still has the capacity (in terms of members, allies, and potential for influence) to ensure that its concerns are not permanently ignored. What is more, Labor's ability to act on the regional level is a critical asset in the fight for regional equity. This is important because actions at the local and regional level will help build public support and pressure for a renewal of the American social contract that globalization and modern business practices have recently shredded.

One need not be an urban geographer to understand the growing importance of regions. In recent years, mainstream corporate leaders across the country have put considerable effort toward building business-oriented civic coalitions capable of exercising power on the regional level. They have done so, in part, because businesses often seek to influence public policies in ways that extend past the municipal boundaries of any one small borough, town, city, or county. In dozens of communities across the country, business executives have responded to this need by forming ad-hoc, cross-industry coalitions (usually misleadingly portrayed as civic-minded groups) that often have a disproportionate influence on political processes in the area in which they operate. In most cases, participation in these groups by citizens without vested interests is greatly limited. Instead, the agenda for these groups are more typically determined by the needs of the businesses that dominate them rather than by the concerns of the local population.

Organized labor is one of the few remaining entities in a position to contest for power with these increasingly influential bastions of corporate influence. Labor has long been organized on a regional basis through Central Labor Councils that are essentially coalitions of the labor unions operating in an area. As such, they can play a valuable role as a counterpoint to the regional influence that has been smartly cultivated by the business community. By acting regionally on behalf of an entire community, Labor has the opportunity to become a spokesperson for the aspirations of its community, which is something that is not possible when Labor focuses only on more narrow concerns, such as negotiating contracts one at a time.

Working Families Excluded from Regional Collaboration

In 1993, the executive board of the South Bay AFL-CIO Labor Council, the Central Labor Council for the Silicon Valley region, went through a planning process to identify strategies to build power for all working people in the Silicon Valley. At the time, organized labor had niche power in Silicon Valley in a limited number of industries. A handful of unions had enough influence with groups such as the chamber of commerce to help their own members. But for the most part, Labor was seen primarily as a special interest, one that represented just 15 percent of the local workforce, with most of its members in so-called old-economy jobs.

Working families had no effective local lobby, particularly on issues of economic development. Instead, local government officials and their agencies were more focused on meeting the demands of the area's well-organized, world-famous, high-tech industry. At the time, most discussions about local economic policy (including taxes, development, and transportation) were dominated by a handful of powerful and influential business-oriented groups. These included the Silicon Valley Manufacturers Group (since renamed the Silicon Valley Leadership Group) and Joint Venture: Silicon Valley Network—a collaborative effort between business and government leaders.

Similar collaborations between business and government that exclude representatives of working families are becoming increasingly common in the United States. Corporate leaders understand that regional competitiveness is affected by public policies related to issues such as land use, taxation, development, and permitting that are determined by often-balkanized groups of local jurisdictions, including cities, towns, and counties. Unfortunately, most of the business-dominated groups formed to address these bureaucratic obstacles to their success often pursue competitiveness for its own sake. Frequently, and despite growing evidence to the contrary, they operate under the mistaken notion that economic competitiveness will translate into a healthy economy. The focus on competitiveness to the exclusion of all other concerns often has very negative consequences on the overall social welfare in a community.

In Silicon Valley, this pattern was well established by the early 1990s. The vast majority of local policy makers were not aware of the most significant issues that local working families confronted, including growing evidence of a lack of upward mobility among the ranks of temporary workers, the area's fastest-growing job sector. "Until the late 1990s, the general public bought into the potent myth that technological and meteoric economic growth had generated universal affluence for the Valley's residents.[1]

Creating a New Agenda for Regional Action in Silicon Valley

In 1995, our South Bay AFL-CIO Labor Council started Working Partnerships USA (WPUSA) with the primary goal of helping organized labor become a more powerful and effective force in our community. In time, the accomplishments of this 501(c)(3) nonprofit revealed a powerful strategy that enabled a small group of motivated activists to shift the debate, increase social and economic equity in a metropolitan area, and enact reforms that have created building blocks for change at the state and federal level.

This organization began as an attempt to increase power for organized labor in Silicon Valley. The model it created is particularly useful for those involved in the movement for metropolitan equity because its inclusive structure and practices build power from the ground up, not the top down. What has made the model particularly powerful is the way it has enabled a shared understanding of the true conditions in a local economy that impact working families, provides effective community education about those issues, and develops motivated civic leaders, arming them with the data they need to highlight and remedy critical unmet needs, most notably, bringing economic development efforts into line with identified community aspirations.

Just prior to its founding, and immediately after its inception, our Central Labor Council was invited by several other groups, including the Silicon Valley Chamber of Commerce, the Silicon Valley Manufacturers Group, and Joint Venture: Silicon Valley Network, to join their economic development efforts. Our leadership team welcomed those invitations and participated whenever we could. It rapidly became clear that none of these other previously existing "insider" groups had placed the concerns of working people at the top of their agendas. One was interested primarily in reducing regulations on business and streamlining local permitting procedures. Another focused its efforts on improving the regional transportation system so workers could commute more easily. The business-dominated groups were also, not surprisingly, always keenly interested in cutting taxes on business.

The only way organized labor would be able to get the concerns of working people at the top of the local public policy agenda would be to put them there ourselves. Our Central Labor Council could not take a seat at someone else's table and have any hope of controlling what would be served. This is true in many other settings where similar business-oriented groups are present. They will often invite token representation from labor and community groups. But we knew we had to get beyond tokenism to have any hope of advancing an agenda for working people in Silicon Valley. What's more, our Silicon Valley organized labor leadership team also knew that we would never be able to demonstrate that organized labor was a capable steward of our local economy if we were just a small subset of some other group. That left organized labor with no other choice: we created Silicon Valley's first public policy research and action group where working people controlled the agenda.

In addition, we knew we needed to develop a strategy to build political power for the long haul. This would enable us to win contests with the insider groups who dominated the local public policy apparatus. That meant we had to develop new capacities and new alliances. To be taken seriously, we had to build a community of activists who would understand the nuances of municipal budgets at least as well, if not better, than the city managers who often dictated economic development policies. Likewise, we wanted to help the same group of activists understand redevelopment regulations better than the experts that big corporations hire. Those experts often create projects that work out well for their already prosperous clients, but not for the community.

A Three-Part, Power-Building Strategy

Federal and state economic researchers, planners, and economists typically don't focus on specific towns or regions. Instead, they usually concentrate on aggregate numbers and much larger population groups. The lack of data about what is happening in specific neighborhoods and communities deprives regional economic planners and policy makers of the most critical data they need, the information that, if assembled, could help galvanize more effective public policies. Instead, municipal planners and policy makers typically rely on aggregate statewide or national figures and statistics. But no one lives in the aggregate. Instead, we all live as individuals and in families and in communities that usually have their own very special, unique, and particular sets of circumstances. This research and information gap often has a very negative impact on the ability of communities to take actions that meet the needs of local residents.

WPUSA developed and implemented a three-part strategy for social change that has helped overcome the damaging effects of the lack of data on local economic conditions. One key result was that, between 1993 and 2003, friends of organized labor and working families managed to capture a majority of city council seats in the two largest cities in the area, San Jose and Sunnyvale. These victories led directly to a number of other policy wins including providing free health-care insurance to all the previously uninsured children in our county, the construction of more affordable housing, the preservation and expansion of transit services in low-income areas, the enactment of a living-wage ordinance, and more detailed and routine analysis of the tangible community benefits—or lack thereof—derived from a wide variety of official economic development programs. Most critically, the work initiated by organized labor in Silicon Valley demonstrated that the success of businesses did not, by itself, mean we had a healthier and more economically secure community.

Policy Analysis and Research
The first part of the strategy, policy analysis and research, helped unmask and highlight economic realities that were formerly overlooked, hidden, or downplayed. For example, our research revealed that many

people were being trained to work in occupations that offered little or no hope of meaningful advancement. We thought the goal of economic development should be to build an economy with greater levels of income security rather than to train people to be pushed into dead-end jobs. But we also knew accomplishing that task meant we would have to bring the community together to understand a complex and interrelated set of issues and then work in common purpose to implement solutions that fit the problems.

Organizing and Coalition Building

The second step was transferring the knowledge, data, information, and ideas we generated to grassroots neighborhood opinion leaders, religious organizations, and community groups and organizing them into powerful new coalitions. We did this through our innovative leadership development institutes, conducted in partnership with San Jose State University. The institutes bring community leaders and concerned residents together with researchers and scholars to learn about, understand, and discuss the economic and political forces that have an impact on the quality of their lives and the health of their neighborhoods. In 1997, WPUSA established its Leadership Institute, which brings together civic leaders, academic experts, union leaders, community activists, clergy, and elected officials in a nine-week course that examines the Silicon Valley economy through a social justice lens. The goal is to create a civic counterpart to business leadership groups that could forge an inclusive, long-term agenda for the well-being of working people and families in poverty.

We wanted to build a public education and outreach program that would bring the results of our research to the attention of a wider audience in our community—in essence, to popularize our work. We knew, from bitter previous experience, that we could not count on the local mainstream media for much help in this respect, given its dependence on big corporate advertising dollars. So, we did our communications the old-fashioned way: person to person. Our goal was to educate others so they understood what was really happening to jobs and industries in our region. We also sensed that an ongoing, grass roots–based public outreach and education campaign would help us cultivate and recruit allies in the community before they were needed for any specific purpose.

Political Action

The third and final step of a power-building strategy is effective political action, including using the electoral process, which research and coalition building facilitates. When a community is successfully informed and organized, the result, not surprisingly, is an informed and organized community that is capable of making real changes. In accordance with the law, WPUSA does not participate directly in politics, but it does provide the research and helps build the leadership capacity of our allied groups to increase civic participation in greater numbers.

From the start, we knew we could not accomplish these tasks on our own. This is why one of our first moves was to put together a group of advisors that included members of groups with which we wanted to develop stronger partnerships in the future, including local religious leaders, community activists, and scholars. In essence, we formalized our preexisting relationships while continually bringing new participants into our circle.

Responding to the Business Community

In 1997, Joint Venture: Silicon Valley Network released a highly publicized and optimistic report that claimed that median household income was rising in the region. The report implied that the local economy was taking care of itself, to the benefit of everyone. We suspected what was really happening was that increasing numbers of working people were living together under the same roof. But how could we possibly make that case?

One of our earliest and most fruitful steps helped us answer that question. Working with our partner organizations and advisors by providing the necessary staffing, we constructed a Community Economic Blueprint, which essentially became a road map for what our community coalition would do. The document helped us achieve consensus and buy-in from others, including a shared agreement about the pressing need to document the real status of working people in Silicon Valley.

We understood, from watching the way the business community had operated, that whoever frames the debate, particularly with solid, unimpeachable research, often gets to take the lead when it comes to suggesting solutions. We wanted to be in the driver's seat when it came to formulating local economic policy, and we accomplished that by

conducting the research and producing the reports that helped us take
the wheel.

In the process, we advanced a new school of public policy that, in the
end, brought together union activists, the religious community, neighborhood groups, and environmentalists, among others, and helped them to
cultivate the power they needed to preserve social progress against the
wars of attrition that often set back progressive causes. In short, organized labor in Silicon Valley realized the need to create a permanent
new local force that could stand up for the needs of our community,
even on short notice. Over time, the members of our coalition also developed a shared understanding that building an ability to rapidly respond
to sudden and unexpected assaults on social, economic, and environmental progress was critical to advancing the goals and preserving improvements made by all the groups who participated.

Creating a Model for Other Communities: Key Reports that Led to Action

In May 1996, WPUSA published "Shock Absorbers in the Flexible Economy," a report that put the issue of nonstandard work on the table
for the first time in Silicon Valley. Before our work was published, the
region's economic boom and its role as the so-called engine of the new
economy were virtually all that anyone talked about. "The document
revealed the hardships for working families associated with temporary
work," notes Bob Brownstein, founding and current director of policy
and research at WPUSA. "It described the low pay, the lack of health
insurance, no pensions, and the often unreasonable administrative practices." The report documented the growing divide between rich and poor
in Silicon Valley and highlighted the growth of poverty-wage and contingent jobs.

The outsourcing section of that first WPUSA report showed clearly,
and for the first time, how the relationship between corporate success
and the well-being of workers was disintegrating during what was supposed to be Silicon Valley's most prosperous period ever. To be sure,
the business pages had it partly right: the area's companies were thriving,
but the fastest growing among them were the outsourcing firms, and
those workers were getting lower pay and few, if any, benefits.

WPUSA's second report, "Growing Together or Drifting Apart?
Working Families and Business in the New Economy," drilled deeper.

Published in 1998, the results revealed that only Silicon Valley's highest-income workers (those at or above the 90th percentile) were earning real wages, adjusted for inflation, above those earned in 1989. Meanwhile, workers on the lower end of the economic spectrum, at the 10th and 25th percentile, had suffered a real wage decline of more than 13 percent since 1989 (Benner 1998, 34). What's more, the data showed that the workers hurt the most during this period tended to be nonwhite and/or those with a high school education or less. A new, more accurate, picture was beginning to emerge: high-tech economy business patterns were deepening rather than reducing the most stubborn social divisions, particularly those related to race, gender, and ethnicity.

Subsequent research and reports by our group focused attention on other critical issues, such as the lack of affordable housing in our region and the problems encountered by immigrants, women, older workers, and families with small children. These reports gave us material we could disseminate and popularize. In his review of our work in 1999, *Sacramento Bee* columnist Peter Schrag wrote that the reports

reinforced the growing evidence that beneath the much-celebrated boom in California's high-tech economy—behind each of those baby-faced Silicon Valley millionaires and every billion dollar initial public offering, there are a hundred low-paid contingent workers with no job security, no benefits and only the most uncertain future. (Schrag 1999, 6B)

Growing Labor Leaders

The Leadership Institute is the glue that holds the three parts of WPUSA together. It has helped our organization increase its influence over time. Our intensive leadership education includes, and moves beyond, previous attempts to create social change that focus almost exclusively on "relationship building." In addition, we built in three value-added components for our training program right from the start.

The first component is strengthening local leaders' understanding and use of policy analysis tools. These tools allow them to ground their organizational strategies in a shared analysis of regional economies and social needs. The second element is offering multiple actors in the social justice movement a shared framework, and, equally important, a common language they can use to speak about disparate issues. By building a common vision of the economy and a shared language in which to talk about economic realities, diverse organizational goals and leadership visions are united into a shared agenda. The third, and perhaps most critical, aspect of this strategy is building ongoing relationships between individuals and groups that transcend single-issue campaigns. By using policy analysis tools to identify issues of mutual concern and potential allies, Working Partnerships USA built new, ongoing broad-based alliances that cut across class, race, ethnicity, and occupational groups within Silicon Valley's highly fragmented economy.

Building Grassroots Power in Various Regions

Like WPUSA, Building Partnerships USA (BPUSA) is a 501(c)(3) non-profit organization. It was founded in collaboration with Los Angeles–based Strategic Concepts in Organizing and Policy Education (SCOPE). BPUSA's primary goal is to help establish more regionally based 501(c)(3) nonprofit organizations with the same capacities as WPUSA and SCOPE. In a natural, national evolution of WPUSA's work, BPUSA takes the concept of peer learning to its next logical step by bringing experienced coalition partners into different communities to help progressive regional leaders become more effective, most notably, by building or strengthening complementary new 501(c)(3)'s.

BPUSA's first institution-building projects got under way in Boston, Atlanta, and Denver in 2005. Similar programs in Milwaukee and the state of Connecticut were scheduled to begin in mid-2006. In each instance, the projects are designed to project and extend organized labor's influence into the broader community to achieve greater measures of equity on a metropolitan scale. The biggest economic problem that faces labor and community groups in Boston, Atlanta, and Denver is rooted in changing demographics and a shift in the way local governments prioritize and then fund public services and economic development.

In Boston, for example, the enormous influx of immigrant workers has presented a challenge to both government and Labor. The main question is how Labor can mobilize and organize these new groups to protect and defend hard-won gains achieved by other workers more than a generation ago.

To build capacity and trust within this new coalition, the Boston organization, Community Labor United, turned to BPUSA for assistance with understanding and using the best practices and models pertinent to their situation. As a result, in 2005, Community Labor United began the process of organizing a series of campaigns designed to require public institutions that receive taxpayer money to align their business practices with community-centered development goals. In 2006, Community Labor United worked to produce the needed research and documents with the assistance of scholars from MIT and the Boston and Amherst campuses of University of Massachusetts.

Likewise, in Atlanta, one of the initial goals of BPUSA's partner organization, Georgia Stand-Up, is to produce high-quality research reports and public education programs that provide more hard data about the devastating impact of the state's probusiness labor laws, including low wages, fewer healthy individuals and families, and environmental damage unevenly centered around urban communities through sprawl and endless miles of subdivision development. The leaders of Georgia Stand-Up report that they see working with BPUSA as an opportunity to build an organization that can educate leaders and communities about these and other inequities while moving strategic policy campaigns forward on the regional and, eventually, the state levels.

"I've been able to bring labor and community groups together on specific campaigns and issues," says Charlie Flemming, president of the Atlanta–North Georgia Central Labor Council. "But the promise that working with BPUSA holds is that we can turn these relationships into a real win that changes the way the State of Georgia treats its working communities."

Organized Labor's Critical Role in the Movement for Equity

The schism in the AFL-CIO that occurred at its 2005 national convention has left Labor embittered and divided, one hopes temporarily. Even so, Labor remains America's largest membership group made up of working families. Only the church and senior citizen groups rival Labor in terms of their capacity to organize and influence the larger society. That puts Labor in an ideal position to partner with others in the fight for social and economic equity. In addition, the Silicon Valley experience teaches us that economic growth, as measured by corporate sales, profits, or productivity, does not necessarily translate into an increase in social or community well-being. Instead, making sure the needs of the whole community are met often requires the intervention of equity actors, most importantly community-based organizations, which operate with an understanding of the importance of re-establishing the link between corporate success and stability and upward mobility for working families.

Many groups are contributing to the drive for metropolitan equity, but few have as much potential power as organized labor to build a grass-roots movement for positive social change. Harnessing that power was, and remains, the main achievement of Working Partnerships USA in Silicon Valley, and it is now the goal of Building Partnerships USA nationally.

Note

1. Barbara Byrd and Nari Rhee describe shifting attitudes among working-class residents of Silicon Valley in a 2004 report for the Labor Studies Institute at Wayne State University, 131–153.

II

Breakthrough Communities: Stories and Strategies in the Quest for Regional Equity

Introduction to Part II

Where part I provides historical background for the regional metropolitan equity movement, part II shares stories of breakthrough communities. These stories document strategies used, lessons learned, continuing challenges, and actions that sustain community and forge bonds among people.

These sixteen stories and strategies are a collective portrait of how leaders and groups confronting local neighborhood and workplace challenges created positive change for sustainable metropolitan regions. The Northeast, the Midwest, the Southeast, and the West are represented. Examples were chosen from the largest metropolitan regions in the country (New York, Chicago, and Los Angeles) as well as less populated areas like Naugatuck Valley, Connecticut, and the lowlands coastal region of Charleston, South Carolina. Stories were included both for the inspiration they offer and the spectrum of strategies employed. The approaches reflected in these stories range from community organizing, community–private sector partnerships, and community development corporations to farm-to-school initiatives, university-based collaborations, and national campaigns.

The leaders you will meet in these stories, many of whom are people of color, are deeply committed to the success and well-being of their communities. Each of the principal actors and organizations has extensive experience working with and on behalf of their constituents. The stories describe how their focus has expanded from improving individual neighborhoods and workplaces to promoting sustainability and structural change at the metropolitan level.

These activists and policy makers are addressing the complexities examined and considered in part I by proposing a new politics of sustainable metropolitan development. Prominent values of this new politics are fairness, participation, democracy, and access to opportunity—in short, metropolitan regional equity.

A Compass for Transformative Leadership

Each successful breakthrough story included in this book reflects four stages common to community organizing efforts. In preparation for mobilization, there is often an initial stage that we have called "waking up," when conditions in the community cross a line or reach a point of critical

Figure II.1
Compass for Transformative Leadership.

mass, which causes an alignment that makes action inevitable. This might be accelerated by external conditions imposed in a hierarchical manner from outside or an internal disintegration that becomes intolerable. For example, a record number of murders in Richmond, California, led Rev. Andre Shumake to organize a tent village protest in one of the public parks. While the Richmond Improvement Association had existed for several years, the escalation of deaths in a short period was the tipping point that led to protest. This stage of waking up often results in one or more leaders in the community giving voice to deeply held values and naming actions or conditions that threaten those values. In some cases, the value is survival. In the background of the Compass for Transformative Leadership (figure II.1) is a spiral, which suggests an expanding awareness that gathers momentum over time, with preparatory actions gathering support in certain sectors, until fragmentation is overcome and there is sufficient clarity and focus to take collective action on a larger scale.

Following this gradual process of people waking up and joining forces is the first stage on the horizontal or action axis of the Compass for Transformative Leadership, "saying no," when a common action identifies and resists the negative forces destroying the community. Common examples are sit-ins, boycotts, demonstrations, or litigation.

The second stage is "getting grounded," stepping back to assess the current situation of the community following an initial mobilization. It is represented as a position at the base of the vertical learning axis of the Compass for Transformative Leadership. In our research we have identified three levels of getting grounded, each with its own tactics and methods unique to the cultural context of the communities: getting grounded in time, getting grounded in place, and getting grounded in community. This inquiry can take months or years, depending on the readiness of the community and the scale of the issue they are taking on. This stage may require historical research of strategies that have been tried or an improved understanding of current legislative and budgetary processes in a wider context that should inform strategy. In several of our breakthrough community stories, the community researched the process by which transit rail stations are established and funded in order to develop a strategy for permanently restructuring and improving access to opportunity in the region. Completing this research and securing funding for the West Garfield station in Chicago took a decade. In Manhattan, the struggle to complete the Second Avenue subway has gone on for more than fifty years. The process of learning and getting organized often catalyzes new leadership, and it provides a new lens through which the community views its history and assets and consolidates its identity. Pride in neighborhood identities that were previously despised often emerges at this stage of the compass.

Having completed this assessment, each successful breakthrough community undertakes a third stage of development: a far-reaching examination of additional possibilities that lie outside the group's conventional wisdom. This learning process often inspires a community to rethink its sense of identity and goals, and to explore more extensively what its members are up against. This stage is at the top position of the vertical learning axis on the Compass for Transformative Leadership and is called "expanding new horizons." At this stage neighborhood entities often find it necessary to ally with new regional partners in order to win their campaigns. In the current breakthrough communities stories, these alliances include black-and-brown coalitions in inner-city neighborhoods that obtain community benefit agreements that expand opportunities for both groups, and coalitions of interfaith denominations spanning inner-city and suburban congregations that create a regional transit plan to serve the entire region.

Finally, in the fourth stage, after consolidating internal power and building external linkages to wider regional institutional support and technical expertise, each successful breakthrough community builds a vision that mobilizes new action and incorporates a wider constituency in their campaign. Their cohesion no longer depends on shared opposition to a common enemy. They have framed a vision and message that consolidates power for a shared future. Breakthrough communities at this stage often utilize strategic communications to consciously frame their message. They are proactive in pursuing their preferred future, rather than reactive. They have "learned how to learn" as a community, and they demonstrate this generative capacity through a variety of flexible responses and practical actions. The capacity to "say yes" through a positive program enables breakthrough communities to consolidate short-term gains as well as longer term policy victories that are part of a vision for sustaining enduring change. During this fourth stage in the Compass for Transformative Leadership process, new allies continue to be enrolled from unexpected places, and a new generation of leaders is consciously cultivated to sustain the action over time.

Each of the breakthrough stories presented in part II documents successful completion of all four stages. For purposes of exposition, the stories are presented in four sections, each focused on a different stage of the process. The stories of these community organizers and visionaries are offered as inspiration to those who face similar challenges in their home communities.

Saying No to Forces Destroying the Community

Numerous studies have documented the traumatic impact on residents who are forced to live in abandoned communities surrounded by concentrated poverty. Issues like public safety, police brutality, poor schools, and the lack of economic opportunity are urgent and relevant to those who live in such neighborhoods. Many of these conditions are the result of globalization and macroeconomic policies. They also result from changes in metropolitan life, decisions being made by jurisdictions and developers far removed from neighborhoods experiencing concentrated poverty (Wilson 2007). On the other hand, residents of poor communities may be displaced by gentrification or large-scale development projects. By the time the community wakes up to the challenge, the forces arrayed against community interests have been so well developed that they are difficult to stop or modify. Although many residents are surprisingly resilient in the face of such conditions, evidence suggests that people exposed to repeated trauma may suffer alienation, loss of self-esteem, feelings of helplessness, paralysis of initiative, and a sense of stigma. For many, joining a social movement to protest environmental neglect or intrusion can be an important step in moving beyond apathy and resignation (Tarrow 2001). Saying no is a way of overcoming silence, discovering one's own voice, and building leadership and agency to transform the world.

Defensive Reaction

The focal point of community reaction is often a highly visible incident, like a toxic release, the closing of a hospital, or police violence. Since a reaction is resistance to a negative force, that force defines the terms and scope of the reaction, and community response is immediate. Task forces are set up to investigate the proximate causes. In some cases, practical remedies are close at hand. A defensive reaction on the part of an aggrieved community may produce a rapid victory for the community. Yet, seldom are such reactions effective in addressing the diffuse daily processes that lead hour by hour and day by day to the isolation and segregation of poor African-American, Laotian, and Latino communities. The civil rights movement of the 1960s, an explosion preceded by decades of grievances, is a case in point (McAdam 1999).

The Power of Resistance

Nevertheless, under the right circumstances, acts of resistance can have impacts of historic proportion. Protesting against negative forces destroying the community is often the first stage in building a social movement. Advocates of direct action, who have exhausted legal remedies such as electoral politics, seek other avenues for effectiveness. The Montgomery bus boycott and the sit-in movement of the late 1950s and the early 1960s that launched the civil rights movement are classic examples of acts of resistance (McAdam 1999). Another example is the 1983 protest by five hundred African-Americans against the siting of a toxic waste dump in Warren County, North Carolina, that triggered the worldwide environmental justice movement (McGurty 2007). In these cases, protesters were able to seize the moment and change an act of resistance into a proactive strategy for transformation. Each of the communities presented in this section illustrate this principle.

Each story demonstrates the capacity of the community to stand up and challenge the patterns of disinvestment that have plagued their regions for decades. In Cleveland, Ohio, citizens saying no to vacant properties confronted the hopelessness and despair that had torn apart their communities. Blight had slowly removed the structures that define "community" and, typically, city leaders had surrendered to resignation in the face of taking on this daunting challenge. Going house by house and block by block, communities worked together to reverse deterioration and reinvent neighborhoods. Our second story is in Rochester, New York, where a visionary mayor empowered community members to rise above the shame of a weak-market city to set the course for their neighborhoods. And in the Naugatuck Valley, Connecticut, by refusing to be trapped by an old industrial identity, new immigrants have joined the community to literally reinvent their region. This is the power of saying no.

Cleveland, Ohio

A port city located on the south shore of Lake Erie, Cleveland was founded in 1796 near the mouth of the Cuyahoga River and grew to become a strong manufacturing center. Referred to as the "Forest City" by Alexis de Tocqueville in his classic work *Democracy in America*, it retained its nickname when then-mayor William Case spearheaded a citywide, fruit-tree planting effort in 1850.

For more than one hundred years, the city enjoyed a reputation as a desirable location, only to experience the gradual collapse of its industrial base coupled with the expansion of suburban development in the decades following World War II. Today, Cleveland, like many midsize cities in America, is struggling to define itself in an increasingly global marketplace. While respected institutions like Case Western University and the Cleveland Clinic point to vital signs of economic stability, the city continues to face the impact of more than 30 percent of its population living in concentrated poverty, as well as the challenge of a staggering 30,000 vacant properties.

Even though some have labeled Cleveland as a city "past the point of no return," David Goldberg and Don Chen of the D.C.-based organization Smart Growth America (SGA) believe there are strategies for "The Comeback City." Locally, statewide, and nationally, SGA focuses on preventing abandonment, redeveloping vacant properties, and supporting more environmentally sound approaches to growth. In their contribution to this volume, the authors detail how SGA has spearheaded campaigns to reverse the trend toward urban ghost towns. As with other regional efforts in the United States, the challenge remains to attract a wide range of stakeholders who are willing to work to change the "throwaway" attitude toward inner cities, while insisting on development that is appropriate for each particular neighborhood. The philosophy that guides SGA is "When every neighborhood is a good neighborhood, the entire region prospers."

Rochester, New York

Resting in the fertile Genesee Valley carved out by huge glaciers in the Cenozoic era, the city of Rochester is surrounded by a lush topography with several waterways coursing through, including the Genesee River. Rochester was America's first boomtown. Colonel Rochester, for whom the city was named, successfully lobbied to have the Erie Canal directed through the center of town in the early 1800s. Toward the end of the twentieth century, Rochester found itself on the brink of collapse. While market forces and shortsighted planning played a significant role in Rochester's decline, the absence of communities of color from the decision-making process further exacerbated the state of affairs.

In the 1990s, Mayor William A. Johnson, Jr., who had been president and CEO of Rochester's Urban League for twenty-one years, instituted an innovative approach focusing on citizen empowerment. Through a participatory planning process, Neighbors Building Neighborhoods, Johnson engaged city residents, businesses, schools, churches, and other community leaders to help chart Rochester's future. In his piece, Mayor Johnson illustrates how, by providing a structure for citizen participation, a community can generate creativity in the planning process and build civic pride. Beginning with opposition to abandonment, Rochester's turnaround provides an example for other cities wrestling with reinventing their future. By remaining highly visible, addressing fearlessly issues of race, and taking risks with his constituents, Mayor Johnson models a way to help achieve the renewal of a community's identity.

Naugatuck River Valley, Connecticut

As the center of brass production in the United States during the nineteenth century, Connecticut's Naugatuck River Valley witnessed an incredible industrial expansion that would eventually rival products from England in terms of both quality and price. Indeed, by midcentury, two thousand *tons* of copper were extracted from the Naugatuck Valley every year and the city of Waterbury was well on its way toward becoming the "Brass Capital of the World." In time, textile mills and furniture manufacturers also prospered, relying on the Naugatuck River for distribution and transportation.

Driving through the region today, one can see its history embedded and inscribed in the landscape—the intersection of its glorious past and its contemporary decline caused by the loss of social and economic infrastructure. The resulting emergence of *brownfield communities*— abandoned or underused commercial or industrial properties—has scarred large stretches of land with environmental stigmas that have been hard to overcome.

Ken Galdston, project director of the InterValley Project (IVP), details how environmentally sound repurposing and economic revitalization can go hand in hand when brownfield transformation is a civic priority. IVP works with a dynamic coalition of labor, immigrants, and community development corporations to reclaim abandoned brick and stone buildings. In addition to helping to create affordable housing and save thousands of jobs in the New England area, IVP has worked to strengthen links among religious organizations, inner-city and suburban neighborhoods, and the various ethnic groups that constitute the region's labor force.

10

Rekindling Hope in Cleveland (Cleveland, Ohio)

David Goldberg

As others pull up their stakes, Julia DiFranco is rooted in Collinwood by her memories. For fifty years, the northeast Cleveland neighborhood just off Lake Erie has been the center of her world. The Collinwood of her memory was a tight-knit, working-class community that seemed the perfect place to raise her four kids. These days, though, Collinwood is split in two by a major interstate, the old industries and their jobs are gone, and lately it seems that more and more occupants of surrounding houses are as well.

"Our neighborhood's gone to pot," Mrs. DiFranco laments. "I had never seen a boarded-up house until about five years ago. Now it seems like they're all over the place. People are moving out to the suburbs for the schools and better police response. I like my neighborhood and I like my house, but sometimes I think it's just a matter of time before everybody leaves. I wish something could be done."

Lifting the Cloud

DiFranco's neighborhood is not unique. Dozens of Cleveland communities face similar challenges, and property abandonment has been an acute and widespread liability for the city. As one observer recently noted, the problem "is a cloud hanging over the city's psyche." Rising to this challenge, a local community development nonprofit, Neighborhood Progress, Inc. (NPI), asked the National Vacant Properties Campaign to analyze Cleveland's vacancy epidemic and offer recommendations for reform. The campaign assembled a team consisting of Alan Mallach, Lisa Mueller Levy, and Joe Schilling, who then conducted nearly fifty interviews with key stakeholders and reviewed a mountain of data, reports, and other materials. By June 2005, the campaign produced

"Cleveland at the Crossroads: Turning Abandonment into Opportunity," a frank, pull-no-punches report not only portraying the scope of the problem but also presenting a pragmatic, step-by-step map to progress.

"Cleveland for many years has been a leader in some community development areas," Schilling said. Responding to the first wave of white flight and abandonment in the late 1960s, Cleveland created one of the first redevelopment land banks in 1976. Also, nearly every neighborhood has its own community development corporation, and the city has been an innovator in addressing blight. "The results of these efforts are visible across the city in ... revitalized neighborhoods with rising market values," the report acknowledges. But for all the progress, vacant properties have again started to spread.

"Most people have a sense that compared to a lot of other Rust Belt cities, Cleveland has had a renaissance," Schilling says. "Our sense is that perhaps Cleveland has leveled off, and some of its institutions aren't working as effectively as they once did."

Taking Stock of Abandoned Buildings

Although the assessment team lacked the information to make a more precise estimate, the report concludes that Cleveland has between 10,000 and 25,000 vacant properties. These include single-family homes, apartment buildings, commercial buildings, and industrial properties. According to the report, housing units continue to be abandoned at a faster rate than new ones are built. Cleveland's home county of Cuyahoga was projected to have more than 10,000 foreclosure cases in its Common Pleas Court in 2005. At the same time, the report notes, "in the absence of aggressive code inspection or the credible threat of abatement or demolition, owners of vacant property in Cleveland have little motivation to repair and maintain their property. As one respondent commented, under the current system 'it's just too easy to own a vacant property in the city of Cleveland.'"

In their interviews with about fifty individuals at thirty agencies, the authors heard complaints that it was hard to get properties in and out of the land bank. "The city was frustrated with the county because of their slowness to act on tax foreclosure," Schilling says. "The communities were frustrated with the housing inspection office because they had too few resources to visit the vacant and abandoned houses. Often they would just do drive-by inspections," missing the serious deterioration that lurked within.

After amassing data and making visits to the area in spring 2004, the assessment team worked closely with local experts to draft and then vet the report. While hoping to make the greatest use of their national expertise, neither the campaign nor NPI wanted to create the impression that a national outfit had come to town and made snap judgments that wouldn't ring true with local observers.

From Analysis to Action

The "Cleveland at the Crossroads" report was released in June 2005, in the thick of a mayoral election. Mayor Jane Campbell's administration moved quickly to respond. As the Cleveland *Plain Dealer* reported, "less than three hours [after the report release], Campbell publicly embraced the study and made it the foundation for a new 'Zero Blight Initiative.' Campbell named Assistant Planning Director Edward Rybka to manage the effort." Rybka, a city councilman for twenty years before taking the planning job weeks before, had made his political name pushing the city to deal with its abandoned properties.

The editorial board of the *Plain Dealer* also weighed in. "For years, critics have complained that there's been a disturbing lack of political leadership on the issue of vacant property. City Hall has not sorted out which places are vacant, why they are empty, or what to do with them. Cleveland at the Crossroads tells how this city, like many others across the nation, can make a dent in the problem." Suddenly, long-standing concerns of community development groups—and citizens—were moved up to front-burner status.

"What was positive about the report was that it quickly got us all on the same page," says Rybka. "Vacant and abandoned properties are an increasing problem in a lot of neighborhoods in Cleveland, despite a lot of work that has been done in the city over the years. We're going to get aggressive about addressing the problems." Already the city is working to bridge communications and procedural gaps between its departments. At the same time, a communitywide coordinating committee is working to improve the interaction between the city and the county, while preparing community organizations to make the most of redevelopment opportunities. "Most people now agree there is more real hope for change than there has been at any time past," says Frank Ford, vice president of NPI.

A Dream Come True

It is all a dream come true for residents like Mrs. DiFranco, as it is for NPI's Ford. "Things are beginning to happen now, and I don't think that would have been possible without this team of experts, the credibility of a national effort, and a well-written report," he says.

The next step, Rybka says, will be to regionalize the effort and link the abandoned properties issue with sprawl development, which is sucking resources and investment out to the suburban fringe. Even suburban leaders are starting to make these connections, especially those leaders in older areas who have banded together as the First Suburbs Consortium to deal with increasing abandonment in their communities. "We want to have the First Suburbs Consortium at the table because they are increasingly seeing the same kinds of issues," Rybka adds. "It's tough for us to compete with suburban development, especially when it is subsidized by state and local policies." Still, he is optimistic that the renewed energy around reversing abandonment will have a positive impact.

"There is no question that these properties represent a huge opportunity," he says. "This is all about making lemonade out of lemons."

The National Vacant Properties Campaign's Recommendations for Cleveland

- Focus code enforcement on key neighborhoods and community development areas instead of responding to complaints one by one.
- Require owners to register vacant properties, and charge them an increasing annual fee.
- Make mortgage companies responsible for maintenance of vacant properties.
- Analyze foreclosures to determine patterns and causes.
- Make it easier for the city's land bank and nonprofit neighborhood development corporations to obtain property.
- Develop a citywide database of property records, including violations, liens and delinquent taxes, and connect it to Cuyahoga County's network.

11

Closing the Gaps: The National Vacant Properties Campaign (Cleveland, Ohio)

Don Chen

The National Vacant Properties Campaign was launched as a response to the devastating property abandonment and hopelessness facing countless American neighborhoods. The campaign seeks to bring about thriving communities that are free from blight and filled with hope and opportunities for their residents. To realize this vision, the campaign is working to create a national movement that helps communities develop systems to prevent abandonment, reclaim vacant properties, and revitalize neighborhoods. The campaign's network of practitioners, policy makers, business leaders, and researchers provides technical assistance to help communities address their vacant property needs. The campaign is also a clearinghouse for the latest information, research, policy innovations, and best practices.

The campaign links citizen advocates, business leaders, urban experts, emergency response personnel, and local officials who have worked for years to stem the abandonment of properties in their cities and towns. It has also connected communities with civic organizations from all over the nation—such as community development corporations, faith-based organizations, nonprofit housing developers, smart growth groups, lenders, realtors, and other key players—to bring properties back into productive use.

What Are Vacant Properties?

While state laws and uniform building codes differ on what constitutes vacant or abandoned properties, the campaign defines them as residential, commercial, and industrial buildings and vacant lots that pose a threat to public safety; or as properties whose owners fail to pay taxes, bills, or mortgages, or carry liens against the property.

Vacant properties can include boarded-up buildings, unused lots that attract trash and debris, vacant or underperforming commercial properties such as empty shopping malls (greyfields), and neglected industrial properties with environmental contamination (brownfields). To date, the campaign has mostly focused on individual homes, lots, and small commercial buildings, because efforts to redevelop brownfields and greyfields are already relatively robust.

In addition to promoting property rehabilitation strategies, the campaign also urges communities to prioritize prevention—mainly through the monitoring of dilapidated single-family homes, apartments with significant housing code violations, and housing that remains vacant for long periods of time.

Why a Vacant Properties Campaign?

Property abandonment has long been a sore issue for countless cities and towns. It is a prominent community development challenge, as well as a major issue for city authorities dealing with tax delinquency and code enforcement. However, what some regard as a rather unglamorous issue has risen in urgency among many unlikely stakeholders, and the National Vacant Properties Campaign has seized on this opportunity.

Smart growth and regional equity advocates have been fighting for years against irresponsible sprawl development in many regions. Their top priority is to ensure that economic development resources—infrastructure dollars, funding for services, jobs—are distributed fairly between older communities (from urban to inner-ring suburban) and newer suburban and exurban places. As Georgetown Law School's Sheryll Cashin says in her book *The Failures of Integration*, "Above all, steering growth inward will contribute mightily to the vitality of existing developed neighborhoods where many people of color live. Coalitions for smarter and more sustainable growth, then, are highly relevant to the project of cultivating successful socioeconomic integration."

Vacant properties represent a tremendous reinvestment opportunity for bringing back underutilized properties in older neighborhoods—an approach akin to smart growth hallmarks such as brownfields redevelopment, transit-oriented development, and infill projects. Such properties also offer opportunities for smart growth advocates who focus on parks and natural areas. Many of them see vacant lots as opportunities for community gardening and agriculture; creating urban parks and greenways; or reestablishing bird sanctuaries, native plant nurseries, and other biodiversity in areas that have recently served as illegal trash dumps or open air drug markets.

Other stakeholders have also expressed a strong desire to get involved. Emergency response workers are concerned about abandoned properties because they account for such a large proportion of crime, arsons, and accidents. Realtors have gotten involved because derelict properties drag down surrounding property values. City governments increasingly recognize that such properties cost money through lost tax revenues, maintenance costs, and a weakened investment climate.

The wide variety of responses to the vacant property challenge has presented a tremendous opportunity for coalition building to solve the problem. It will take such a coalition—one that is broad-based and powerful—to transform the system by which vacant properties are handled by local jurisdictions.

Obstacles Blocking the Path to Reclamation

The causes of property abandonment are fairly simple. Some property owners fall on hard times and lose their homes to tax foreclosures. Others lack the insurance necessary to rebuild after a fire or other damage. The presence of vacant and deteriorating buildings drives property values down, which motivates other people to consider leaving.

By contrast, the barriers to reclamation are complex. Many people assume that such problems simply result from poverty, declining population, and weak real estate markets—nothing a strong market couldn't solve. They would be only partially right. The other major challenge is that in most jurisdictions, a newly abandoned property wallows in a process that virtually ensures that it will remain empty or cycle through a long succession of slumlords and tax liens. Communities typically handle their vacant properties through a process designed to maximize short-term tax revenues (e.g., foreclosures, tax lien auctions), which misses opportunities to support long-term community development and the stabilization of the real estate market. Misguided policies, entrenched industry practices, and noncreative thinking present a procedural morass that thwarts even the most ambitious communities that have tried to create new reform policies, codes, government programs, and citizen campaigns.

Breaking this cycle is one of the campaign's primary goals. "A lot of practitioners tend to focus on the technical aspects of vacant property revitalization," says Joe Schilling of Virginia Tech and cofounder of the National Vacant Properties Campaign. "Fortunately, others look at the problem through the lens of broader policies related to smart growth, community development, and regional equity. Part of our mission is to provide the big picture and bring together all of the different visions into a cohesive blueprint for revitalization."

Better Information: Laying the Foundation for Effective Action

One of the first steps in developing a better property reclamation system is to gain a clear understanding of where vacant properties are and what they do to communities. Yet, most communities cannot say for sure just how many vacant properties exist within their jurisdictional boundaries. The campaign is working directly with cities and counties to help them improve their documentation. Innovative approaches are leading to better information systems, including tracking and inventory tools, high-tech geographic information systems (GIS), and speedier ways to contact potential claimants to clear the title on a property.

Another critical need is to quantify the costs of abandoned and vacant properties—in crime, lost tax revenue, fire danger, and environmental hazards, among other problems. The campaign has compiled and disseminated the best available information in a monograph, *Vacant Properties: The True Costs to Communities*, published in spring 2005. Some of the report's findings include:

Vacant properties dramatically drag down the value of surrounding properties.
Blocks with unsecured vacant properties generate much more drug-related activity, thefts, and incidences of violent crime than otherwise comparable neighborhoods.
More than 12,000 fires flare up in vacant structures each year, most the result of arson, costing $73 million in annual property damage, firefighter injuries, and even deaths.
Vacant lots cost $3 billion to $4 billion in lost revenues to local governments and school districts annually.

The Need for Jurisdictional Coordination

In addition to lacking information, jurisdictions lack strong coordination. Indeed, some cities have as many as fifteen separate agencies charged with different aspects of the same problem, often with differing missions and perspectives. Although cities have many tools—including code enforcement, tax collection, recording of liens, and condemnation—to date there has been little focus on the bigger picture.

Effective approaches to vacant and neglected properties require wide intergovernmental cooperation and efficient intra-agency coordination. Elected officials could appoint high-level coordinators who would have the authority to lead multiple agency efforts to address vacant property reclamation. The campaign is working with many cities, counties, and states to spread the good news about leadership models and solutions that work.

This coordination also needs to take place among public officials and other community stakeholders. City development officials and their allies in the community—including nonprofit development corporations, housing groups, realtors, and lenders—are typically uncoordinated, and sometimes conflict occurs when attempting to turn vacant properties back to effective use. The campaign is working to introduce players to one another and broker lasting partnerships for policy reform and new procedures.

Communitywide Responses to Vacant Properties

The challenge for communities is: where to begin, and how to do it? The campaign provides technical assistance and model practices to local officials, civic groups, and business leaders to help them design more holistic policies and implement more effective programs to address the blight of vacant properties. By working closely with local hosts and advisory committees, we have forged trust among diverse stakeholders and have enhanced their understanding of the causes of abandonment and the need for communitywide strategies.

When selecting an assessment team, we reach out to former and current practitioners. Many communities have pioneered innovative strategies to get their vacant properties under control, yet opportunities for peer-to-peer learning have been practically nonexistent, so most people working on these issues spend years toiling away in isolation. Those who actually have done this work in other cities carry immense credibility with the practitioners who are seeking help. They speak the same language and share a similar culture.

Ohio has been a particular focus of our recent technical assistance work. In June 2005, the campaign released its Cleveland assessment report through a series of town hall meetings. More than four hundred practitioners, policy makers, community leaders, and neighborhood residents attended these sessions on the different facets of an effective vacant property initiative—comprehensive information systems, aggressive code enforcement, and efficient land banking. The mayor responded with her own Zero Blight initiative and appointed a former city councilmember as the city's first vacant properties coordinator. Several members from the advisory group, along with the official sponsor of the assessment, Neighborhood Progress, Inc., continue to work with local foundations on a series of strategies that will translate the assessment team's recommendations into action.

Later in the summer of 2005, the campaign released another assessment report in Dayton, Ohio, sparking similar interest and results. Nearly one hundred business, government, and community leaders participated in a briefing presented by the campaign's assessment team. The Dayton/Miami Valley assessment report included more than twenty-five policy recommendations and sixty specific action items. The Miami Valley Regional Planning Commission is now starting to convene working groups to adopt the most compelling of the campaign's recommendations.

The campaign is building on its work in Cleveland and Dayton to develop a statewide action agenda on vacant properties encompassing public education, policy proposals, legal reforms, and business leadership. In October 2005, the campaign was the lead sponsor of Ohio's first vacant properties conference—the largest conference of its kind ever held in the nation. At the conference, a wide array of public officials, citizen activists, academics, and nonprofit groups joined together to start developing an action agenda for 2006 and beyond.

Reaching Out to Particular Neighborhoods

A recent Federal Reserve Bank study of neighborhood indicators provides powerful evidence that targeting resources is an effective strategy for combating vacant properties. For example, the City of Richmond, Virginia, established the Neighborhoods in Bloom program in 1999 to focus its limited neighborhood reinvestment funds in six neighborhoods suffering from crime, disinvestments, and vacancies. An important attribute of these neighborhoods was their rich historic character, which was threatened by vandalism, crime, and increasing neglect by absentee landlords. Within two years, the program had resolved 810 code violations, made 130 home repair loans, and rehabilitated or repaired 118 properties. Today, crime in the six neighborhoods is down and property values are up.

Next Steps

Abandoned properties can be found in virtually every American community, from Rust Belt cities that have fallen on hard times, to Sun Belt boom towns that are letting their old commercial districts fall apart. In every community that faces this problem, reclaiming these properties has proven to be a quagmire. The National Vacant Properties Campaign sees the possibility of minimizing blight and bringing these properties back to life—even for places with the worst economic conditions.

The solution is to focus as much on people as we do on properties. The campaign's growing network of experts, practitioners, and civic leaders has started to develop a strong sense of shared purpose: to create a community—no less a movement—of people who see abandoned properties not as liabilities, but as opportunities for revitalization. Policy approaches focus on individuals and families as well. Recommended strategies include not only the most efficient ways of doing things but also more humane approaches to difficult issues such as tax foreclosure. For example, Michigan and New Jersey's new tax foreclosure laws both include new grace periods that can provide more time to families who have fallen behind on their payments and are struggling to get by. Finally, our set of best practices includes numerous tools to help practitioners and localities work together better—both within their borders and within their larger regions—to achieve common goals.

The campaign seeks to ensure that the real estate transactions, development projects, and legal decisions that drive vacant property issues are undertaken with awareness of the needs of the people and communities that such efforts are supposed to serve. In prosperous times, our communities' homes and storefronts represent the best of what our nation has to offer—a fair shot at the American Dream, a roof over our heads, and a place to belong. When we leave these places behind, the loss is equally profound—family tragedies, failed careers, and discarded dreams. In recovering lost properties, we restore hope and strengthen community.

12

Neighbors Building Neighborhoods: Community Stewardship to Revitalize Midsize Cities (Rochester, New York)

Mayor William A. Johnson, Jr.

Rochester, like other cities, was created by human interaction, not government. Since its incorporation, Rochester has had four kinds of government: a trustee system, a weak mayor system, a city manager system, and a strong mayor system. Every forty years the form of government changed, but it never really mattered. What always kept the city going was the positive interaction of its residents.[1]

Often elected officials are threatened by anything they cannot control. Citizen-based planning can be unpredictable, messy, and difficult to manage. Perhaps I feel comfortable with an empowered citizenry because I did not follow the traditional political route. I did not rise through the ranks of the local Democratic Party to be anointed mayor. I was head of the Urban League of Rochester for twenty-one years. I saw how citizens' views were rarely respected, and this as much as anything prompted me to run for mayor. In 1993, I entered the Democratic primary as a challenger to the political establishment. I won the primary and general elections with a slogan of "We, not me." That slogan became the credo for my administration.

The Rochester story has been about *partnerships and building community*. Contrary to what you may have read, there is no urban renaissance going on in most older American cities. Viewed from a spatial context, American society consists of poor, primarily black cities surrounded by sprawling, primarily white suburbs. The backdrop for economic and racial segregation is the virtual abandonment of American cities by the federal government, and often by their state governments as well. Federal aid to American cities has essentially stopped. For most cities, this has meant a loss of about 20 percent of their budgets since President Ronald Reagan began his conservative revolution in 1981.

A Legacy of Home Rule

New York is what is called a home rule state. Each municipality in the state is fully independent, with complete authority over land use planning and economic development within its borders. Annexation and amalgamation are practically impossible, which means that municipal borders never change. No county, regional, or federated local government systems exist of the sort that can be found in Canada and in many southern and western U.S. states.

This has resulted in intense competition between neighboring municipalities for people and businesses that pay the property taxes that fund schools and other public services. Naturally, newer suburbs with an abundance of undeveloped land will be the big winners, while cities with no room to expand will be the big losers. Since 1950, the population of the city of Rochester has declined 34 percent, while that of the suburbs has increased 320 percent.

Home rule also means that municipalities can use their zoning powers to keep out lower-income individuals. Roughly 80 percent of the entire metropolitan area's poor live in the city of Rochester. Racial segregation accompanies economic segregation, and 90 percent of the region's people of color live in the city.

Let's take a closer look at the city of Rochester. The city's official poverty rate is 25.9 percent. Twenty of the city's ninety-one census tracts are classified as high poverty. We refer to this as our "crescent of concentrated poverty." This is the only concentrated poverty in the entire metro area. However, these numbers are misleading. "Extremely poor" census tracts in the United States are defined as those with 40 percent or more residents living below the poverty line. If, instead, we applied Canada's threshold of 26 percent, then forty-three of Rochester's ninety-one census tracts would be classified as "concentrated poverty tracts." In the United States, the official poverty threshold for a family of four is a low $18,900. If we used Canada's more reasonable low-income cutoff line—the point at which individuals or families must devote 60 percent of their annual income to basic necessities—then Rochester's poverty rate would not be 25.9 percent, but 62.5 percent.

This scenario is by no means unique to Rochester. For example, Rochester has 3,300 vacant residential and commercial buildings, a relatively modest number. Cleveland has more than 33,000, Baltimore has nearly 50,000 vacant properties, and larger cities like Philadelphia and Detroit are in the 100,000 range. All eastern and midwestern U.S. cities are in the same boat, with wide swatches of poverty and abandonment even though many of their downtowns have been revitalized in recent years.

A City Rich in Assets

Every city has its problems, but every city also has tremendous assets, and Rochester is no exception. We are located on three great waterways: Lake Ontario, the Erie Canal, and the Genesee River. We also have twenty-six miles of publicly owned waterfront. Our three-thousand-acre park system, designed by Frederick Law Olmsted, is one of the most beautiful in the country.

Some of the region's proudest history is also linked to Rochester. The city was the home of Susan B. Anthony and Frederick Douglass, and became a center of the antislavery movement and the women's rights movement. The city also has more buildings listed on the National Register of Historic Places than any other city its size in the country. We have one of the finest collections of museums in the country. We are also the center of a dense concentration of universities, with nearly 70,000 college students residing in the county. Most of the region's art and cultural institutions are in the city.

Most of our neighborhoods are quite attractive. We have strictly enforced regulations against nuisances such as graffiti, high grass, vehicles parked in yards, and poorly maintained exteriors. Our city is filled with great people. Rochester residents will roll up their sleeves and work for their neighborhoods when given a chance. They are the salt of the earth.

Neighbors Building Neighborhoods: Citizen Empowerment as a Central Strategy

In Rochester, our continual challenge is to revitalize the city, which despite its unique assets, is poor and has limited public resources to work with. As a primary strategy, we decided to tap into the energy and knowledge of our citizens. When I became mayor in 1994, one of the first initiatives of my administration was a process called Neighbors Building Neighborhoods (NBN).

With the consent of our neighborhoods, the city was organized into ten geographical sectors for planning purposes, respecting existing neighborhood associations' boundaries. Organizations and residents in each sector were encouraged to envision a future for their neighborhood, the result being ten action plans with a total of 1,450 action steps. The plans were not created by a professional planner, but by the citizens themselves.

We want people to dream when they create their plans, but not with an anything-goes attitude, because we also want them to manage the implementation of their plans. To make sure the plans are grounded in reality, the city sets an eighteen-month horizon, after which each sector plan is revisited, revised, or redone. Also, proposals are allowed only if the citizen-planners can identify a partner or partners who will sign a form in support of the project. If they cannot identify resources to fund an idea, and if they don't have a committed partnership to help, then the idea doesn't go into the action plan. This forces the entire NBN process to focus on *outcomes*, not general goals. Since the inception of Neighbors Building Neighborhoods, an average of 76 percent of each plan's projects have been completed.

The City of Rochester does not fund the NBN process for more than incidental expenses. Also, the city doesn't choose NBN leadership, and, generally, city staff do not even go to neighborhood meetings. The city acts as a partner, and provides every tool that it can to help the involved citizens implement the projects in their plans. It is focused on outcomes rather than on control of the details of how to get there. All the city wants to know is what residents are trying to get out of the process, and whether the plans reflect the consensus priorities of the sector.

The NBN sectors are linked by a NeighborLink computer network, giving them access to such resources as city databases, GIS mapping software, and 3-D virtual planning tools, as well as secure e-mail and an information management system. A team of volunteers—our community technology leaders—works with the city to maintain and update the network.

Perhaps the most powerful tool that residents have is a voice in the city's budget. When I was president and CEO of the Urban League of Rochester, neighborhoods had a say only in the neighborhood development allocation (4 percent) of the approximately $18 million that the city received each year from the federal government for community development. Small as it seems, this level of citizen involvement was actually quite progressive for an American city. But I wondered what would happen if citizens were involved in establishing spending priorities for the entire city budget. That's exactly what we have now—the entire $450 million annual capital and operating budget is supported by the citizen planning process.

The Impact of Zoning

As people got into the nitty-gritty of planning, they began to feel restricted by zoning. They identified a number of common concerns focused on the existing zoning code. In response, beginning in 2000, we held more than 120 public forums over a three-year period to rewrite our zoning code.

In the United States, almost all zoning ordinances are based on what is called *Euclidian zoning*. Development in a given area is restricted to single uses, such as residential-only, commercial-only, or industrial-only. This makes it difficult to create the kinds of healthy, mixed-use and mixed-income urban neighborhoods envisioned in the NBN plans. Our new zoning code includes standards for higher density mixed-use development, delineation of urban village areas, pedestrian-oriented development, and flexible adaptive reuse, as well as a streamlined permitting process, more objective codes enforcement, and incentives for public art and creative design.

The new zoning also helps residents deal with some thorny issues. For example, many older U.S. cities face the huge problem of *nonconforming uses*—the vacant, obsolete factory in a residential neighborhood, or the abandoned gas station in a historic neighborhood. Rochester's zoning code enables the creative reuse of these properties. It is a powerful tool that has led to the adaptation of many vacant buildings in our neighborhoods to small-scale commercial and mixed uses. We also borrowed from Toronto's King's regeneration initiative, such that we now have no use restrictions for self-enclosed buildings in our city center. We don't care what use a building has—provided it is legal—as long as it meets strict design criteria. This flexibility allows us to fill vacant buildings.

Outcomes of Citizen Empowerment

NBN sectors have been able to leverage private partnerships and funding to accomplish many things the city otherwise might not be able to accomplish. Examples include new playgrounds, street furniture, public art, neighborhood resource centers, neighborhood festivals and street cleanups, neighborhood marketing initiatives, three hundred new community gardens, neighborhood business incubators, and a land trust that purchases and rehabilitates run-down properties.

The land trust received a $1 million foundation grant to establish an urban agriculture program. They operate a vineyard and a farm in the city. The urban farm also has been an outdoor classroom for more than three hundred college students in eight disciplines at the Rochester Institute of Technology. These are only a few examples of literally hundreds of small but important projects that give a neighborhood a sense of social and economic vitality.

All the NBN sectors now have community development corporations, formed to expand revitalization capacity. For years they have been involved in the construction of affordable and market-rate housing. Now they are beginning to move into commercial development.

One project of which I am particularly proud is our supermarket initiative. Through the NBN process, the city and some of our poorest residents developed comprehensive reinvestment strategies for their neighborhoods. Then we presented these plans to retailers. To make a long story short, seven new full-service supermarkets have opened in city neighborhoods. The first store opened in our poorest neighborhood, where median household income is only $14,000. More than 100,000 square feet of additional retail space was also opened by merchants who used the supermarkets as anchors.

Another example of the NBN process is our extensive program to rebuild major streets as gateways. The NBN sectors have secured several major state grants for amenities. Besides new sidewalks, lighting, landscaping, and pavement width, neighborhood charettes look at issues such as strengthening commercial areas along the street and turning vacant land into redevelopment sites. As the streets are rebuilt, the Rochester Housing Authority—which is responsible for public housing—rebuilds all of their units on the street. The housing authority, which is independent of the city and manages nine thousand units of public housing, now evaluates its investments according to NBN plans.

The Neighbors Building Neighborhoods process allows us not only to leverage more resources but also to target them in ways that create a real synergy—a total that's greater than the sum of its parts. Like many other cities, we are rebuilding our waterways. Several hundred million dollars of private and public construction are under way or planned, including a thirteen-mile pedestrian and bicycle trail to link our established neighborhoods with the new developments. In these and virtually every other project in which the city is involved—at every step of the way, from beginning to end—citizens are at the table in a decision-making capacity.

Now, the United Way allocates its funding based on the NBN plans. In cities like Rochester, if you don't have enough money to do something yourself, you will be working with several nonprofits. Our nonprofits are getting the message that they don't have to go out and do their own planning. Residents have already said what they want and need. In the case of the United Way, that means $22 million each year spent the way residents want it spent.

Assessing the Challenges

I have given a taste of the benefits of the Neighbors Building Neighborhoods process, but what are some of the drawbacks?

One is that it can take a lot longer to get a project done. A public works project that used to take a year to complete now takes two or three years by the time all stakeholders are properly heard. This can add to costs, but the result is a better project.

Special area needs like downtown development also becomes problematic. Sectors naturally focus on issues within their neighborhoods, with the result that unique zones, like downtown, do not rise to the top as priorities. So while many of the neighborhoods surrounding our central business district revitalized quite impressively, the commercial core was decaying. The city's position was that the heart of downtown would come back only when the entire region wanted it to come back, only when the broader community would recognize downtown as the region's core and a shared asset. The city and its residents by themselves wouldn't—and shouldn't—do all the heavy lifting.

I am very proud of our citizen empowerment process. It has been emulated by large cities such as Houston and Miami, by midsize cities such as Des Moines, Iowa, and Newark, New Jersey, and by small cities such as Erie, Pennsylvania, and Corning, New York. We were even asked to give a few presentations in New York City, including one to their city council. But the reality is that cities in much of the United States are caught in a dilemma. They are starved for resources, yet they are expected to provide more services to an increasingly poorer constituency. How much can you realistically expect from empowered citizens when a quarter or more of them are living in concentrated poverty, and many of the issues they face are strongly influenced by policies and actions and jurisdictions outside their sectors and the city?

In truth, I wish that NBN were supplemented with urban growth boundaries as in Portland, Oregon, or a metro government like Toronto has—some way for the city to capture a share of the regional growth. Such a system would get everyone from the neighborhood level to the regional level pointing in the same direction. But there is no political will for regionalism in upstate New York. NBN is the best we can do with the hand we have been dealt.

The Spirit of Urban Communities

Today, every politician in America champions "community" and "citizenship." However, as we know, these terms can mean very different things. On the right, they symbolize the pledge of allegiance, respect for authority and religion, and the replacement of the welfare state by private agencies that appeal to the spirit of voluntary cooperation. For people on the left, a revival of citizenship and community seems to require economic as well as political decentralization.

Words like *citizenship* and *community* make us feel good, which no doubt explains their popularity. Yet, if you listen carefully to the speeches of people such as George Bush or Hillary Clinton when they talk about community—as they often do—they are talking less about community and more about their fear of forces that destroy community, such as social fragmentation, hypercompetitive capitalism, and self-seeking individualism. They do not use the latter terms, of course, but that is what's beneath the rhetoric.

Most politicians bemoan the disintegration of the traditional family. A few even bemoan the disintegration of traditional neighborhoods. But it appears beyond the scope of their imaginations to seriously consider the reconstruction of face-to-face communities as a way of restoring a sense of connection.

In Rochester, we assume that virtue lies in self respect by empowered men and women, not in powerful, bureaucratic institutions. Institutions work by leveling people and cities to a common type. This is why reforms such as urban renewal failed. All cities were categorized according to a set of standardized pathologies such as "blight" or "poverty," and standardized "solutions" were then imposed on them. Then when urban renewal inevitably failed, people in damaged neighborhoods were expected to pull themselves up by their bootstraps. This is hardly a way to inspire people.

In Rochester, we believe that community, in its deepest and richest sense, must always remain a matter of face-to-face interaction. That this idea can seem radical in America is a sign of how much we view people as objects, rather than the subjects, of community. Historian Lewis Mumford put it well when he said, "the best economy of cities is the care and culture of human beings."

Planning an Urban Renaissance: Rochester 2010

In 1997, we began a very public three-year process to incorporate the ten sector plans into a new comprehensive master plan called "Rochester 2010: The Renaissance Plan." Building from the base of NBN visions and priorities, thousands of metro area residents at nearly a hundred public forums decided how they wanted the city to look, and what role they wanted it to play in the region by the year 2010. This gave residents a citywide and regionwide focus as well as a neighborhood focus, and increased suburbanites' stake in the city. At that time, we aligned our budget with the goals of the 2010 plan. In fact, the goals of the 2010 plan have become the performance measures for which department heads are accountable. If a proposal doesn't support the goals of the citizen-created 2010 plan, it doesn't get into the budget. There are no exceptions.

Because the city puts its money where its mouth is, people remain energized and engaged. Citizens do not have to wonder if the city will support them. In particular, they do not have to wait for the city's permission to act. They have the tools, knowledge, confidence, and support to go out and form partnerships and leverage resources for the projects they design.

Note

1. The material this chapter is based on has been drawn from two speeches delivered by Mayor Johnson: "Rochester: The Path Less Traveled" (June 21, 2001) and "New Times, New Visions, New Challenges" (September 15, 2005).

13

Transforming Brownfield Communities: The Naugatuck Valley Project (Connecticut)

Kenneth Galdston

The Naugatuck Valley of Connecticut contains 189 brownfields (abandoned, heavily polluted former industrial sites) along 45 miles of highway. This is the story of how people in the Valley are organizing to clean up these sites, redevelop them for productive use, and engage local residents in this work through the Naugatuck Valley Project (NVP) brownfields remediation/jobs campaign.[1]

The Naugatuck Valley and Manufacturing Jobs

From the late 1700s to the late 1900s, the Naugatuck Valley was the center of the world's brass industry and the site of some of the nation's most intense industrialization. The Valley was the location of Seth Thomas's clockworks, Charles Goodyear's first experiments in vulcanizing rubber (and the subsequent founding of the U.S. Rubber Company), and Gail Borden's first successful efforts in condensing milk.

The Valley had waterpower and an increasing supply of potential workers, including farmers whose farms were failing from overfarming of the soil and stiffer competition from farms being started in western territories. It also had a base of skilled workers who had arrived on their own from Great Britain or who had been recruited to begin work in home-based manufacturing.

Brass manufacturing in the Valley began in 1790 in the homes of British immigrants who made brass buttons. Yankee peddlers—roving salesmen with a variety of goods—helped these businesses grow by widening the market for the items they produced. The development of rolled brass sheets for button production improved productivity. By the 1820s, the surplus stock from the brass-rolling operations was being sold on the open market, establishing the roots of what became the Valley's chief industry—brass production on a massive scale (Brecher, Lombardi, Stackhouse 1982).

The demand for military goods made in the Valley, used in a dozen wars and military campaigns, gave the region staying power as an industrial hub from the early 1800s through the Vietnam War. The resulting intense industrialization transformed farmland and small settlements into cities and towns whose focus was factory production and whose wealth came from manufacturing. The largest of these was Waterbury, with a current population of 110,000.

Today, almost all development along Connecticut Route 8 consists of factory sites. Route 8 runs along the Naugatuck River, beginning in Bridgeport in the south and ending in Winsted at the northern end of the Valley, with no major east-west highways until U.S. Route 84 in the middle of the Valley at Waterbury. This gives the Valley an insular feeling, and the red brick factory buildings that dominate Valley towns compete visually with rugged forested landscape and high swirled rock walls.

During World War II, these factories had fifty thousand jobs, most of which were skilled and union-represented. While the Valley's economy thrived, especially in times of war, the by-product of intensive manufacturing was pollution of the air, water, and land. Smoke darkened the sky. The river, a source of power, was also an industrial sewer. The land in and around the factories was a dumping ground for oils, chemicals, and other waste products.

The environmental consequences for those living in the Valley were daunting. When Trout Unlimited, an organization devoted to restoring aquatic life in the Naugatuck River, won the removal of factory dams from the river in the 1990s, they described their success in the following terms:

The Naugatuck [was] long regarded as one of Connecticut's most polluted rivers. Throughout the 19th century, factories and municipalities along the river openly dumped sewage and factory waste into the river. By the 20th century, pollution and dams had taken their toll on the river's health. The Naugatuck historically supported migratory fish runs. However, conditions started to deteriorate during the late 1700s, when the Naugatuck became one of the first rivers in Connecticut to be extensively dammed. Dam construction continued during the industrial revolution of the 1800s. The dams provided power, cooling water, rinse water, and boiler water for industries, including brass and rubber manufacturing. These industries and the municipalities that grew up around them discharged untreated wastewater into the river. The fisheries were decimated. (Trout Unlimited 2000)

Longtime area residents recall visitors coming into the Valley in the 1950s to purchase Keds sneakers at factory-store prices from the U.S. Rubber plant. The visitors commented as much about the smell and the air pollution as about the bargain prices for their Keds.

By 1980, as the result of compliance with new environmental regulations, visible air pollution had been diminished, while efforts to address river pollution were just beginning. But relatively few people considered the pollution to be a major issue, until factories started closing because of global competition and lack of investment by new conglomerate owners.

Organizing the Naugatuck Valley Project

The Naugatuck Valley Project was created in 1983 as a broad-based, multi-issue coalition of church congregations, labor unions, and community and tenant organizations. NVP is a member of the InterValley Project (IVP), a New England organizing network focused on organizing and leadership training for sustainable and equitable development.[2]

At the time of NVP's founding, many companies whose roots went back to the early 1800s had been bought out and were now owned by outside conglomerates, and scores of factories were at risk of closing. Still, more than ten thousand skilled, union-represented jobs remained in the early 1980s. But typically the new owners of the factories used their acquisitions for short-term cash flow, disposing of them after just a few years. Within less than fifteen years, there was a wave of plant closings and many jobs were lost.

NVP challenged the conglomerate owners to invest in the plants, to give advance notice of any sale of the firm, and to cooperate in feasibility studies of employee buyouts of the firms. Between 1983 and 1993, NVP helped to bring about such feasibility studies at more than thirty plants. In three cases where employee ownership was feasible, NVP helped the employees become owners. The most important of these was Seymour Specialty Wire, the nation's largest employee-owned and -controlled industrial firm between 1984 and 1991.

Where a buyout by employees was not feasible, NVP focused on sales of firms to other owners who would keep jobs in the Valley. If that goal turned out not to be achievable, NVP focused on severance benefits. NVP helped save more than three thousand jobs during this time, for periods lasting from one to seven years or more (Brecher 1990, 93–105).

By the time the wave of closings had subsided, NVP had already turned to job training and creating good jobs to replace those being lost to the factory closings. NVP organized Valley Care Cooperative, a worker-owned home health care company employing seventy-five previously low-income women. NVP also helped to create the Multi-Metals Training Center, which trains and places an average of eighty students per year in positions in the surviving firms, often family-owned, which continue precision metal manufacturing in the Valley.

Yet, even as NVP enjoyed these successes, many other jobs were being lost. Companies continued to be moved or shut down, and the number of empty industrial sites grew.

Brownfield Remediation and Jobs Organizing

By 1995, NVP organizers had begun to hear from residents about their concerns about living next to abandoned factories. As NVP leaders surveyed the Valley's closed mills, they began to learn about "brownfield" development. Brownfields—previously developed sites with some level of pollution—are so named in contrast to greenfields, the open, never-developed land favored by developers. Most of today's sprawling development is greenfield development, in which shopping malls, office parks, or housing developments are built on vacant, often verdant, land—former farm fields or woodlands—changing its character permanently.

Brownfield development seeks to clean up, or "remediate," older properties and reclaim them as worksites, stores, warehouses, and even residential and public-use buildings. The benefits are increased property values, safer and healthier neighborhoods, new jobs, and better public or commercial services. Reuse of sites can also alleviate the concentrated poverty in cities, and provides an alternative to new development in outlying areas, thus reducing sprawl and traffic.[3]

Together with key allies, NVP worked to bring the Valley's brownfield sites to the attention of public officials, bank presidents, and planning agencies, ultimately achieving EPA designation of the Valley as a regional remediation pilot area, and bringing in $800,000 for the Valley Brownfields Pilot Program. The Pilot Program identified 189 brownfields, and as of 2005, a dozen sites have been remediated and have begun to be redeveloped.

Local Jobs and Federal Dollars

Many Valley brownfield sites were being cleaned up by outside contractors using workers from outside the Valley. NVP sought to secure a training site for environmental remediation technicians in the Valley, and encouraged members of the community to seek this training. The two Workforce Investment Boards that together serve the Valley communities agreed to establish an Environmental Remediation Technician Training Program at Naugatuck Valley Community College. This program trained fifty-seven Valley residents in its first two years of operation and placed forty of them in well-paying environmental remediation jobs. The NVP Jobs Committee is also seeking a local hiring preference in publicly funded remediation projects.

Vision and Necessities

Nearly two hundred brownfield sites have been abandoned by corporations who profited from the labor of the Valley in good times and then left. The resulting heavily polluted sites create both a burden and an opportunity for Valley leaders.

NVP's leaders have learned to think creatively, to envision ways that built spaces can be refashioned to be "eco-effective"—to use the term of William McDonough and Michael Braungart, from their book, *Cradle to Cradle* (2002, 68–91). Their work challenges the belief that human industry must damage the natural world, and describes buildings designed to conserve energy using natural models, and manufacturing processes that return material to nature for reuse.

In developing its vision of the Valley, NVP has the chance to heal more than the brownfield sites. They have the chance to restore balance to the regional development pattern by opening for reuse millions of square feet of center-city space, thereby reducing the pressure for development in the remaining open land in communities at the edge of the Valley region.

NVP's work in the Valley has focused for more than twenty years on a key Valley resource: the factories and the incredible intellectual, creative, and physical energy that went into creating the firms whose work they housed. NVP is fully applying their own intellectual, creative, and physical energies to this campaign, and they're in it for the long haul.

Brownfield Organizing: Strategies and Solutions

Through its campaign to remediate and redevelop scores of brownfield sites, Naugatuck Valley Project leaders have learned a number of lessons that could be useful to communities and regions dealing with toxic waste or brownfield issues. A summary of key learnings follows.

"Why Are We Organizing?"

In organizing for brownfield remediation and redevelopment, most leaders already know a number of answers to this first "Why" question:

They don't like living next to something large and potentially unhealthy or unsafe.

They can imagine many uses for the site once it is cleaned up, ranging from returning it to a productive workspace to seeing it used for commercial or even residential purposes.

They know that cleaning up the site will make their neighborhood a better place to live, and, for property owners, it may increase the value of their property.

Analyzing Self-Interests

Some key interests underlying brownfields campaigns:

- an end to neighborhood blight
- protection of human health, especially children's health
- creation of, and training for, skilled, better-paying jobs with promising career paths
- reinvestment in and reuse of key center-city resources
- reduction of sprawl
- improved income for neighbors of these sites

Such campaigns also reflect longer-term interests such as having a voice in shaping one's future, exercising one's political rights, and acting on one's core values.

While an organization including religious congregations such as NVP thinks about values in a self-conscious way, any group organizing to take on this task can benefit from asking its members "Why are you stepping forward to do something about this? What is at stake here for you?" After the most likely initial answers identified above, it is valuable to keep asking each other "Why?"

The answers that come from continuing to raise the "Why" question lead to consideration of even more fundamental values, such as the following:

Self-determination Those who live and work in a neighborhood have a right to shape its future.

Democratic organization and participation Acting collectively in a democratic process is a right, and leads to outcomes that are fairer for all.

Building community Building community through a campaign leads to strength.

Faith and justice values Such core values call upon us to uphold the dignity and worth of each individual, and advocate for fair and equal access to resources regardless of race and ethnicity.

An obligation to nature and to successive generations The earth and succeeding generations endure beyond our own time. We must act as good stewards of this land and of nature.

Building a Successful Issues Campaign

Strong campaigns generally include the following elements: First, after getting clear on self-interests and underlying values, *engage diverse individuals and institutions* that have a stake in the campaign's issue. One way to achieve this is to launch the campaign through a community-based organization such as NVP, which organizes in inner-city, blue-collar, and suburban communities. It is useful to gain the involvement and backing of institutions ranging from religious congregations and labor union locals to neighborhood, ethnic, and service organizations.

Building a strong campaign also means taking the time to *build community* among campaign participants. NVP leaders hold brief one-on-one conversations with each other at the start of virtually every large meeting to create and strengthen relationships among themselves, especially across the lines of race, ethnicity, religion, and income. Often a focus question helps the process, such as "Why are you concerned about this issue?" or "How does this affect you and your household?"

Effective campaigns are led by a *broad and diverse leadership team*. A conscious focus is placed on *developing the leadership skills* of a variety of leaders through formal training and sharing of responsibilities, so that new leaders can exercise their new skills.

Learning to *listen to one another* is a truly indispensable skill. We listen especially to those directly affected by the issues, so the campaign engages the energy and vision of those with the most at stake. *Good questions* and *good listening* lead to *good stories* that crystallize what people are facing and what they want from the situation, and energize members and allies.

It is important to *develop an accurate power analysis* before moving into action, to discern the history and trends related to the issue, as well as the key decision makers central to achieving your goals.

A good power analysis also looks at the question of who might be on your side: those with an immediate stake in the issue, such as local congregations; those whose stake might be more tangential but still powerful, such as a suburban congregation that feels called to act in solidarity with the poor; a small business organization that wants the neighborhood economy improved; or a health care organization concerned with environmental factors that affect health.

When developing a campaign, *research models or strategies* for addressing the issues and borrow from among the best of these. For example, David Rusk's book, *The Inside Game/Outside Game* (1999), puts forward a dual strategy of working locally while also fighting for larger policy reform that helps local efforts. NVP works to get particularly important brownfield sites cleaned up immediately, while also organizing to get more federal resources into the region for long-term remediation and redevelopment.

In pursuing particular models, *develop working relationships with technical assistance providers* who know those models well. This results in more organizational credibility with the range of players one may need in the campaign, and can also lead to learning relationships with their allies—for example, a loan fund that supports environmental cleanup.

It is important to *see a larger strategic framework* for thinking about and planning the organization's work. NVP benefited greatly from seeing the relationship between center-city revitalization and the reduction of sprawl. This allowed NVP to bring suburban and small-town congregations together with center-city congregations. NVP leaders were exposed to this idea through IVP workshops conducted by john powell of the Kirwan Institute and Myron Orfield of the Institute on Race and Poverty. This helps highlight the importance of *participating in the national conversation* about these issues.

With a power analysis and models in hand, it is important to *begin campaigns with small steps that help build momentum*—immediate, actionable goals within the reach of campaign participants.

Well-planned first steps lead to initial successes that give focus to the work, dramatize a particular story emblematic of the issue as a whole, and give a sense of possibility that something can be done in a situation that many might have considered impossible to improve.

Addressing large-scale environmental damage requires *multisector collaboration*. A key to winning complex issue campaigns is *building alliances* with a broad range of organizations and individuals, including not-for-profit, public, and private sector players.

In its brownfields remediation campaign, NVP built a broad alliance of member groups, ranging from center-city Pentecostal and African-American churches to center-city and suburban Catholic and Protestant churches, as well as labor unions and tenant organizations. NVP then went on to bring together chambers of commerce, Valley banks, local and state government officials, and a regional planning agency to launch a campaign to make the Valley the site of a regional EPA Brownfields Pilot Program. NVP is now developing a collaboration with the Valley's congressional delegation, state legislators, the Archdiocese of Hartford, the Yale School of Public Health, and private developers who specialize in brownfield redevelopment.

When each party stands to gain through an overall success, small differences are easier to live with. It also helps to have the group leading the collaboration be headed by a *strong coordinating team* that has the trust of the various partners.

Ten Tips for Community Organizing

1. Engage diverse individuals and institutions.
2. Build community while organizing for change.
3. Build a broad and diverse leadership team.
4. Listen closely to one another.
5. Develop an accurate power analysis.
6. Research models or strategies.
7. Develop working relationships with technical assistance providers.
8. See a larger strategic framework.
9. Participate in the national conversation.
10. Build alliances and engage in multisector collaboration.

Effective Strategies for Communicating Your Vision

The campaign should be buttressed by a well-considered message that conveys the organization's agenda, the benefits for each party and the community at large, and the underlying values at stake. The most effective strategy for communicating the message is face-to-face contact with stories and testimonies that convey the human consequences of action or inaction, supported by facts and figures grounded in local history and area social and economic trends.

While people generally recognize that they should not have to live with pollution that affects their health or safety, it can be challenging to link this perspective with the complex message that it's also necessary to remediate, restore, and develop specific sites. When communicating this message, NVP begins by addressing the health and safety risks in a given situation, documenting the dangers, and making sure people know about them. The next step is highlighting the benefits of a revived neighborhood or locality. The message is further strengthened by appealing to the broader vision of being stewards of the earth by remediating environmental damage and restoring a natural resource. The final step is appealing to social justice—the most powerless should not disproportionately end up living closest to brownfield sites—which helps to communicate more forcefully still.

Presentations to member groups or potential allies incorporate the following elements: 1) an overview and history of the issue, 2) stories or testimonies, 3) the national context, 4) examples of what others facing this issue have done, 5) a vision of what can be done together, 6) questions and answers, 7) the current status of the campaign, 8) a proposed first—or next—step, 9) open discussion, and 10) an invitation to join in the campaign.

As the campaign moves ahead, leaders organize public action meetings to dramatize the problem and focus media attention on the issues being addressed at any given moment in the campaign. Public actions cause more people to become aware of and involved in the campaign, either directly as members of the campaign, or as members of their organizing team in their NVP member group (Pierce 1984).

The Larger Context for Campaigns

Campaigns have the greatest chance of achieving good results for those who live near these sites *only* if these people have a seat at the table and a voice in the decision making. These stakeholders must be represented through a powerful community-based organization.

As campaigns grow to demand center-city brownfield remediation for an entire region, as well as reduction of sprawl, the affected constituencies grow more diverse. NVP's experience demonstrates that diverse, well-organized, and well-prepared constituencies have the best chance of succeeding. The development of the EPA Brownfields Pilot Program in the 1990s, as well as the creation of state funds for brownfield remediation, have created a platform for organizing around these issues.

Finally, the rise of Evangelical Christian organizations and growing numbers of Catholic, Jewish, and Protestant organizations that embrace biblical admonishments to the faithful to be stewards of the earth, could hasten an emerging consensus that sees brownfield remediation and re-use as part of earthly stewardship.

World As It Is; World As It Should Be

NVP is working to create the world as it should be from the ruins of the world as it is. Remnants of a fading industrial history haunt their present and future, but by fighting for cleanup, redevelopment, and jobs, they are transforming the future not only for themselves and their communities but also for generations beyond their own.

Notes

1. The NVP was organized by the Roman Catholic Archdiocese of Hartford, the Episcopal Diocese of Connecticut, the United Auto Workers, and the Connecticut Citizens Action Group.

2. See the InterValley Project Web site for more information. Available at www.intervalleyproject.org.

3. See the National Brownfield Association Web site for more information. Available at www.brownfieldassociation.org.

Getting Grounded in Place, Time, and Community

In Chicago, Bethel New Life took on the challenge of transforming abandoned buildings into active, vibrant community spaces and world-class destinations. The Atlanta Neighborhood Development Partnership, Inc. (ANDP) applied innovative approaches to build mixed-income housing that supports the needs of low-income families while combating sprawl and other regional challenges. In South Carolina, communities learned the "rules of the game" and applied legal instruments, both to secure rights and to plan their futures. Each of these stories illustrates how groups gain benefits for the community by building strength through knowledge.

As part of standing against destructive forces and practices, often communities and social movements need to step back and assess their capacity and options. Marginalized groups may lack stable resources that mainstream organizations take for granted, such as money, organization, and access to influence. An emerging social movement must discover its own assets, build solidarity around shared goals, and overcome obstacles and challenges from major opponents. Creating an effective social movement requires the firm foundation of a community learning process that brings people together as a unified force for change.

The term *collective identity* refers to the distinguishing characteristics formed through the shared experience of a group of people. Such experiences may be positive or negative. For instance, African-Americans and other people of color have encountered poverty, abuse, discrimination, exploitation, and a legacy of denial by the established institutions of society. Yet, these communities also draw upon positive traditions of struggle to overcome hardship, as do organizations and institutions engaged in political mobilization for racial and economic justice.

Participants in the civil rights movement of the 1960s—whether white, black, or brown, elite or working class—experienced a strong sense of collective identity. However, since the mid-1960s, Martin Luther King Jr.'s vision of a "beloved community" has grown dim. Toward the end of the 1960s, the black power movement resulted in a deeper sense of collective identity among African-Americans, while at the same time fragmenting the earlier civil rights coalition. Paradoxically, however, the civil rights and black power movements became the model for ethnicity-based social movements in Latino, Native American, and Asian-American communities, as well as other social movements based on identity (such as women's liberation and gay rights).

The 1970s and 1980s saw a period of reaction against social movements. A right-wing politics emerged in which communities of color and new identity-based social movements were cast as either causing or exacerbating a range of social problems. This shift, along with other society-wide dynamics of change in the post–civil rights era, undermined the cohesiveness and solidarity of progressive social movements.

Rebuilding a new and positive sense of collective identity is integral to successful social change. Within the regional equity movement, the job of strengthening collective identity often falls to churches and other community organizations. This can be achieved through community learning efforts such as framing new and important issues, developing shared narratives of the past, exposing community members to inspirational stories, and initiating collective action.

The construction of collective identities across the boundaries of race, class, and jurisdiction is a prerequisite for achieving other important goals. With a strong collective identity, a community can transform its social and physical space into an environment that supports the realization of community goals. Centrally located vacant property can be seen as an asset rather than a liability, valuable to existing residents as well as attractive to new mixed-income populations.

Getting "grounded in place" means learning about community lands and how they can benefit residents and others throughout the metropolitan region. This requires working with local stakeholders and government officials to inventory sites that can be put to good use. By learning about legal and administrative dimensions of land acquisition and development and becoming familiar with the natural and historic assets of the community, smart strategic decisions can be made. It is also vital to know how the market works and to make sure that public decision-making is efficient, effective, and transparent. Large-scale public subsidies for infrastructure—such as schools, transportation facilities, and hospitals—can benefit communities rather than displacing them. As middle-class populations return to the city, new strategies can ensure that the community's existing low-income population benefits from smart growth.

Chicago, Illinois

Bethel New Life, a faith-based community development corporation, exemplifies the commitment to place that can help to transform, even reinvent, a community's basic institutions. Led for much of its existence by Mary Nelson, it has been active in Chicago's economically challenged West Side for more than forty years.

Revitalizing community treasures (such as the hundred-year-old Garfield Park Conservatory, saved from closure in the mid-1990s) is but one example of how Bethel's approach of working person by person, block by block, has turned problems into assets for disenfranchised neighborhoods on the West Side. For Nelson, civic challenges are portals to untapped opportunities such as transit-oriented development, community-based job training, and child care programs. Nelson's work demonstrates the value of patience and tenacity, coupled with creative reuse of existing structures and learning new skills.

Atlanta, Georgia

Over the past three decades, the greater Atlanta region has surged with development and now faces the fiscal, social, and environmental consequences of unplanned development and sprawl. With no natural boundaries to slow its sprawl, the metropolitan region now extends out from the city center to suburbs and exurbs, almost sixty miles away.

One of sprawl's insidious effects in this region over the years has been to lock people of color out of opportunity, limiting their access to jobs and housing. Under the direction of Hattie Dorsey, the Atlanta Neighborhood Development Partnership, Inc., works with community development corporations to create economically viable mixed-income communities so that families with a wide range of incomes can live near their workplaces in the central city and its surrounding inner-ring suburbs. Through initiatives in development, advocacy, and housing financing, ANDP helps families facing economic challenges to benefit from the region's opportunities. Since 1991, ANDP has developed more than six thousand mixed-income housing units in Atlanta's metropolitan region.

Coastal South Carolina

Getting grounded in place, time, and community is particularly challenging when the land where one's family has lived since the Reconstruction changes hands due to a lack of the full "bundle of rights" of private ownership. African-American land loss in the southeastern United States is a classic illustration of the nation's racial blind spot that is part of the challenge in fostering equitable metropolitan development.

In the Sea Islands off the South Carolina coast, a unique community has been under pressure by market forces for several decades. Living in the shadow of the region's embedded racism, many long-term residents have been left without benefit of legal protection. Conventional representations often separate urban and rural development, viewing African-American relationships to the land as being of marginal importance. In a collection of stories about regional equity, this chapter may appear to be an idiosyncratic offering. Yet, this loss of land in South Carolina's coastal region, given its origins in the failures of post–Civil War Reconstruction and in the ensuing institutionalization of racist practices, has significant regional equity implications. Developers of retirement homes and resort areas in South Carolina and other places in the Southeast have realized windfall profits from speculating in these valuable coastal properties. As a consequence, large extended families have been forced to relinquish their land for pennies on the dollar.

The chapter by Faith R. Rivers, assistant professor of law at Vermont Law School and former executive director of the South Carolina Bar Foundation, and Jennie Stephens of the Coastal Community Foundation of South Carolina tells the story of how heirs' property owners have been helped to retain their property in the face of developers' determined efforts, through education about their property rights and through direct legal support. Supporting residents to secure the rights to their land carries with it the additional benefit of helping to preserve the region's besieged Gulla-Geechee culture.

14

Community Activism for Creative Rebuilding of Neighborhoods (Chicago, Illinois)

Mary Nelson and Steven McCullough

"The Green Line Is Closing"

In late 1991, news that the Chicago Transit Authority was closing a one-hundred-year-old transit line linking a low-income minority West Side Chicago community with jobs and services in the suburbs galvanized the Bethel New Life community organization into action. Closure of the elevated transit line would be another blow, on top of deindustrialization, brownfields, and being a credit-starved community.[1]

The Green Line cuts across low-income communities of both the South Side and West Side of Chicago, as well as serving seven western suburbs. To preserve the transit line, Bethel New Life searched for allies and found not only South Side community groups just as concerned about the line's closure, but also Greenpeace, the Transit Riders Association, the Neighborhood Capital Budget Group, and some suburban townships and business communities. Together these groups formed the Lake Street El Coalition in 1992. Local political representatives from the city and suburbs were also prodded to join in the effort to save the line from closure.

Thus began a long, slow process of organizing for fairness in public transportation and for citizen participation in transit decision making.

From Riots to Rebuilding

The circumstances that led to the founding of Bethel New Life had roots in the summer of 1966, when a series of riots in the West Side neighborhoods of Chicago caused widespread damage. This sparked white flight to the suburbs and increased the apathy of absentee landlords. The West

Side swiftly declined as businesses closed, housing values plummeted, investment dropped, and banks redlined neighborhoods.

By 1979, when Bethel New Life began as a housing ministry of the Bethel Lutheran Church, other churches had been gradually leaving the West Side. The faith-based corporation got its start by putting up $35,000 to renovate a nearby dilapidated, turn-of-the century building— investing in housing development for a community others had all but ignored.

Bethel's leaders soon realized that they could never subsidize housing enough—this alone was not sufficient. People had to have jobs—decent, living-wage jobs. With a majority of the employment opportunities moving to the suburbs, a spatial mismatch was created. This led to a growing awareness of the need for mass transportation linking city residents to jobs in the suburbs.

From Protest to Possibilities

Over many years as a community catalyst and developer for the West Garfield and Austin neighborhoods, Bethel New Life has followed a strategy that has worked time and again: *Identify an issue, form a coalition of allies, hold a press conference, put pressure on public officials, keep the issue in the public eye, devise solutions, and find funding.*

In the face of the threatened closure of the Green Line, Bethel linked with disparate groups who understood that organizing needed to be regional, and who were united by a common understanding of the need for accessible public transportation. Hearings and meetings with political representatives highlighted the gross federal spending disparities and unfairness between highways and mass transportation. Joint meetings of city and suburban groups with local and federal transit officials were held to discuss the possibilities. The American Civil Liberties Union became interested.

After demonstrations, protests, and representations at transit board meetings, the Chicago Transit Authority ultimately committed $380 million in redirected funds to rehabilitate and renew the one-hundred-year-old line. The project was completed in 1998. A joyous victory celebration was held once the commitments were made. However, it was evident that monitoring would be vital to ensure that the community benefited from all aspects of the project.

Bethel Center: A "Smart, Green" Building

During the transit line rehabilitation process, training sessions offered by the Center for Neighborhood Technology (CNT), a nonprofit technical assistance organization, helped community groups like Bethel New Life see new possibilities for transit-oriented development. When Bethel's staff looked with fresh eyes at the dark, dilapidated intersection of Lake and Pulaski, they discovered that almost three thousand people got on and off at this stop each day. This constituted a major community "asset" and a previously undiscovered opportunity. An effort was launched to develop a commercial center at the transit stop.

Ten years after the protests around the line closure began, not only was the line renewed but Bethel also dedicated a "smart, green" 23,000-square-foot commercial building, the Bethel Center, at the Lake Pulaski intersection. Bethel's new facility is a model "smart" building, with photovoltaic cells, a "green roof" of insulating soil and wild flowers, automatic light dimmers, and recycled materials. A bridge joins Bethel's building to the Lake Street Elevated (El) Train platform that now links community residents with the western suburbs, the location of many of the jobs in the Chicago metropolitan region.

This $4.5 million building was made possible through creative financing. It took more than five years to piece together the financing and grants, partly because the process was hampered by turndowns from three banks before the final loan was in place.

Bethel Center anchors the redevelopment of the area near the transit stop. It houses Bethel's employment and community technology center, a child-development center serving more than one hundred clients, and six commercial storefronts that have given local businesses a chance to expand—creating more than seventy new jobs for local residents.

The new building is also the home of a pilot model of a financial services center, in partnership with a local bank and Thrivent Financial for Lutherans. The goal of this center is to move low-income people into the financial mainstream, and to put predatory payday lending out of business in the community.

The commercial center complements the work by Bethel and other groups to stabilize the surrounding area. Bethel has built more than fifty new homes in walking distance of the center, with more planned. Rev. Evan Hines, pastor of Keystone Baptist Church in the Lake Pulaski neighborhood, notes a growing sense of hope: "People see their community being transformed, a community being restored." He gives much of the credit to Bethel New Life.

Reversing the Flow of People—A Regional Attraction

Bethel also played a major role in establishing the alliance dedicated to rehabilitating the Garfield Park Conservatory, a group of greenhouses and gardens extending over a city block. "Bethel looks at a need and they go right to where the need is," says Eunita Rushing, president of the Garfield Park Conservatory Alliance. Built in 1907, and once Chicago's botanical crown jewel, the conservatory had fallen on hard times by the late 1980s. "Physically, it was in bad shape and it was kind of a ghost town," says Lisa Roberts, director of the conservatory. "People weren't coming anymore."

About fifteen years ago on a wintry Saturday, Bethel received a phone call. The conservatory's decrepit heating system had failed and the institution's future was in jeopardy. Bethel got the alderman involved, brought in Friends of the Park, and held a press conference in the cold conservatory saying that it is a national treasure and that it cannot go to rot and ruin.

While local support was sought, Bethel was also able to draw on resources well beyond the West Garfield area by understanding the importance of the conservatory to the region and making the wider connections. First, they helped to link the high school across the street into the efforts. Then, they followed up by identifying allies, pressuring public officials, devising solutions, and helping to find the money to make things happen.

With an infusion of almost $8 million of public and private money, the Chicago Park District committed to rehabilitating the conservatory, and people came back—from 74,000 in 1998 to more than 600,000 in 2002, when the conservatory hosted a special exhibit by Dale Chihuly, a nationally renowned glass artist. Chihuly's striking and fanciful glass flowers, lily pads, and other plants displayed among the conservatory's specimens were an unexpected blockbuster. Seeing the crowds of people was "like winning the lottery," says Roberts.

With the conservatory officially back in business, the Chicago Transit Authority spent $7 million to open a transit stop next door—again with pressure from Bethel, the Conservatory Alliance, and other groups. The horse stables adjacent to the conservatory have been transformed into an April–October market, with garden shops, citywide artisans, light food, and even entertainment on a regular basis. The conservatory has become a regional destination and a tourist attraction. People look with new eyes not only on the renewed conservatory but also on the developments in the surrounding community.

What Do You Do with a Closed-down Inner-City Hospital Campus?

Bethel's other centerpiece is the Beth-Anne Life Center, the ambitious renovation of the former 437-bed, 7-building St. Anne's Hospital site. Completed in many stages, the 9-acre campus is now home to 125 units of independent seniors' housing, 85 units of assisted living, a cultural and performing arts center, a small business center, a children's day care center, and Bethel's administrative offices.

The Urban Land Institute has cited the center as a national model of adaptive reuse. Like the transit building that came much later, this development took many years, many different government and private sector grants, strong community support, and a great willingness on the part of banks and the Beth-Anne Life Center board of directors to take a risk. "It will take a miracle," headlined a Chicago business newspaper. Ultimately, it took twelve years and a lot of push and pull.

A Green Development Agenda

After engaging in a community-planning process and taking a look at the ongoing opportunities, Bethel has adopted a "green agenda" that emphasizes mass transit, adaptive reuse of existing buildings, addressing environmental concerns, and an approach to neighborhood problems that links them with the problems of the larger metropolitan region. The plan includes not only transit-oriented development but also greening elements like pocket parks, tree planting, traffic-calming strategies (traffic circles and cul-de-sacs), pedestrian-friendly designs, and energy efficiency. Indeed, Bethel's new commercial building cuts energy costs by approximately 50 percent.

As a start to bringing a green approach into its housing efforts, Bethel New Life built fourteen single-family homes with the guarantee that heating bills will not exceed $200 a year. All their affordable homes currently under construction have solar roofs, and Bethel will continue working to ensure that green, affordable housing is part of all future development projects.

Informed by these examples, other churches in the community as well as private developers are also developing green housing. Chicago mayor Richard Daley, who has become interested in the greening of the city, has supported building code changes and zoning amendments that push developers to include similar components.

Transforming a Community; Living a Vision

Bethel is committed to helping the community move forward in a way that does not leave anyone behind. During twenty-five years dedicated to championing a "smart growth" approach to inner-city development, Bethel New Life has played a key role in the construction of more than 1,100 units of affordable housing; helped some 7,000 people find living-wage jobs; created quality, low-cost day care; helped ex-convicts reenter society; sponsored clinics; and helped addicts get off drugs. It has also had remarkable success in its fight to preserve and improve public spaces and facilities.

Bethel has evolved into a diverse, faith-based community development organization—a local organization with a national reputation for cutting-edge initiatives. Local residents form the backbone of the board of directors, and a strong network of relationships with local churches, block clubs, peace beat groups, and community institutions keep Bethel grounded with diverse input and accountability.

The source of Bethel's energy and vision is their faith—God's call to be about the work of justice and community. Weekly church gatherings provide a strong sense of support for the long haul and serve as reminders of God's vision of people thriving in community.

Not everything Bethel has tried has worked—but at least they have tried. One needs to have the courage to try, and to hope that it will work. Strong connections with other churches—both local and suburban—have made all the difference in mobilizing the moral and political leverage to move things forward in such challenging times.

Bethel's development work and the Chicago real estate boom caused Bethel to shift gears. After spending years trying to lure development money to the West Side, Bethel began to concentrate on making sure that money didn't flow in so fast and so indiscriminately that longtime residents were pushed out.

Lessons Learned

Bethel has learned a lot through its many efforts toward "Sustainable Communities, Equitable Growth." It identifies six important ingredients for such work:

1. *Participatory planning*, involving community residents in all aspects of planning and priority setting. This can be messy and difficult, especially when delays in plans make people lose hope, yet it is essential.

2. *A transit-oriented, walkable city*, including housing, transit, commercial, and play spaces.

3. *Energy-efficient, "green" construction and design*, creating affordable homes with reduced operating costs.

4. *Environmentally friendly economic development*, converting brownfield sites into jobs and other uses.

5. *Community improvements*, including "greening" with trees, placement of homes to reduce energy costs, safe play spaces for children, and traffic-calming strategies that enhance pedestrian safety and ease.

6. *Advocacy for equitable growth in the regional context*, especially for equitable allocation of infrastructure improvements, affordable housing, and public transportation.

Issues like equitable spending on infrastructure, land use planning, equitable government funding for mass transit, affordable housing, and clean air all depend on a regional strategy. These are high on Bethel's current agenda as it seeks to achieve impact on a wider scale. To do this requires regional and statewide strategies that combine a wider range of citizen groups and interests. Here again, faith-based connections and relationships help cross the borders of confined communities through a common commitment to fairness and the common good, and an opportunity to communicate with one another around the regular weekly religious services where people gather.

Community development projects are not easy; often they take many years to bring to fruition. Community groups can take on different roles in the development process—from being developers themselves, to agitating for development that benefits the community, to collaborative and partnership efforts.

It is important that a clear set of "community benefits" is incorporated into community development projects. For example, such benefits might include local and minority hiring commitments, patronage of local small businesses and vendors, opportunities for local investment, or development standards that include pedestrian-friendly designs. Experience indicates that these community benefit commitments need to be spelled out in specific and measurable terms with monitoring devices in place, then established in a written agreement and publicized to the wider world to gain public support and encouragement for the developer (whether private or government) to meet these goals. For example, first-source hiring agreements require a developer to fill a certain percentage of the jobs from the community (as the first source). Usually a local group helps recruit such potential employees and verifies community residence.

Bethel advises others working on similar issues to begin by looking at what you have to start with. Look at the assets in your community and see how you can use them. Then launch the visioning process with that as a foundation and build from there.

A key strategy for success in the management of community-based organizations is to focus on the outcomes that you want to achieve, and to be specific about what you want to achieve. When you have the end in mind, you can work more skillfully on getting there.

As you proceed, it may be useful to keep the following five things in mind: 1) do your homework and know your "stuff"; 2) create a lot of allies; 3) be willing to share the limelight; 4) cultivate "staying power"; and 5) don't burn bridges—you may need them later.

Another essential thing to keep in mind is that bringing out other people's skills is an essential aspect of leadership. You do not have to do it all yourself—in fact, you cannot do it all yourself. Get clear on the help you need—then put out the call for help, and it will come.

A Strategic Vision

In a recent presentation at a national conference, Bethel offered the following recommendations to community groups and government and private sector developers:

Be sure the local community is at the table early and in all aspects. Although this may be a bit unruly and take longer, inclusion saves a lot of grief later.
Get acquainted with the community and the people—not just with the politicians. Such relationships go a long way in mitigating misinformation.
Figure out regular and effective communication channels to keep people informed. Secrets ruin many a development; they break the fragile sense of trust needed to work together.

Build on capacities and opportunities, not on needs. Instead of identifying all the needs and deficits of a low-income urban community, have residents identify the positives and strengths of the community, such as access to public transportation or proximity to the downtown.

Capacities include the residents, too—such as a viable work force, lots of elderly, etc. Once you have identified capacities and opportunities, build on these.

Government should provide incentives and/or requirements for private sector developers to involve community partners. The prime opportunity to require community participation and local hiring is when a private developer is seeking zoning changes, easements, infrastructure assistance, or even funding. In return for this assistance from government, developers can be asked to include community groups at the table, and to make commitments for local hiring.

Bethel encourages community groups to discover the "hidden assets" of their communities, such as transit stops that link one community to another, institutions that can be revived, and people who can be empowered to see the possibilities and then work to make them real. The following lessons are offered to community groups facing challenging projects:

Large projects are a long, hard struggle. You need staying power for the long haul. It is often better to break large projects down into phases, so that results can be celebrated and courage gathered for the next part of the challenge.

"Keep your eyes on the prize." Stay focused on the outcome you have hoped for and do not get sidetracked.

You need champions and allies. Search out even the most unlikely ones who share some common interest.

Find the win-win, but make sure the community benefit is foremost, not just your own organizational good.

Find resources for hanging in there. Keep in mind that delaying is a tactic that may be used to get you off their backs.

Share the glory, and a good word, even with the people who you find most difficult to work with.

Miles to Go Before We Sleep

The journey from protest to possibilities changed Bethel and the whole community, one step at a time. The process included transforming vacant lots and dark commercial corners into new housing and community development, using public transportation as a major "connector" between racially and economically separated urban/suburban communities, and breaking down the walls of isolation of a "ghetto" community to help it become a regional attraction. It took persistence and hard work—fueled by faith and the support of the faith community—and the involvement of government, citizens, and the private sector.

Community transformation can happen even in the most difficult situations. It is a process that requires a vision of community that brings people together in action, and that links a neighborhood to the wider "community." It is an exciting, challenging, difficult, rewarding journey—one that starts with the first step.

There is still much to do. The journey continues.

Note

1. Portions of this chapter have been adapted from Rob Blezard's article, "Chicago Hope," in *Ford Foundation Report*, (Summer 2004): 14–19. Available at http://fordfound.org/pdfs/impact/ford_reports_summer_2004.pdf.

15

Opportunity-based Housing in Atlanta (Atlanta, Georgia)

Steve Lerner

"I just wanted it to be possible for a secretary to walk to work from her downtown apartment," said Hattie Dorsey, president and CEO of the Atlanta Neighborhood Development Partnership, Inc. (ANDP), one of the frontline nonprofits working to make the way development is done in this country more equitable to low-income residents.

Rather than building segregated enclaves where the rich and poor live separated from each other, Dorsey works to create mixed-income developments that permit working-class residents of Atlanta to live in good neighborhoods with decent schools close to where they work. Founded by Dorsey in 1991, ANDP now has a budget of $4.3 million and a staff of seventeen.

To date, ANDP has been remarkably successful at building affordable housing, owning 773 mixed-income units. In addition, it operates a $12 million revolving loan fund for nonprofit and for-profit developers of affordable and mixed-income housing. Over the years it has loaned $21 million that has helped finance 3,364 units.[1]

However, none of this came easily, Dorsey recalled in an interview. Initially, some of the power brokers in Atlanta were resistant to the idea of mixed-income housing. "Let's face it. The assumption is that if black people move into a neighborhood they will bring undesirables with them. There is no getting around it. We bring race and class issues to the [negotiating] table and insert them into the conversation," she said. "Are we always welcomed to do that? The answer is no. We advocate for inclusion, and we are not always invited. Sometimes we just show up." Dorsey, who often heard sermons about boycotts and protest marches at her father's church, and knew Martin Luther King Jr. and Sr. as frequent visitors to her home, plans to title her autobiography, if she ever writes one, "Elbowing My Way to the Table."

Exclusionary Zoning Causes Jobs-Housing Mismatch

"Making the Case for Mixed-Income and Mixed-Use Communities," published by ANDP in 2004, documents the causes of the crisis in affordable housing in Atlanta and points to a number of possible solutions (Goldberg 2004). One central cause of the lack of affordable housing in the metro region is the promulgation of exclusionary zoning rules by affluent suburban jurisdictions, which effectively exclude low-income residents from moving in. As a result, teachers, policemen, firemen, secretaries, bank clerks, gas station attendants, restaurant employees, and hospital orderlies are unable to live near where they work in the suburbs. This exacerbates a variety of urban problems including unemployment, long commutes, traffic congestion, and worsening air quality.

Under exclusionary zoning rules, whole counties have forbidden the construction of multifamily buildings, making it impossible to build affordable rental units. Douglas County set minimum lot and house sizes that jack up housing costs to the point that working-class people cannot afford them. In some jurisdictions, there are even zoning provisions that dictate what kind of exterior finish is required on houses, further increasing construction costs and housing prices.

"It costs two to three times what a teacher makes in salary to live in some of our suburbs," Dorsey commented. To compound the problem, many affluent suburban jurisdictions forbade the Metropolitan Atlanta Transit Authority from extending their mass transit lines out to job centers in the suburbs (Bullard, Johnson, and Torres 2001). Exclusionary zoning rules reduce the stock of affordable housing in the affluent communities where most jobs are being generated. "Just do the math," Dorsey suggested. Two-thirds of the jobs in metro Atlanta and one-third of the family incomes are less than $40,000 a year; with the average housing price at $190,000, most working-class residents can't afford to buy a home. One in four families now pay more than a third of their income for housing. To economize, many families are doubling or tripling up in substandard rental units. For Atlantans in the bottom tenth of the income scale, who make less than $19,000 for a family of four, only 6 percent of the housing in the region is affordable, and most of that is in desperately poor, inner-city neighborhoods or in older working-class suburbs.

The unwillingness of officials in affluent suburbs to permit the construction of affordable housing, combined with limited transit routes, leaves working-class residents with difficult choices. Some choose to take several buses and then walk to work in suburban areas. This trek can take hours and require them to get up before the sun rises and get home well after dark, leaving little time to spend with their families. As a result, many urban children see little of their parents and are left to their own with little mentoring or parental supervision.

In areas where there are no buses or trains out to the suburban jobs, workers must buy and maintain a car. With the annual cost of transportation per family ranging from $5,446 in the city to $8,000 in the suburbs, some people simply can't afford a car, and are effectively cut out of the job market.

Those who do manage to buy cars find their ability to buy a home dramatically decreased. The expense of owning enough cars so the average family can get to school and work is estimated to cut $75,000 off the size of the mortgage for which they qualify, thus diminishing their ability to buy a first home. Recognizing that families that live near their work or can take transit to work have lower transportation costs, some lenders have created Location Efficient Mortgages (LEMs) that provide higher mortgages to families that locate near where they work.[2]

Costs to Business

Exclusionary zoning rules that create the jobs-housing mismatch also hurt local business. "Some businesses in Atlanta are beginning to recognize that they are suffering financially because their workers have to commute over long distances on traffic-clogged highways," observed Dorsey. Employers who make large investments in worker training find it difficult to retain workers who must make a long commute. This has brought some of the big employers to argue for mixed-income housing projects, Dorsey said. Some of the largest businesses in the city, such as Varian and Arcada National, guaranteed the early loans ANDP made for mixed-income housing construction because they wanted to bring employees closer to where they worked.

Despite the obvious merits of mixed-income housing, convincing suburban county officials to increase the diversity of types of housing they permit is challenging. Many of them are convinced that keeping out working-class residents will keep property values (and property tax revenues) high. This perception, however, is contradicted by a study from the Joint Center for Housing Studies at Harvard University and the

Neighborhood Reinvestment Corporation. The study looked at property values in "working communities" in metropolitan areas, where 56 percent of that U.S. population lives, and where their household income is 60 to 100 percent of the area median income. It found that communities with a diversity of housing types and costs (including multifamily housing) actually have higher property values than similar communities with less diversity of housing types (few multifamily units), and a smaller range of housing costs.[3]

Despite this evidence, many suburban officials are unwilling to lower their zoning barriers to the construction of affordable housing in their jurisdiction. There is some hope that this trend may change as suburban residents recognize their own life-cycle housing needs. "Grandparents are getting fed up with the fact that their children and grandchildren cannot afford to live near them," Dorsey noted. People need different kinds of housing for different phases of their lives, she explained. For example, when college graduates return home they find there are no units they can afford to rent in the neighborhoods where they grew up. And older residents who want to move into a smaller condo or apartment find none is available in their community. This forces them to choose between staying in a house that is too big or expensive for them or leaving a community and a network of friends that they love.

Olympic Push for Mixed-Income Housing

The decision to bring the 1995 Summer Olympics to Atlanta created an impetus to re-examine the city's housing stock, particularly in the area where the games would be held. "The city fathers didn't want the world to come to Atlanta and see all our pimples," Dorsey observed bluntly.

At a gathering of city decision makers discussing what kind of housing should be built around Centennial Olympic Park in downtown Atlanta, some participants advocated a high-end development that would look like the apartment buildings around Central Park in Manhattan. But Dorsey argued that it should be designed as a mixed-income neighborhood that included both market-rate and affordable units.

After a brief silence, the meeting room erupted into a heated debate. A number of participants told Dorsey there was plenty of low-income housing in Atlanta and asked why the city should build more in this desirable location. "They just looked at the messenger and that is what they heard," she said. When further challenged, she said, "Your problem is that because I am a black woman you thought I was saying we should bring in all low-income people, but I said mixed-income!" In the end, Dorsey's vision of building mixed-income housing around Centennial Park prevailed.

Centennial House

Near Centennial Park, ANDP partnered with the Sun Trust bank to build Centennial House, a 101-unit condo complex that included 26 affordable units that secretaries, schoolteachers, firemen, policemen, and young professionals could afford. By bringing $1.2 million in equity to the table, ANDP was able to lower the cost of the subsidized units by $50,000 each so that they sold for $120,000 for a one-bedroom unit to $160,000 for a two-bedroom unit. By contrast, the market-rate units sold for about $200,000 to $300,000.

Today, Dorsey's vision of helping a secretary live near her work in downtown has become a reality for April Simon, thirty-two, a hard-working young woman from North Carolina. Simon moved to Atlanta and landed a job at Invesco where she has risen to associate in sales and marketing. Four years ago, in 2001, she was faced with a stark choice: she could either pay high rent to live in the city or find an apartment in a distant suburb with a long commute.

Seeking a better solution, Simon took a seminar on how to buy a house, and when she heard about the affordable condos on sale at Centennial House she jumped at the chance. With little savings and a job as a secretary, she barely qualified for the $120,000 mortgage. The deal was made possible with $6,000 in a down-payment-assistance loan from the Atlanta Development Authority. In all, Simon counts herself lucky. "We have a good mix here: blacks, whites, Asians, homosexuals, heterosexuals, people who drive fancy cars, and those who take MARTA," she said, "and you can usually find someone with the skills to fix just about anything by placing a notice on the bulletin board."

Clustering Mixed-Income Developments in Transitional Neighborhoods

ANDP seeks to cluster its infill developments in the "transitional neighborhoods" that are undergoing economic and demographic change so that their projects will have a positive impact on the community and encourage others to invest in the area. One such neighborhood is the Martin Luther King Jr. Historic District located near King's birthplace. Previously known as the Sweet Auburn District, the area was a center of black commerce from 1860 to 1940. Then, in the 1960s, a highway project bisected the district, exacerbating economic disinvestment that was already plaguing the community. Businesses closed up, houses were abandoned, vacant lots went to seed, and residents who could afford to decamped.

Recently, ANDP helped prime the development pump in the neighborhood by investing $3 million from its loan fund and raising $18 million for the construction of Studioplex, a 153-rental-unit complex built in a 104-year-old cotton warehouse that had been converted into a Sears storage facility. The renovated building now contains 112 live/work units, 24 of which are affordable and rent for $482 to $575 depending on size. The rest are market-rate units that rent for $760 to $1,200.

Farther out at the periphery of the city, in an area known as Adamsville, ANDP is completing the first phase of a massive, thirty-two-acre project where they purchased two troubled apartment complexes, relocated families to other housing within a three-mile area, and demolished the buildings. In their place, ANDP is putting the finishing touches on a 152-unit, four-story senior living complex, complete with its own doctor's office and examining room, dispensary, movie theatre, library, pool, and herb garden. Adjacent to this complex, they plan to build 150 town houses and single-family homes.

Mitigating the Impact of Gentrification

"At the outset we had such economically deteriorated neighborhoods that we were glad if anyone moved back into town," Dorsey recalled. Later Atlanta became a "hot" real estate market, and $400,000 homes were going up next to homes that previously sold for $75,000. With real estate prices soaring, some low-income homeowners found their property taxes rising to the point where they were forced to sell. Worst hit in such situations are families with children, elderly people on fixed incomes, and renters evicted as landlords turn apartments into condos. "We didn't put protections for these people in place in time," notes Dorsey.

One partial solution is to ensure that homes are reassessed and property values go up only when they are sold, thus protecting long-term, low-income homeowners from being displaced. Further, community development corporations (CDCs) can be funded so they can continue to build and maintain affordable rental properties. Local jurisdictions can help by donating land they own for this purpose, or by funding CDCs and local land trusts to buy up properties while they are still cheap as sites for future affordable housing.

Slow growth cities, such as Baltimore and Philadelphia, can learn from Atlanta's experience and secure land while it is cheap so that there will be a place to build affordable housing once the real estate market heats up. "Here in Atlanta, we didn't adequately plan for our own success," concludes Dorsey.

Although the gentrifying real estate market in Atlanta brought challenges, the work of Dorsey and ANDP to spread the construction of mixed-income housing is at the cutting edge of the "equitable development" movement, and is likely to gain favor with urban planners in the coming years. This is not simply because mixed-income developments will give low-income residents of Atlanta better access to jobs, good schools, and safer streets—although it will certainly do this. Mixed-income developments help solve a number of problems that afflict the middle class and the affluent as well as the poor, like traffic congestion and air pollution—two of the most pressing problems that Atlanta's urban planners face.

Finally, the mixed-income development strategy advocated by ANDP has the potential to reduce the concentration of poverty that creates areas where few are tempted to invest, and to convince business leaders that it is an important way to attract middle-class residents and business investors back to all corners of the Atlanta metro region.

Notes

1. ANDP's Community Redevelopment Loan and Investment Fund, Inc. (CRLIF) helps fund both its own mixed-income projects as well as those of other local CDCs involved in the same type of work. In addition to a $1.5 million contribution from the U.S. Treasury, CRLIF has also been capitalized by loans and grants from the Bank of America, GE Capital, Wachovia Bank, the Robert Woodruff Foundation, and others.

2. See S. Bernstein, Center for Neighborhood Technology (Chicago). Bernstein pioneered the Location Efficient Mortgage.

3. A. Von Hoffman, E. Belsky, J. DeNormandie, and R. Bratt, "America's Working Communities and the Impact of Multifamily Housing" (Neighborhood Reinvestment Corporation and the Joint Center for Housing Studies of Harvard University, 2004, p. 2). Available at www.jchs.harvard.edu/publications/communitydevelopment/w04-5.pdf. Note that this study looks at "working communities" where the median household income is between 60 and 100 percent of the area median income. Working communities are host to a majority of Americans and contain some 157 million Americans or 56 percent of the population.

16

A Regional Approach to Affordable Housing (Atlanta, Georgia)

Hattie Dorsey

Sprawl and Race

"Hot-lanta," as it is sometimes called, is the capital of and largest city in Georgia. Today's Atlanta has so many great things to offer: amazing nightlife, beautiful museums, four sports teams, Coca-Cola's headquarters, successful music artists, wonderful festivals, a rich and controversial history, and a true Southern hospitality without compare. It has also gained notoriety for being home to the Confederacy, the setting for *Gone with the Wind*, a seat of the civil rights movement—and the poster child for rapid urban sprawl.

The related problems of sprawl and inner city abandonment have been evolving over the past three decades. Although the city's former mayor, William B. Hartsfield, proclaimed Atlanta to the world in the 1960s as "The City Too Busy to Hate," a polarizing racism was still the reality within its borders. As a result of white flight in the 1970s and 1980s, the city of Atlanta had one of the highest poverty rates of any city of its size in the 1990s. A total lack of investment in inner-city, largely African-American neighborhoods left the city's urban core highly undesirable. Dilapidated housing and overgrown weeds plagued many inner-city neighborhoods, and crime had become the order of the day.

Prior to the emergence of Atlanta Neighborhood Development Partnership (ANDP) in 1991, community development efforts in Atlanta were minimal and the city was ill equipped to address its large-scale and complex socioeconomic needs. The city had largely overlooked these issues, the philanthropic community barely recognized them as a concern, and activist organizations from the past (such as the Urban League) were not at the table. No energy was going into rebuilding the city's core communities, and no one was talking about the fundamental issue underpinning the problem: race.

The Creation of ANDP

In 1989, *The Atlanta Journal-Constitution* won a Pulitzer Prize for a series of articles called "The Color of Money" (Dedman 1988), highlighting the discriminatory practices of lending institutions toward African-Americans across the economic spectrum. If a well-educated black man with a six-figure salary could not get a loan to make improvements to his house, what hope did the rest of black Atlanta have? Shortly after "The Color of Money" was published, a *Wall Street Journal* article that referred to Atlanta's "myth and hype" angered some of the city's leaders who had proclaimed that Atlanta had moved beyond race. The article drew attention to our fatal flaws and advertised that we weren't nearly as progressive as we claimed. While the message was hard to take, issues that had been ignored for years were finally surfacing. At this time, the concept of ANDP began to take form in discussion groups.

Simultaneously, Atlanta began its bid for the 1996 Olympics, portraying itself as a city of Southern hospitality and racial harmony. Once it was announced that Atlanta had won the bid, the business community's anxiety increased when all of the locations chosen to hold stadiums for the games turned out to be in blighted communities. Now the entire world was going to be in our backyard to see firsthand the dirty laundry we had desperately tried to keep hidden.

This predicament put the creation of ANDP on a *very* fast track. The "powers that be" remembered talk of an organization like this forming, and they were suddenly quite interested in making it happen. ANDP ended up playing an integral role in reshaping targeted neighborhoods through an on-the-ground approach, working closely with newly formed neighborhood community development corporations (CDCs) and their residents. This marked the beginning of a change in Atlanta, and by the late 1990s, the city finally saw itself on an upswing.

Atlanta in the New Century

Greater Atlanta finds itself not unlike other metropolitan regions that are prospering outside the central city. Its diverse economic base includes rapidly growing white-collar industries that are increasing per capita wealth and portend continuing regional affluence in the future. Population and job growth show no signs of slowing, with projections indicating an increase of 2.3 million residents in the next twenty-five years.

Atlanta's reputation as an inexpensive place to live and do business has spread far and wide—and so has the region. This rapid expansion has been accompanied by inadequate planning, by policies that discourage mixed-income and mixed-use developments, and by the lack of core connectivity in a region with too many separate municipalities and no main entity thinking about the big picture. As a result, Atlanta now has serious land use issues, accompanied by problems with air and water quality, traffic congestion, and the financial burden of rising housing and transportation costs for the average wage earner.

Atlanta has become a place of economic opportunities for both whites and African-Americans, and has attracted new immigrants from Latin America and Asia. Between 1990 and 2000, the metropolitan Atlanta region saw the white population increase by 10 percent, the African-American population by 70 percent, and the Asian population by 255 percent. Most notable, however, is the Hispanic population's explosive 461 percent increase. Much of this new population resides in three of Atlanta's twenty-eight counties.

The Atlanta region has embraced the cheap labor force that these new residents represent, without being prepared to address the underlying issues that come with a largely non-English-speaking pool of workers. The majority of these immigrants have low-income jobs such as food service, yard care, hotel maid service, construction, and child care, and a variety of challenges prevent them from attaining basic needs such as education, transportation, and housing. The necessary social infrastructure, transportation, and housing policies are not in place to ensure safe housing near the jobs where they are employed. Housing discrimination is still a huge problem, although many people have been unwilling to consider the issue as part of the regional agenda.

A Crisis in Housing

Mounting evidence indicates that our region is losing its ability to provide modestly priced housing for families. Far too many Atlantans no longer enjoy the option of living in the same neighborhood throughout the course of their life. The people who keep our basic systems running—those who teach our children, police our streets, and staff our offices and call centers—find it nearly impossible to afford housing in the jurisdictions that employ them, and the news is worst for the third of our families who earn a total of $40,000 or less per year. While Atlanta's job mix is shifting toward lower-paying service and retail jobs, the choices for affordable homes near job centers are growing narrower, since almost no one is building housing for the people who fill those jobs.

After fourteen years, ANDP has come to realize that affordable housing and neighborhood revitalization issues reach beyond the boundaries of neighborhoods, cities, and counties and beyond the capabilities of any single organization to manage. If the Atlanta region is to preserve the quality of life for which it is known, we must find a way for our civic, political, and business leaders to make it a priority to provide affordable housing in reasonable proximity to job centers. Developers, both for-profit and nonprofit, need to work together to address this issue.

Model sustainable communities include affordable housing and economic development so that moderate- and low-income families can live near jobs, services, shopping, and other life amenities to which those of greater means have access. Turning this understanding into action has involved the cooperative efforts of the Atlanta business community, national and local foundations, and numerous partners such as CDCs, the nonprofit and for-profit development community, and the public sector.

ANDP has brought the term *mixed income* to the forefront in Atlanta by speaking about it at every table. Indeed, the way people regard the pattern of Atlanta's development is beginning to shift. Awareness is growing of the essential role mixed-income development can play in challenging sprawl.

ANDP's commissioned report, "Making the Case for Mixed-Income and Mixed-Use Communities: An Executive Summary" (Goldberg 2004, 28), describes the challenges arising from the practice of strictly segregating land uses and incomes. Large concentrations of lower-income families, isolated from the opportunities available to other income groups, breed hopelessness and crime. In contrast, mixing incomes within jurisdictions accommodates people in various stages of life. In a successful "choice" community, both an executive and his seventh-grade daughter's teacher could live and shop in the same district, while the child could ride her bike to see her grandmother or her brother who recently graduated from college.

Forty years ago, Jane Jacobs elaborated on this topic in her book, *The Death and Life of Great American Cities*. She wrote of integrated, manageable communities with a diversity of people, transportation, architecture, and commerce. The holistic approach that she espoused is at the core of all of ANDP's efforts today.

Sound Public Policy Is the Key

The biggest current obstacles to getting our work accomplished are the subtle exclusionary policies that keep lower-income individuals and families out of more affluent jurisdictions. These include policies that mandate large minimum-lot and house sizes, place moratoria and long-term bans on multifamily construction, require fewer homes per acre than local comprehensive plans call for, and require costly exterior finishes.

Various federal, state, and local policies have contributed to the mismatched availability of housing, transportation, and employment in the region. The promotion of suburbanization through incentives such as the exclusive use of state gasoline revenue for the building of roads and highways has been a major cause of this imbalance. Other such incentives, combined with prosuburban development policies, have included exclusionary land use practices, tax incentives for the wealthiest Americans (e.g., mortgage deductions, capital gains tax laws, and the largest federal housing subsidy program in the country—the mortgage interest rate deduction), public subsidies that attract businesses and developers to the suburban fringe, and investment in suburban public infrastructure. Together, these policies and incentives set the stage for inequity in the region's distribution of housing, employment, and amenities.

Up to this point, most local policy makers have focused their attention on only transportation as the means of addressing Atlanta's issues. What they have failed to realize is that the region's intolerable traffic can be traced to the use of land—specifically, single-family units on large lots—resulting in low-density development reinforced by exclusionary zoning practices. The availability of quality, affordable housing closer to job centers and to goods and services would be a significant step in addressing this problem.

ANDP has spent significant time and energy working with local politicians and regional leaders to advocate for policies that will shape a sustainable community for the future. ANDP's study of housing affordability in metro Atlanta made it clear that the issues cannot be solved by any single level of government or by jurisdictions acting alone. The first and most important step is to establish a broad consensus and spirit of collaboration for solving our issues, applying common rules of the development game at the state and local levels.

Tools for Success: Policy and Communication Strategies

In November 2003, ANDP's Public Policy Subcommittee developed a set of public policies to advocate within local, county, and state government, as well as the private sector, through 2010. Some of the policies can be realized in the short term, while others will take a number of years, and the help of our partners, to accomplish. In the near term, the three most important policies in the view of ANDP are inclusionary zoning, regional fair-share plan, and zoning/land use plan alignment.

Inclusionary Zoning Policy
To increase the availability of affordable housing, we support encouraging municipalities to adopt an inclusionary zoning policy that would reward developers for making 15 to 20 percent of new developments meet affordable housing rates. In return, developers would receive the right to build more units, faster approvals and permits, tax or impact fee abatement, or other considerations. As a result, communities diversify and strengthen their tax base, retain a competitive advantage in attracting employers, and aid in reshaping economically and environmentally damaging commuting patterns.

Regional Fair-Share Mixed-Income Housing Policy
We have proposed a regional plan that would correct the growing imbalance between affordable housing locations and job centers by distributing housing for people of all income levels equally throughout the region. Fair-share efforts, which must be enacted by local governments, are also designed to combat zoning conditions that lead to economic stratification of housing, limited housing options, and traffic congestion.

Zoning/Land Use Plan Alignment
The Georgia Planning Act of 1989 requires local governments to develop comprehensive plans that include the housing and transportation needs of all its citizens. Unfortunately, local zoning maps and day-to-day rezoning decisions often do not reflect this intent. We believe local governments need to change zoning maps and development policies so that

they are consistent with their own plans, and their housing and transportation goals.

Improving Communications

We want to see ANDP become known in the region as championing housing affordability for everyone. We know the necessity of good communication and education, but putting these insights into practice has posed some significant challenges. One has been the stigma attached to the terms *mixed-income housing* and *affordable housing*, assigning them to the poor, indigent, and nonworkers, and people on public assistance. Many people also believe that affordable housing will look cheap, will be poorly maintained, and will reduce property values. A race component also factors into the resistance.

ANDP has worked to develop messages that will change beliefs and attitudes—and, ultimately, behaviors—of policy makers, business leaders, and the general public around affordable mixed-income housing. We do this by humanizing the issue and by drawing attention to the benefits and opportunities of mixed-income communities, as well as the problems of sprawl and polarization.

We are developing a campaign that connects the dots between more affordable mixed-income housing and improvements in traffic problems, air quality, and other basic quality-of-life issues. The long-term viability of the Atlanta region is dependent upon housing diversity that brings with it a diversity of workers, providing a broad economic base that will strengthen business and create better communities.

We intend to reach each of these constituents via direct one-on-one engagement; small group discussions in the form of sector-specific roundtables; and media outreach in the form of newspaper ads, op-ed pieces, billboards, articles, and radio and television spots.

We have learned that in the Atlanta region, we need to use several angles to tackle metro equity issues. The business community is mainly concerned about economic factors, so language that refers to "competitive advantage," "workforce housing," and "workforce development" is instrumental in communicating our goals. The core message to policy makers addresses "quality of life" for their constituents.

Because much of the general public holds the misconception that affordable housing is unattractive, attracts crime, and reduces property values, politicians often have a difficult time moving this issue to the forefront of their agendas unless their constituents are holding them accountable. Our public communications campaign focuses on dispelling these myths through careful education and demonstration of what a successful mixed-income community looks like.

A Vision for Affordable Housing

We want to develop communities where people can live, work, play, learn, and age. We foresee a regional approach to housing that combats sprawl, emphasizes balanced growth and mixed-income housing, and preserves green space. Our overall goal is to ensure that jobs, transportation, and land use are always a part of the fundamental strategy and are not discussed as separate issues.

Meeting the challenge of affordable housing in our region may be our greatest test yet. It will require the commitment of dozens of local governments, as well as support from business, state government, and citizens at large, to apply the tools we have determined will help us show how a modern metropolis can provide more equitable housing and keep its competitive edge.

17

Preserving Heirs' Property in Coastal South Carolina (Charleston, South Carolina)

Faith R. Rivers and Jennie Stephens

The Low Country of South Carolina is known for its moss-covered trees, marshland, beaches, creeks and rivers, Southern cuisine, historic buildings and plantations, and the Gullah-Geechee culture. The Low Country—spanning nine counties, from Horry in the north to Beaufort in the south—is a landscape rich in heritage and tradition. The growing appreciation of this heritage and culture is changing the landscape of these communities. What once was primarily an area of bedroom rural communities is now becoming populated with transplanted retirees, upscale resorts, gated communities, and exponential growth.

Nestled among all this growth and development is heirs' property.

The Evolution of Heirs' Property

Heirs' property ownership is a significant problem facing the African-American community in coastal South Carolina. Heirs' property refers to real property held without clear title. The property is typically owned and inhabited by indigenous families, a significant number of whom can trace their ownership back to purchases by former slaves during the Civil War and Reconstruction. The vast majority of heirs' property owners are black.

Typically, isolated African-American communities like Wando, Huger, and Wadmalaw Island are highly concentrated with heirs' property. At one time, these communities were, in the words of a community leader, located "behind God's back"—isolated from major neighboring cities such as Charleston and Beaufort. However, this started to change in the mid-1900s when governments began investing in bridge infrastructure and flood insurance. Up until this time, these properties were generally inaccessible and presented substantial investment risks. Now that this infrastructure is in place, coastal land values are on the rise and outsiders are interested in obtaining heirs' property. Many landowners have already lost their land to developers and have no idea how it happened. Others are signing away their interests without fully understanding the consequences of the simple act of signing on a dotted line. Still others are being threatened by land use ordinances and zoning plans.

Heirs' property owners understand that as heirs die, the interest in the property is transferred to other heirs, but they fail to realize that this new interest makes the entire tract vulnerable to tax sales, partitions, liquidations, and "courthouse steps" sales. In major cities in the North, many family members whose ancestors left the South during the Great Migration hold interests in land to which they have no intention of returning. They become easy targets for outsiders, and many sell their interests to developers, who then initiate partition suits to force public sale of the property at auction for much less than the actual value received by developers in subsequent sales. The family members are rarely able to raise sufficient funds to outbid developers in order to keep from losing their land. The distant relative may be totally unaware of the impact this will have on the rest of the heirs when the developer who has purchased a share turns around and forces a sale of the property.

For many heirs' property land parcels that have been part of the family since Reconstruction, the land and the land deed have remained unchanged since the original conveyance. For generations, the original purchasers of the land and their descendants died without wills, and their estates were never properly probated. The ancestors simply wanted the property to remain in the family. With the deed to the land registered to a deceased family member, the land has been handed down from generation to generation through the intestacy laws that apply when there is no will. It is then owned by a group of relatives who possess fractionated interests as tenants in common. The only changes taking place have been family members coming and going, setting up permanent and mobile homes on the land.

Consequently, small family villages have sprung up on these lands. This process seemed quite adequate for heirs' property owners. They were never educated as to how they could best protect their most prized possession.

Providing a place for family members to call home, a place to live, to keep a garden, and to remain in close proximity with the family is exactly how many ancestors intended their property to be used. However, the method that heirs' property owners have used to preserve their property for future generations has made them vulnerable.

"From the River to the Road"

During interviews with heirs' property owners, the full import of this situation became clearer. A middle-aged grandmother's simple phrase, spoken without bitterness or anger but with a hint of profound regret, summed it up. While looking toward Bohicket Creek on Wadmalaw Island, she told us, "We used to own all of this land from the river to the road." This grandmother's mournful description of how her inheritance has been reduced from one hundred acres of land to a mere one-acre plot—*systematically and without just compensation*—says it all. Although these individuals did not create the heirs' property problem, it has reached across the ages to meet them face-to-face on the cusp of the twenty-first century. Despite their best efforts, they are slowly losing the land. As land values escalate and cities such as Charleston and Beaufort expand, and as developers continue to capitalize upon the coast as a lucrative venture, this class of property owners will continue to shrink. These landowners have become victims of their cultural practices—practices that conflict with the property laws of South Carolina.

The Bundle of Rights Is Half-Empty

Many African-American property owners are unaware of the problems associated with heirs' property, the disadvantages of owning such property, and the means by which they can protect their most prized possession. Furthermore, misinformation and lack of understanding about property ownership cause numerous problems for low-income landowners.

Heirs' property is perpetuated by the lack of estate planning. By operation of de jure and de facto discrimination, coupled with economic limitations, throughout most of the twentieth century, generations of African-Americans did not enjoy meaningful access to legal counsel. In the absence of a community understanding of the importance of legal documentation of clear title, and without meaningful access to the legal system, heirs' property ownership continued from generation to generation, with heirs' property owners often unaware of legal avenues available to them to retain ownership and control of the land.

In part, this is because heirs' property owners may distrust strangers who come into their community showing an interest in or asking questions about their property. Often heirs' property owners have seen valuable property taken by legal means from their friends, family, and community members by land speculators, developers, and unscrupulous businessmen. For this reason, landowners are distrustful of private attorneys and are reluctant to seek legal assistance. Moreover, many heirs' property owners cannot afford representation when legal problems do arise that involve their land. Through the operation of legal and administrative procedures like partition orders and tax sales, heirs' property owners are in constant danger of losing their inheritance. These threats relegate to a disadvantaged class of property ownership all African-Americans who have inherited their land through intestacy.

Losing the Land Legacy: A Historical Perspective

Despite the disappointments of wartime and Reconstruction land distribution policies, land acquisition was an important objective for African-Americans who had just gained the right to own property after the Civil War. Nowhere was this truer than the Low Country of South Carolina. More than 16,000 African-American families acquired at least 50,000 acres through persistent efforts to shape Reconstruction-era land distribution policies. Though these coastal lands were low and often unproductive for agriculture, families treasured the land and made great sacrifices to hold on to their property.

This was no small feat during Reconstruction and the subsequent Jim Crow era. African-Americans who acquired land did so mostly by private market purchases, often in the teeth of threatened violence, limited access to credit, and overt discrimination. Some refused to record their deeds officially for fear of reprisal (Lewan and Barclay 2001a). Moreover, access to legal assistance was extremely limited during the postwar period.[1] Even today, African-American landowners may shun the assistance of lawyers because of their own past involvement, or their friends', with lawyers and the legal system that led to the loss of land (Lewan and Barclay 2001b).

During the twentieth century, land loss depleted the land and economic resources of African-Americans throughout the United States.[2] The decline in black landownership had a number of causes, including the migration of blacks from the rural South. However, land-takings also contributed to the decline in property holdings.

In an eighteen-month investigation, the Associated Press documented a pattern of African-Americans being cheated out of their land or driven from it through intimidation, violence, and even murder (Lewan and Barclay 2001a). According to Ray Winbush, the director of Fisk University's Race Relations Institute, "[I]f you are looking for stolen black land, just follow the lynching trail." These acts of violence were concentrated in the rural South—including a substantial number of such murders in the Low Country of South Carolina.[3] Land losses linked to lynching were doubly devastating to families struggling to overcome the legacy of slavery.

The Associated Press reported that the true extent of land-takings from black families would be difficult to determine because of significant gaps in public records; they found many deed books with pages torn from them and many records that had been crudely altered. The AP also found that about a third of the county courthouses in Southern and border states have burned—some more than once—since the Civil War. In some cases, the fires were deliberately set; in every case, key land documents were destroyed.

When Heirs' Property Is Not Protected

Can you imagine waking up one morning to be told by the local sheriff that you, along with your wife, children, and grandchildren, must vacate the premises on which you were born and have lived for seventy years? Mr. Waters (name changed), living in Wando, experienced just this.

In 1890, Mr. Waters's grandfather, a former slave, purchased seventeen acres of mosquito-infested marshland next to the plantation on which he worked. At his death, this gentleman left strict oral instructions that all family members have a place to live on this land. This oral instruction was passed down from generation to generation. Each member of the family honored this wish until 2002, when Mr. Waters and his sister, who no longer lived on the property, were unable to agree upon an appropriate division of the land, and his sister filed an action to sell the property.

Although the governing statute expressed a preference for land to be divided rather than sold, the judge indicated that he could not decide how to divide the land between the siblings. After hearing from experts who testified to the value of the property, the judge ordered that the land be sold and proceeds shared in proportion to each heir's interest. The property was sold for $900,000 to investors, who within three months in turn sold the property for almost $3 million.

Mr. Waters did not fare well by this transaction. The only winners were the lawyers who were the first to be paid from the proceeds of the sale, and the developers who purchased the property. Family members shared only what was left. When all was said and done, Mr. Waters and fourteen of his family members were escorted by police off the property that he considered to be "his gold." His share of the proceeds from this sale netted him almost $27,000—just enough to purchase a mobile home. South Carolina law considers mobile homes personal property, which depreciates in value, rather than real property, which appreciates. So while Mr. Waters's new homestead would only lose value over time, the new owners of "his gold" would go on to create a subdivision of large waterfront homes and make a multimillion-dollar return on their initial investment.

The Heirs' Property Preservation Coalition: Making a Seat at the Table

The need for heirs' property services is pervasive. In November 2000, eleven governmental, nonprofit, and faith-based organizations in the Low Country area determined that more than five hundred people needed heirs' property services. The purpose of the Heirs' Property Preservation Project is to support efforts by low-income heirs' property owners to use, retain, or preserve their property in the face of rising development and governmental pressures in the Low Country area. The effort seeks to halt proliferation of heirs' property, stem the tide of heirs' property loss, and eliminate the disadvantaged status of heirs' property under the law. The Heirs' Project offers legal education, as well as legal and mediation services, to heirs' property owners throughout the Low Country. More than one thousand family members have been educated about their options, and mediation services are provided to help families come to agreements on the future of their commonly held lands. In the mediation process, members of the family have a neutral, third-party volunteer assigned by the Heirs' Project who helps them work through any disputes they may have about ownership or other issues related to the property or family. The objective is to empower families to make informed choices about their property and to come to an amicable resolution. Families may wish to conserve their property and maintain the family homestead, leverage their interests to build homes, access capital for key investments, establish family developments, or explore other uses. These choices are part and parcel of the bundle of rights that property owners possess—but unfortunately, heirs' property owners have not had these options.

As a coalition of organizations and service providers dedicated to preservation of heirs' property, the Heirs' Property Preservation Project took a three-pronged approach to addressing the issues presented by this form of property ownership. The project empowers heirs' property owners to preserve their land inheritance through the three primary strategies of *law-related education, legal services and policy analysis,* and *legal reform efforts* that aim to restore their full bundle of property rights.

Creating a broad coalition and involving partners in the program's development and implementation empowered each group with a voice "at the table"—without ceding control of the project's agenda, goals, or strategies to contrarian forces. Giving public recognition to all partners provided value that increased each group's willingness to support the endeavor. Most important, involving heirs' property owners and community representatives in the project's governance, planning, and implementation helped to ensure that the project was rooted in a community-based orientation. In addition, the participation of community leaders helped the project to enhance its credibility, which in turn contributed to its successful community outreach and education efforts.

The South Carolina Bar Foundation[4] and South Carolina Bar Media Services Division collaboratively produced an educational video on heirs' property.[5] The video has been distributed to hundreds of heirs' property owners throughout the Low Country and has been presented at trainings for the probate court, Low Country Masters-In-Equity, and the pro bono panel of attorneys. Lawyers from across the state and nation have thereby recognized the important work of the Heirs' Property Preservation Project.

Exploring Strategies to Restore the Bundle of Rights

Alternative Legislative Approaches
Through the Heirs' Property Preservation Project, the South Carolina Appleseed Legal Justice Center reviewed the partition statutes of

Southern states that have attempted to address the challenges presented in partition sales.[6] Two major strategies have been taken by sister states. North Carolina imposes an enhanced burden of proof. The party seeking sale must prove that he or she would suffer "substantial injury" if the property is divided rather than sold.[7] Georgia and Alabama provide opportunity for heirs to buy another family member's interest through a right of first refusal and a private sale requirement, respectively.[8]

While these measures provide additional protections for heirs' property owners, several equity issues remain. None of these provisions ensures compensation for heirs' property owners who have made substantial improvements to the property or who have assumed responsibility for maintaining the property and paying property taxes. Further, no resources are made available unless the property is sold. Moreover, right-of-first-refusal or private sale options are of no effect to low-income heirs' property owners who do not have the financial means to purchase other heirs' interests.

Tax Sale Procedures and Land Loss

Through the Heirs' Property Preservation Project, the South Carolina Appleseed Legal Justice Center investigated tax assessment and sales procedures in coastal counties and reviewed potential solutions to the problems presented by tax sales.[9] Where the delinquent tax amount is relatively small compared with the property value, and the tract is suitable, heirs can request division so that the amount of property sold is only enough to satisfy the amount of delinquent taxes. In *Folk v. Thomas*,[10] the South Carolina Supreme Court declined to require counties to make a divisibility determination in all tax sales absent specific direction from the state legislature.[11] However, the court did indicate that a landowner could request a divisibility study, and must affirmatively do so to obtain this relief.[12] Unfortunately, most heirs' property owners are unaware that such a request can be made.

Bridging the Black-Green-White Divide to Balance Preservation and Growth

Safeguarding heirs' property ownership in rapidly developing coastal areas presents a particularly complex issue that demands a thoughtful and comprehensive set of solutions, including comprehensive land use strategies that respect and protect land holdings that are part of traditional settlement patterns. Through the Heirs' Property Preservation

Project, the South Carolina Coastal Conservation League identified the challenge as meeting needs for infrastructure that protects the health and safety of residents in rural areas, and accommodating the settlement patterns of heirs' property owners, while enabling families to preserve their lands in the face of increasing tax assessments on developable property.[13]

As no grassroots planning organization takes a consistent, active role in planning and development in the Low Country, low-wealth communities rarely have input in the development of comprehensive plans, while developers and high-wealth communities are routinely included in these processes. Efforts must be made to bridge the "black-green-white" divide that often places antigrowth environmentalists at odds with the interests of heirs' property owners and community leaders seeking to support economic development measures that enhance the standard of living in these communities. Working together, these interests can develop new strategies that promote equitable growth without penalizing low-income families seeking to enhance their economic standing, and that do not come at the expense of heirs' property owners. Project leaders made a conscious decision that the Heirs' Project would not position its work as "antidevelopment"—but rather would seek to promote equitable development that empowers heirs to make informed decisions about their property.[14] Moreover, the communications strategy has emphasized the basic property rights aspect of the heirs' property challenge. While that strategy has met objection from various property rights groups opposed to zoning and land use planning restrictions, it should be noted that on the whole, real estate attorneys that represent developers have not broadly attacked the Heirs' Project.

Where Do We Go From Here?

The current state of heirs' property ownership and the challenges this form of ownership presents took more than a century to develop. Its genesis and proliferation are due in large part to the denial of access to the legal system to a broad class of people, particularly African-Americans. In many social and economic arenas, the nation and the state of South Carolina have not been able to formulate remedies to inequities that stem from this period in our nation's history.

In reflecting upon the Associated Press investigation of African-American land takings, Congressman James Clyburn, who represents significant portions of the Low Country and a majority of African-Americans who likely own heirs' property, challenged the nation to consider how to stop the tragedy of land loss.

Just like many blacks with roots in the South, I grew up hearing stories of land lost by relatives and family friends. These stories were so commonplace and pervasive that I worked with Penn Community Center for many years before I came to the Congress, studying these land takings. To date, Penn Center has collected reports of 2,000 similar cases that remain uninvestigated. And there are other institutions around the South collecting the same kind of information. The question now is where do we go from here?[15]

The creation of the Heirs' Property Preservation Project is an important first step in the effort to utilize evolving legal, social, and political systems to restore the bundle of rights to this class of property owners. What started as a coalition of concerned individuals to address an issue that greatly impacted a vulnerable group of property owners has now grown into a separate nonprofit entity that hopes to work itself out of existence because heirs' property owners will eventually just be known as landowners—period—as a result of clearing titles or legislative reform within the state of South Carolina.

Armed with the knowledge and power to manage their property, heirs' property owners can continue to sustain the rich cultural heritage and land legacy of the Low Country.

Notes

1. No African-Americans were admitted to the South Carolina Bar until 1868, at which time only three members were sworn in. Through the remainder of the century, only forty-six African-Americans became members of the bar. By the turn of the twentieth century, the number of African-American lawyers had decreased dramatically to twenty-nine. See J. R. Oldfield, "Black Lawyers in South Carolina," *Journal of American Studies*, 23 (1989).

2. The U.S. Department of Agriculture Agricultural Census revealed that African-American farmland holdings declined from 15 million acres in 1910 to only 1.1 million acres in full ownership and 1.07 million acres in part ownership in 2001. While the number of white farmers has declined during this period, African-American ownership was reported to have declined two and a half times faster than white ownership.

3. See T. Lewan and D. Barclay, "Torn from the Land: Landownership Made Blacks Targets of Violence and Murder," in the Associated Press, 2001, citing S. E. Tolnay, E. M. Beck, and F. Brundage, *A Festival of Violence: An Analysis of Southern Lynchings, 1882–1930*, Urbana/Chicago: University of Illinois Press, 1995.

4. The South Carolina Bar Foundation serves as the charitable arm of the South Carolina Bar and administers the state Interest on Lawyer Trust Accounts (IOLTA) program.

5. Faith Rivers (producer) and Warren Holland (director), *Heirs' Property* [Videotape] South Carolina Bar Foundation (2003).

6. For a full discussion of statutory options, see Danielle Metoyer, "Heirs' Property: A Survey of Neighboring State Partition Statutes," Appleseed Legal Justice Center (2003). This report was developed for the Heirs' Property Preservation Project under the direction of this chapter's coauthor, Faith Rivers.

7. See N.C. Code 46-22.

8. See GA. Code Ann. Sec. 44-6-116.1; AL Code Sections 35-6-100 through 35-6-104.

9. For a full discussion of tax issues, see Danielle Metoyer, "Heirs' Property: Potential Solutions to the Problems Tax Sales Present," Appleseed Legal Justice Center (2004). This report was developed for the Heirs' Property Preservation Project under the direction of this chapter's coauthor, Faith Rivers.

10. 543 S.E.2d 556 (S.C. 2001).

11. Ibid., 558.

12. Ibid., 558, 559.

13. For a full discussion of tax issues, see Michelle Sinkler, "Heirs' Property: Planning & Zoning," South Carolina Coastal Conservation League (2004). This report was developed for the Heirs' Property Preservation Project under the direction of this chapter's coauthor, Faith Rivers.

14. One HP 102 Seminar featured Tom Barnwell, a native of Hilton Head Island, who has successfully developed family heirs' property, providing one of the few sources of affordable rental and single family housing on the island.

15. Congressman James E. Clyburn, press statement, "Black Land Loss Must Be Addressed," February 7, 2002.

Exploring New Horizons: Connecting Local Struggles to Global and Regional Stories

The civil rights movement of the 1960s worked to address problems of racial segregation and urban poverty in the context of an *industrial* vision of society. In the opening decade of the twenty-first century, solutions to these continuing problems must be developed in a *postindustrial* context, the emerging larger story of our historic epoch (Korten 2006). Case studies in this section illustrate three ways that community organizing efforts have attempted to build bridges to this larger story: 1) by new collaborative leadership roles for working people and communities of color, 2) by linking organizing to the concept of "just sustainability," and 3) by connecting local efforts to a metropolitan regional context.

In an increasingly multicultural, multiracial world, everyone should have the opportunity of learning new and collaborative ways to lead. Responding to persistent racism, prejudice, and bigotry in American society, a multicultural movement has arisen during the past three decades. This movement has had a healthy impact on society as a whole. In the twenty-first century, we are challenged to build on the success of multiculturalism, supporting leadership to create a new sense of the human community as a whole, greater than the sum of its parts. Communities of color and working people have responsibilities to bring forward the issues of their particular constituencies, in ways that improve the quality of life for this larger human community, acknowledging our interdependence while respecting the health of the planet.

Sustainable development is widely defined as development that meets today's needs without compromising the ability of future generations to meet their needs. "Our Common Future," the report of the World Commission on Environment and Development, explicitly referred to goals of reducing poverty and inequality as central to sustainable development. Yet, many environmental organizations in industrialized countries have overlooked these core components of sustainability (Portney 2003). True sustainability cannot be pursued separately from a quest for social and economic justice (Agyeman 2005; Baland, Bardhan, and Bowles 2007).

The quest for regional equity is also an outgrowth of the environmental justice movement (Bullard 2007b). Grassroots organizations in the environmental justice movement have made strong connections between sustainability and the quest for social justice. Many advocates of social justice recognize the possibilities in aligning with the growing worldwide interest in combating global warming, reducing pollution, and protecting biodiversity. Such sentiment can be harnessed in support of incentives for economic change. The imperative of sustainable development also offers opportunities for synergy with issues such as transit-oriented development, mixed-income neighborhoods, affordable housing in close proximity to emerging job centers, reinventing the food system, and the greening of cities.

During much of the twentieth century, advocates thought of cities as compact urban places contained within municipal boundaries. Such a perspective is no longer adequate. Urbanized places are now spread out across the landscape in cities, suburbs, and rural areas (Gottdiener and Hutchson 1999). Immigrants arriving in the United States in the early twentieth century typically settled in inner-city enclaves, while in the twenty-first century many immigrant populations are bypassing older cities altogether and moving directly to the suburbs, where poverty is spreading (Puentes and Warren 2006). Addressing concentrated poverty in the United States in the twenty-first century requires a shift of geographic imagination and consciousness. As David Rusk points out in his influential book, *Cities without Suburbs*, the *city* is the region (Rusk 2003). Another study of fifteen metro regions by the Institute of Race and Poverty found that by 2000, roughly half of the African-American population—and more than 60 percent of Latinos—lived in financially stressed suburban areas (Orfield and Luce 2005).

Developers who arrange the shape of our metropolitan regions—including the location and physical design of housing, workplaces, shopping areas, and schools—are in effect committing the whole society to the decisions they make. When these decisions are made without the input of those who are most affected, there are devastating consequences for marginalized populations. The chapters in this section document ways that social justice advocates have been innovative and effective in addressing environmental issues within the new spatial reality of metropolitan landscapes.

Los Angeles, California

In addition to meeting the needs of future generations, and integrating the environment, the economy, and social equity, sustainable development must also address needs at different spatial scales: global, regional, and local. The struggle for a community benefits agreement (CBA) at Los Angeles International Airport (LAX) presents a promising model for addressing conflicts at these different levels. The airport development represented a countywide response to the challenge of expanded global air transport, while the agreement resolved local economic, health, and environmental concerns.

With a yearly total of more than 50 million passengers arriving and departing on scheduled carriers, LAX is one of the world's busiest transportation hubs. Like many major airports in the United States, the early landing strip that would become LAX was once farmland. Wheat, barley, and lima beans were the principal crops of that area before the historic Hanger No. 1 was constructed in the 1920s.

For communities living in the shadow of LAX, the proposal to modernize the airport offered by developers to the City of Los Angeles in 2000 included little to ensure that their quality of life would not be degraded. Yet, by forming a multiracial, multigroup coalition—and working effectively with city officials and business interests—these communities secured an agreement benefiting thousands of people, while avoiding a lengthy battle in the courts. Danny Feingold, director of communications for the Los Angeles Alliance for a New Economy (LAANE), tells the story of a legally binding community benefits agreement, which set a historic national precedent for community improvements around large-scale projects.

In a companion piece, Greg LeRoy looks at the efficacy of community benefits agreements that have been forged throughout the United States. Founder and executive director of Good Jobs First, LeRoy has been called "the leading national watchdog of state and local economic development subsidies" and "God's witness to corporate welfare."

New York Metro Region

Following the devastating impact of the September 11 attacks in Lower Manhattan, a unified vision for rebuilding the region was needed. Because a strong network of community groups was already engaged in regional planning, a community-based response was organized for the disaster and rebuilding. Civic collaboration in the New York metropolitan area has resulted in plans that reflect the visions and values of a broad constituency.

The Regional Plan Association (RPA) in New York City facilitated a dialogue between the disparate factions in Lower Manhattan following 9/11. An article coauthored by RPA's Robert Yaro, Chris Jones, Petra Todorovich, and Nicolas Ronderos tells how a regional approach using a community-based process responded to the changing needs of one of the world's largest metropolitan areas.

Detroit, Michigan

Detroit is a weak-market city with a long history of abandonment. It is also the most racially segregated metropolitan area in the United States. Victoria Kovari, winner of the prestigious Leaders for a Changing World award, tells the story of how the Detroit-based organization, MOSES, worked with the Gamaliel Foundation, an interfaith group, on land use and transportation equity issues. Such bold work by a faith-based group demonstrates that communities of worship can be a source of sustenance and moral leverage, engaging the policies and structures that shape people's lives. Following Kovari's chapter, three perspectives are offered by the Gamaliel Foundation's Greg Galluzzo, Mike Kruglik, and Rev. Cheryl Rivera, reflecting on the values and vision that underlie a faith-based approach to social change.

18

LAX Rising (Los Angeles, California)

Danny Feingold

To a first-time visitor, the classrooms at the Oak Street Elementary School in Inglewood, California, could easily be mistaken for holding cells at a minimum-security penitentiary. The windows are sealed shut to lessen the roar of the constant plane traffic overhead. No natural light makes its way into the space. There is no air-conditioning, so on hot days the rooms are sweltering. Outside, there is no escape from the planes, whose unrelenting aural assault serves as a menacing sound track for lunch and gym classes.

The appalling conditions at Oak Street Elementary extend far beyond the gates. For decades, communities adjacent to Los Angeles International Airport (LAX) have endured an array of environmental hazards and quality-of-life nightmares, from rattling homes to asthma-inducing diesel emissions to the constant din of truck traffic. The worst of these impacts are felt in the poor, largely minority communities east of the airport: Inglewood, Lennox, and South Los Angeles.

All this is poised to change, however, thanks to a groundbreaking agreement reached in late 2004. The pact between the City of Los Angeles and a coalition of more than twenty labor, environmental, and community-based organizations allocated $500 million to improving conditions for residents living in the shadow of LAX. Under the agreement, homes and schools are to be soundproofed, emissions and truck traffic dramatically reduced, and millions of dollars earmarked to provide local residents with training for living-wage jobs, among other provisions.

By any measure, the LAX community benefits agreement (CBA) was a landmark achievement. Not only was it by far the largest and most comprehensive such agreement in the country, it was the first to be negotiated with a governmental agency, setting a precedent that can be emulated and replicated across the United States. The LAX CBA also represented a giant step forward in the growing effort to forge coalitions between labor and environmental groups, which have often found themselves on opposing sides of land use and development battles. Just as important, the agreement was a huge victory for low-income Latinos and African-Americans, who overcame decades of disenfranchisement, as well as a history of racial tensions, to create a formidable alliance that secured previously unimaginable guarantees from city officials.

Creating a New Paradigm for Economic Development

Traditionally, airport development in Los Angeles—as in most cities—has been imposed upon, rather than negotiated with, the adjacent communities. Airlines, construction firms, land use experts, and elected officials have all had a role in shaping the future of the world's sixth-busiest airport. However, despite the far-reaching environmental and economic impacts of airport projects, residents in surrounding neighborhoods had been left out of the process.

The decision by Los Angeles mayor James K. Hahn to undertake a massive airport modernization offered an opportunity to change this paradigm. Hahn's $11 billion proposal stirred immediate controversy, and it was clear that he would need community support to win approval. The nonprofit Los Angeles Alliance for a New Economy (LAANE), which pioneered the concept of community benefits agreements, recognized that Hahn's ambitious LAX plan opened the door to a community-organizing campaign—and to an agreement exponentially bigger than anything that had come before.

Since the beginning of 2000, LAANE has helped negotiate community benefits agreements for half a dozen private developments, including the expansion of the downtown Staples Center. These agreements have ensured a range of benefits for the mostly low-income communities affected by development projects, from the creation of living-wage jobs, local hiring, and affordable housing to providing green space, child care centers, and traffic mitigations. They also have brought about a fundamental change in the dynamics of development, giving residents a real voice in shaping the future of their communities. This new model has put parents, clergy, activists, and other neighborhood leaders across the table from developers, rather than relegating them to the sidelines or leaving them with no option but litigation to oppose harmful projects.

While previous community benefits agreements provided a foundation, it was clear from the start that the LAX campaign presented some daunting challenges. First, there was the need to build a broad, cohesive coalition that would include groups with differing interests and a history of conflict. Equally formidable was the vast scope of both Mayor Hahn's proposal and the impact of LAX on adjacent communities. Then there was the controversy surrounding Hahn's plan, which many doubted would survive the political minefield that had doomed earlier efforts at airport expansion.

"We knew that there were major hurdles, and that the whole process could disintegrate," says Danny Tabor, a former Inglewood city councilmember and longtime activist who lives under the LAX flight path and played a pivotal role in bringing the community benefits agreement to fruition. "But the potential gains were too great to let the opportunity pass. Those of us who had suffered the consequences of living near LAX were determined to seize the moment and create a better future for our children."

Building a Movement for the Community

In August 2003, the first meeting of the LAX coalition for Economic, Environmental & Educational Justice was held. With more than twenty-five member groups, the coalition represented a broad cross section of the residents most directly affected by the multiple impacts of LAX.

The first seeds of the coalition had been planted six months earlier when LAANE met with Environmental Defense, a leading environmental group that had been working for several years on issues related to LAX. The idea of pursuing a community benefits agreement appealed to Environmental Defense, which was dubious about the prospects for a legal challenge to LAX expansion. Other environmental groups liked the concept as well.

Meetings with a range of labor and community-based organizations yielded the same results—everyone agreed that a community benefits agreement was the best hope for ensuring that any major LAX development plan addressed the real needs of residents near the airport. This view was shared by the Lennox and Inglewood school districts, which saw a chance to improve conditions dramatically for thousands of current students and generations of children to follow.

The convergence of labor, environmental, community, and educational groups marked a breakthrough in coalition building. While unions had generally supported LAX development as a source of new jobs, environmental and community organizations as well as the two school districts had opposed the airport expansion, fearing more health and environmental problems. Indeed, in communities around the country, the perceived tension between economic and environmental interests has stood as a key roadblock to the formation of larger, more powerful social justice alliances.

With the community benefits agreement, however, all of the parties involved recognized an opportunity to transcend these divisions. When Mayor Hahn's proposal and the associated Environmental Impact Report (EIR) were released in the summer of 2003, the nascent coalition sprang into action, analyzing the plan and disseminating it to potential coalition members. This process helped clarify the issues and produced the initial outline of an agreement.

Finding a Place at the Table

Once the coalition was formally established, three committees were formed—one on jobs and small business, one on education, and one on community and the environment. Grassroots groups worked side by side with policy and research organizations to craft an extensive list of demands reflecting the very real needs of their constituencies. "Everyone understood that we weren't going to get everything," explained Maria Verduzco Smith, chairperson of the Lennox Coordinating Council, who played a key role in the coalition. "At the same time, the problems faced by the community are so pressing that it was difficult to pare down our proposal. Ultimately we decided to push for far more than any of us thought we would ever get."

This turned out to be a good strategy—and a lesson that activists in other regions may want to keep in mind. The strength of the coalition, combined with the political difficulties encountered by the mayor's plan, lent an unforeseen viability to the most expansive hopes for the community benefits agreement. "Even as prospects for the mayor's plan grew more questionable, our coalition continued to grow and become more unified," said William D. Smart Jr., LAANE senior organizer and chief negotiator for the coalition. "We were well positioned to negotiate a strong agreement."

Political realities notwithstanding, the mayor and airport officials demonstrated a genuine commitment to the community benefits concept. Negotiations with the coalition weren't just a nod to expediency, but rather reflected the officials' understanding that an inclusive process was the right approach for all parties involved. Nevertheless, talks with the city pressed on for nearly a year, punctuated by stalemates that at times appeared to threaten the agreement. "We had some bottom-line demands that were nonnegotiable," added Smart. "And the city had its interests to protect. This made for some pretty high drama."

In the end, though, the two sides reached an agreement whose size and scope far exceeded what many in the coalition had thought possible. The two school districts stand to receive nearly $230 million to soundproof classrooms and install windows, improvements that will benefit thousands of students. Tens of millions of dollars are to be spent to reduce diesel emissions and to electrify airport gates. Another $15 million is to go toward job-training programs for local residents, who will be guaranteed first-shot at jobs created by the modernization project. (Los Angeles's current mayor and then-councilmember Antonio Villaraigosa pushed hard to increase the job-training allocation, which originally was significantly less). The agreement also includes provisions to help minority- and women-owned businesses win LAX contracts.

Facing a Brighter Future

While the most immediate effects of the LAX community benefits agreement are to be felt by residents living near the airport, the pact also promises to have a major impact well beyond the homes and schools near LAX flight paths. In particular, the environmental mitigations—including emissions controls and the conversion of trucks, buses, and shuttles to alternative fuel—will bring lasting environmental benefits not only to impacted communities such as Lennox and the cities of Westchester and Inglewood but also to the millions of tourists who visit LAX annually. In the wake of September 11, cities across the country have been drawing up plans to upgrade their airports, and as they move forward, many of the same environmental and economic concerns will be raised.

Savvy community leaders and farsighted elected officials will look to Los Angeles as a model for airport development. Those who do will discover that an effective approach to projects of this scale must balance the needs of communities, businesses, and the general public. More broadly, the community benefits model is one that can be adopted to transform economic development throughout the country, so that the interests of business and government are no longer pitted against those of communities.

"The success of the community benefits model underscores that, for too long, we have been presented with false choices—business growth versus good wages, and jobs versus a healthy environment," says Tabor. "We weren't willing to accept that anymore, because we knew there was a better way."

19

Community Benefits Agreements: A Strategy for Renewing Our Cities (Los Angeles, California)

Greg LeRoy

Community benefits agreements (CBAs)—like those won in Los Angeles —are now also being won in other cities such as Milwaukee and San Jose. These project-specific contracts between developers and community coalitions ensure that redevelopment projects benefit local neighborhood residents with first-source access to jobs, living wages, affordable housing, and other benefits.

Word of the Los Angeles victories spread rapidly, catching the attention of activists in many other cities, and prompting a how-to handbook first published in 2002 and reissued in 2005 (Gross, LeRoy, and Janis-Aparicio 2005). The combination of demonstrable results, enthusiasm among community action coalitions, and well-established trends in the urban real estate market, seems to guarantee that CBAs are here to stay—and that they will become more common across the country.

A nascent back-to-the-city trend finds many older areas growing in population for the first time in decades. This city-thickening trend is being driven by family demographics, by some cities' spatial limits, by housing market forces, and by an emerging corporate consensus that sprawling development patterns make metropolitan areas dysfunctional and less economically competitive. At the same time, the resulting surge in redevelopment of already inhabited areas means growing friction between the needs of incumbent residents and the desires of newcomers—friction that CBAs are perfectly suited to help resolve.

National political trends also bode well for CBAs. Declining federal aid for urban areas means cities are being pressed to leverage more private funding and more public benefits from fewer government dollars. Resurgent labor organizing, especially in the low-wage service sector, is increasingly built upon coalitions with community groups—using procurement and economic development rules to promote living wages, accountability in job subsidies, and project-specific CBAs.

Finally, the underlying ideology of CBAs—that taxpayer dollars and corporate accountability are rightful levers to promote racial and economic justice—is a long-established tradition in community organizing and civil rights advocacy in America. CBAs are an echo of lessons from grassroots pioneers such as A. Philip Randolph and Saul Alinsky, as embodied in some of the most creative and promising organizing in America today.

Back to the City

Many of the nation's urban areas are experiencing a distinct back-to-the-city trend, reversing cities' decades-long decline in population. Eight of the nation's ten largest cities experienced population growth in the 1990s; only Detroit and Philadelphia did not. While the overall movement of people to the suburbs still outpaces growth in urban cores, many cities have been attracting a steady influx of residents, particularly young professionals and empty nesters. For some cities, newly arriving immigrants are also a boon.

Urban observers attribute this trend to several factors. Some baby boom generation parents have become empty nesters and therefore no longer choose their address based on school quality. They want shorter commutes and more access to leisure time amenities. More young adults are also interested in living in urban settings. Raised on television shows like *Friends* and *Seinfeld* instead of boomer fare such as *Father Knows Best* or *Leave It to Beaver*, members of generations X and Y more often consider dense urban places to be hip. In most big cities, newly arriving immigrants are moving into older neighborhoods, continuing a long-standing American tradition.

Besides demographic and pop culture trends, some metro areas are changing because of physical limitations. Some urban areas have natural barriers (such as coastal California cities abutting mountains or landslide areas) or growth management policies (such as Oregon's now-endangered urban growth boundary system) that have slowed their geographic growth, forcing development back inward to infill, rehabilitation, and even demolition and redevelopment (Katz and Altman 2005, 32–33).

Partly as a result of this trend, most urban cores also experienced a noticeable decline in the concentration of poverty, as more middle-income people have moved into older areas. It is certainly desirable for many reasons that cities become more economically diverse—including social stability and a healthier tax base—but this trend also inevitably means growing friction associated with gentrification and displacement of low-income residents. This highlights one of the most desirable aspects of community benefits agreements: they enable longtime city residents who endured the hard years to harness the back-to-the-city trend and channel it to their benefit. In the same way that the Community Reinvestment Act enables city residents to gain home equity and thereby benefit from rising home values, CBAs help core-area residents benefit from job creation in older areas (Jargowsky 2003).

Emerging Corporate Consensus Against Sprawl

Several notable corporate-sponsored efforts have been launched to curb sprawl and promote smart growth, including fulfilling two key components of many CBAs: transit choice for commuters and a greater supply of affordable housing. Similarly, many employers see smart growth planning in their self-interest, as evidenced by groups such as Chicago Metropolis 2020, the Silicon Valley Manufacturers Group, the Regional Business Coalition of Metro Atlanta, and the many corporate-sponsored efforts chronicled by the National Association of Local Government Environmental Professionals.[1]

The "Metropolis Principles" promoted by Chicago Metropolis 2020 are perhaps the most explicit statement of such beliefs. More than one hundred major Chicago-area employers publicly embraced the principles, stating that in the future when they decide where to expand or relocate, they will give major consideration to whether the jobs are accessible by public transportation and close to an adequate supply of housing affordable to typical working families.[2] Some observers read this declaration as a not-so-subtle message to exclusionary suburbs: mend your ways or lose your tax base.

The analysis central to this Chicago work has been clearly stated by 2020's leaders: if the Chicago region continues to grow in the same sprawling way, it will become ever more dysfunctional (for commuting and labor-market access) and polluted—further degrading quality of life and causing the region to lose both existing employers and bids for desirable new businesses. That is a sure recipe for economic decline.

Declining Public Resources for Urban Development

Strong CBAs make it possible to use public funds to leverage more benefits from private dollars. In that sense, CBAs are a terrific vehicle for government efficiency and cost-effectiveness. This is especially critical given that federal aid to cities has generally been declining since the Carter administration (with temporary improvement in the Clinton years). Given the ballooning federal deficit and many costly new priorities—such as the War in Iraq, Homeland Security, and Hurricane Katrina rebuilding —it is very likely that federally funded job programs will suffer long-term funding cuts.

State aid to local governments has been the second-largest victim of state budget cuts in recent years, topped only by cuts of funding to higher education (Greenblatt 2005). There have been deep cuts in federal programs such as housing assistance, with states forced to cut services or scramble to make up the difference from their own budgets. Consequently, many cities are strapped for resources. For such cities, CBAs offer significant promise of solutions, both for economic development efforts and for the types of community services and programs that CBAs typically include.

With the decline in federal funding to cities, it is hard to question the argument that when precious public dollars are spent on a project, they should leverage as many private dollars as possible. CBAs are an appealing way to achieve that goal, because they enable community groups to make demands of developers that city governments won't make for fear of gaining a reputation for having a "bad business climate." Pro-CBA community groups have in a sense bypassed hesitant city councils (in some cases with individual council members' blessings), negotiating enhancements that city officials couldn't or wouldn't seek.

CBAs are a win-win situation for city council members and mayors. Elected officials have always touted the arrival of big new projects in their communities; using that development to leverage even more benefits and amenities that directly serve their constituents' needs makes the officials look even better. Despite widespread budget crunches on all levels of government, CBAs provide a way to fund projects and services that would not happen otherwise. Experience is proving that, far from spooking developers, CBAs are helping businesses to succeed. Indeed, in a few cases, developers are even seeking out CBAs to help facilitate the approval process for their projects.[3]

Organized Labor Getting Back to Its Roots

In their efforts to revitalize new-member organizing, some unions are returning to their roots, to a neighborhood-based form of organizing that readily lends itself to community benefits agreements and other smart growth policies. Churches, ethnic associations, and other community groups were the foundation of multiracial organizing drives that birthed the nation's industrial unions in the 1930s. Indeed, job-related issues were key issues in some of Saul Alinsky's most famous community campaigns. In his Kodak shareholder campaign with a Rochester, New York, community group for example, the demand was for local hiring of African-American workers.

In the 1970s and early 1980s, Latino community groups in Chicago, mentored by the National Training and Information Center (NTIC), ran local-hiring campaigns to bring pressure to bear on companies that had received Industrial Revenue Bonds (tax-free, low-interest loans). NTIC was founded by Gale Cincotta, a third-generation Alinskyite, and a leading advocate for the Community Reinvestment Act. Local-hiring campaigns were part and parcel of her neighborhood reinvestment agenda.

Using public money as a lever against private-sector discrimination is also a long-established civil rights strategy benefiting urban workers. When defense contractors discriminated against African-Americans early in World War II, A. Philip Randolph's Brotherhood of Sleeping Car Porters organized the March on Washington. That movement forced President Franklin Roosevelt to create the Fair Employment Practices Commission—a precursor to the Equal Employment Opportunity Commission—to desegregate the nation's defense plants.

Today's movements for a living wage and for accountable development are an echo and a reaffirmation of these proud organizing traditions, leveraging public money—either for procurement or for economic development—to win local benefits. Returning to their roots to survive and rebuild, unions are targeting those urban workers who are most exploited: women, people of color, immigrants, and low-wage workers who cannot afford a car and are hurt the most by housing inflation and the related necessity for long commutes. The AFL-CIO's Union Cities program and the activist network Jobs with Justice both target such workers, as do fast-growing unions such as the Service Employees International Union. These same workers are the primary intended beneficiaries of regional growth policies that improve transit, build more affordable housing, and promote reinvestment in low-income areas. Economic development dollars are integral to all such efforts, providing organizing "hooks" at every turn.

Recovering Self-Confidence in Our Urban Centers

After half a century of being told that their cities are undesirable—that they are dirty, dangerous, and full of unqualified workers—the leaders of many of America's municipalities have a depressingly low level of civic self-worth. This lack of confidence has blinded many urban leaders to the market realities that are making cities increasingly profitable investment opportunities.

Our nation's capital is a case in point. After decades of sprawl-related decline, Washington, D.C., suffered a famine of capital investment during the 1990s administration of Marion Barry. But after a federal control board righted D.C.'s finances and the board's CFO, Anthony Williams, became mayor, an enormous amount of pent-up demand surged back in, and the city's skyline grew crowded with cranes.

Misjudging the market, Williams's administration made a series of fundamental economic development errors typical of a city with low self-image. It launched a costly Tax Increment Financing (TIF) program that included no real targeting of needy areas. It enacted a form of subsidy for venture capital that has proven costly and ineffective in other states. It spent more than $700 million to build an enormous convention center that is operating far below capacity and running at a large loss. It has failed to create any sort of linkage mechanism to ensure that its most distressed neighborhoods benefit from the resurgent central business district. It has also committed to building a costly new baseball stadium despite an overwhelming body of evidence that the D.C. Nationals team will not benefit the city's economy.

Internalizing negative stereotypes about themselves and feeling desperate about their futures, many urban leaders have become easy marks for developers seeking massive subsidies. City officials dare not demand quid pro quos for fear of losing deals. The community-labor coalitions that are forming to demand CBAs are critical ingredients to cities in recovering their self-worth and commanding the respect they deserve. Savvy elected officials who want to revitalize their cities and strengthen their civil societies would be wise to school themselves on CBAs and the forces driving their popularity.

The Challenges Ahead

Seeking to help coalitions in other cities win CBAs, the Los Angeles Alliance for a New Economy and three other California groups have formed the Partnership for Working Families (PWF). PWF provides technical assistance and is targeting cities with strong labor movements and promising redevelopment targets.

But even the best technical advice cannot replace the good old-fashioned community organizing necessary to build strong coalitions. Coalition builders must be inclusive, creative, and continually aware of developing trends in government and the marketplace. CBAs hold enormous promise, when pursued tenaciously and with constant attention to the need to make sure they benefit all parties. Organizers must protect the process against potential abuse—as one New York City case already suggests, some developers are likely to seize upon CBAs as a new and improved "divide and conquer" device to grant concessions to some members of a community while excluding others.

The challenge for organizers and those supporting them will be to remain ever vigilant about trends in the market and how those trends can best be channeled for community benefits. Organizers can use the CBA model to harness the same kind of entrepreneurial vision and drive that have powered successful businesses in America, and link them with the social justice and environmental commitments that have fueled our best community-improvement campaigns.

Notes

1. See "Profiles of Business Leadership on Smart Growth: New Partnerships Demonstrate the Economic Benefits of Reducing Sprawl," National Association of Local Government Environmental Professionals (1999) and "Smart Growth Is Smart Business: Boosting the Bottom Line & Community Prosperity," NALGEP and Smart Growth Leadership Institute (2004).

2. "The Metropolis Principles: A Commitment by Chicago Area Business Leaders to Bring Housing, Transportation and Jobs Closer Together," Chicago Metropolis 2020 brochure, available at http://www.chicagometropolis2020.org/documents/metropolisprinciples.pdf.

3. For example, the National Association of Industrial and Office Parks' research division has commissioned research on developer-community relations that include CBAs.

20

Reshaping a Region After September 11 (New York Metro Region)

Robert Yaro, Chris Jones, Petra Todorovich, and Nicolas Ronderos

Introduction

Ground Zero, the place most associated with the September 11 terrorist attacks, is a mere sixteen acres. Yet, the site bears the weight of one of the most far-reaching, complex sets of expectations the nation has ever seen. Grieving family members, community residents, regional workers and businesses, national citizens, and world leaders all have a stake in what gets built there and the policies that guide the recovery effort.

Ground Zero lies within the physical, historic, and economic center of the New York metropolitan region, which spans much of New Jersey, Connecticut, and New York and is home to 23 million people. The aftermath of 9/11 has affected both community and regional planning, and has revealed the web of human, physical, and institutional connections that bind the tristate metropolitan area, dramatizing the challenges of maintaining constructive collaboration and overcoming long-standing divisions among a broad range of constituencies.

This story will be told through the lens of two places that could hardly be more different, but whose fates are linked both by long-term patterns of metropolitan development and by specific responses to 9/11. Lower Manhattan has the third-largest concentration of office space in the country and a diverse range of affluent and low-income residential neighborhoods. East Harlem is one of the poorest communities in the region and subject to increasing gentrification. As this article demonstrates, the lessons learned here after 9/11 are relevant to anyplace dealing with redevelopment or major infrastructure developments on a metropolitan or regional scale.

September 11 and the Region's Communities

The death toll of almost three thousand lives in the 9/11 attacks cut across geographic, economic, racial, and social lines. The interconnectedness of Lower Manhattan to the region—and of the region to the world—became painfully and abundantly clear.

In the commuter parking lots of New Jersey, Connecticut, and New York suburbs, cars were left by owners who never returned home from work. Immigrant families, whose members had found good jobs at Windows on the World, the celebrated restaurant at the 110th floor of World Trade Center One, mourned their lost loved ones, as did the families of bond traders and insurance brokers from Cantor Fitzgerald and Marsh & McLennan.

The physical and economic destruction was also profound. The attacks destroyed 13 million square feet and damaged an additional 17 million square feet of office space. More than seven hundred small businesses closed as a result of the attacks and more than one hundred thousand jobs were lost throughout the region. The attacks knocked out or damaged tunnels serving New York City subway lines, costing $855 million in damage.

Neighborhood Impacts in Lower Manhattan

Residents of Battery Park City, the Financial District, and Tribeca were prevented from returning to their homes for days, weeks, and, for those closest to Ground Zero, more than a year. Security restrictions on Lower Manhattan streets cut off the main access point to Chinatown, a community of more than eighty thousand residents. Just as it began to recover, in the spring of 2003, Chinatown was dealt a second blow when the SARS epidemic scared away tourists and caused a second decline in retail and restaurant patronage.

The way the tragedy of 9/11 played out for different communities of Lower Manhattan underscored the diversity of this district, which was defined for the purpose of federal rebuilding funds as all of Lower Manhattan south of Houston Street. Even before 9/11, Lower Manhattan had begun to change, with rapid growth of residential population in Battery Park City, the Financial District, and Tribeca. Despite the disruption caused by 9/11, these three communities grew their housing stock by almost half between 2000 and 2004 by adding 6,560 new units to an existing supply of 13,835.[1] Many of the new units were created by converting old office buildings into condominiums and apartments, almost all of them market rate. In the 2000 Census, the median income for these three neighborhoods ranged from $90,000 to $137,000. The median annual incomes for Chinatown and the Lower East Side were $24,000 and $25,000, respectively.

The growth of Lower Manhattan's residential population attracted more services, restaurants, and amenities to the Financial District, but not enough to make it competitive with the mixed-use character of Midtown. Nelson Rockefeller first championed the creation of the World Trade Center in the 1960s to stem the exodus of corporate tenants from Lower Manhattan's historic, outdated office buildings to the modern office buildings rising to the north. But the strategy was only partially successful because Lower Manhattan faced transportation disadvantages compared to Midtown. Midtown has easy access to the suburbs—Long Island, Connecticut, and Downstate New York—home to many of the corporate executives who make location decisions for their companies.

Neighborhood Impacts throughout the Region: The Case of East Harlem

East Harlem, a neighborhood long defined by its Puerto Rican residents, had begun to change in the years leading up to 9/11. Its Latino population has increased even as this group has diversified from its Puerto Rican base to include a much wider array of Dominicans, Mexicans, and others. It has also retained a large African-American population, but even here a larger number are foreign born. Over the past decade, its income has increased and it has gained population, including an increasing white population.

The burden of layoffs and downsizings in the economic downturn following 9/11 hit low-wage workers particularly hard. A study released shortly after 9/11 by the Fiscal Policy Institute estimated that 60 percent of those laid off following the attacks earned $11 per hour on average. Workers who lost their jobs were not only in downtown offices, restaurants, and garment factories; they were also airline workers, hotel employees, and others who worked and lived in neighborhoods far from Ground Zero.

In recent years, new development has begun in East Harlem after decades of inactivity, spurred in part by the area's location near

Manhattan's affluent core, which makes it attractive to revitalization and reinvestment but vulnerable to the threat of rising housing costs for existing residents. When the City of New York rezoned East Harlem in 2003 to allow increased density and development, the number of permits issued for new residential units demonstrated its attractiveness to private investment. Indeed, East Harlem exhibits the contrasts within so many neighborhoods faced with the anomaly of a booming housing market and a weak job market.

The Power and Challenges of Coalitions

Regional Plan Association played a central role in the rebuilding process by convening in October 2001 the Civic Alliance to Rebuild Downtown New York, engaging broad participation of civic and business leaders, professional planners and architects, environmental and transportation advocates, active community residents, and academic institutions. The coalition built consensus around planning principles and objectives for the rebuilding process, while city, state, and federal agencies were still occupied with the urgent task of rescue, recovery, and restoration of critical services.

A key principle emerging from that planning period was that the Lower Manhattan rebuilding process must be public, participatory, and transparent. The Civic Alliance sponsored a public forum for six hundred people at the South Street Seaport in February 2002 to identify broad principles and vision for the rebuilding process. The forum attracted the attention of the City of New York and the Lower Manhattan Development Corporation (LMDC), the newly formed state agency tasked with overseeing the rebuilding process.

Listening to the City

Partnering with the LMDC, the Civic Alliance planned an even more ambitious public forum in the summer of 2002 to coincide with the release of the LMDC's preliminary concept plans for the rebuilt Word Trade Center complex.

On July 20, 2002, forty-five hundred people crowded the Jacob Javits Center for "Listening to the City," an electronic town hall meeting sponsored by the Civic Alliance. Participants sat face-to-face at ten-person roundtables and were led through discussion topics by professional volunteer facilitators. Each table used a wireless laptop computer to enter major points of consensus on each topic they discussed. Themes from the entire room were displayed on large screens, giving participants instant feedback on their views from the other people in the room, and the opportunity to vote on questions with wireless keypad voting devices.

Shortly before lunch, the participants reviewed six concept plans for the World Trade Center site that the LMDC had prepared. Each image showed a different approach to arranging 11 million square feet of office space on the site around a memorial. But the participants rejected the plans. Finding the plans "too dense" and "not ambitious enough," participants called for the integration of nonoffice uses. In terms of a setting for a memorial, a majority of participants rated almost all of the plans "poor or unacceptable."

The results of the meeting made front-page news, multiplying its effect. Public officials pledged to go back to the drawing board—and indeed they did, launching a new design process that resulted in the selection of Daniel Libeskind as master planner for the site. Never before in New York had a public meeting turned the planning process around in such a way.

The result is a greater expectation that the public be actively engaged in decisions about planning and development. The attention to architecture and design resulting from the meeting continues to influence New York planning projects, with big-name architects like Santiago Calatrava and Frank Gehry selected for major projects like the WTC PATH station and the Brooklyn Nets Arena.

Dealing with Internal Tensions

In addition to the Civic Alliance, a variety of coalitions emerged after 9/11 to focus on different aspects of the rebuilding process. New York New Visions, a group of architecture and design professionals, donated their time to planning and advocacy work, with an emphasis on physical planning and design. Rebuild Downtown Our Town (R.Dot), a coalition of Lower Manhattan residents, business owners, and artists, produced papers incorporating maps and design techniques to illustrate solutions for revitalizing Lower Manhattan. The Labor Community Advocacy Network (LCAN) was formed in early 2002 to address the plight of low-income people, immigrants, and people who lost their jobs as a result of 9/11. LCAN distinguished itself by expanding its scope to all of

New York City and the economic effects of 9/11 in all five boroughs. Finally, a project by the Municipal Art Society entitled "Imagine New York" provided public outreach through a series of more than two hundred public workshops in communities throughout the region.

In addition, several established institutions focused on the rebuilding effort. Most notably, these included Manhattan Community Board One, representing Downtown residents and businesses, and the Downtown-Lower Manhattan Association (DLMA), representing the business community. The Civic Alliance engaged all of these organizations through the spring of 2002 in developing recommendations for the rebuilding process.

One of the recommendations was that rebuilding and improving the transportation infrastructure of Lower Manhattan was a top priority. There was almost universal recognition by community groups, planners, and business groups that Lower Manhattan was at a transportation disadvantage to Midtown. Correcting this would require immediate investments in restoring the severed transit links, as well as improving Lower Manhattan's access to the suburbs.

This unanimity of the civic community resulted in the allocation of a $4.55 billion earmark of federal funds for transportation projects in Lower Manhattan—a huge victory. But tensions emerged regarding the details of which transportation projects to fund after the disrupted links were restored. RPA and some transportation advocates recognized that the funds for transportation could be channeled to speeding construction of the Second Avenue Subway—a project that had stalled due to lack of funding. Yet, despite the project's merits for Lower Manhattan and the entire city, the idea of advancing the Second Avenue Subway with rebuilding dollars never caught on.

Business leaders in Lower Manhattan advocated a $6 billion rail link from Downtown to John F. Kennedy Airport and the Long Island Rail Road, arguing that this "quick" and highly visible project was necessary to restore business confidence in Lower Manhattan. This was an ongoing source of tension within the Civic Alliance, in which one group believed the project would benefit relatively few suburban and airport riders. The difference caused a fissure within the Civic Alliance in 2003, when downtown business groups proposed using Community Development Block Grants toward the JFK/LIRR rail link. The Civic Alliance's refusal to support the project resulted in some downtown business groups dropping out. Partly as a result, the Civic Alliance underwent an intense reevaluation of its rules for membership, voting, and governance. This has helped the Civic Alliance to deal with later controversies, and membership has been largely stable since then.

A Model of Urban Transit-Oriented Development

In East Harlem, before 9/11, RPA was already working with the community to build support for the Second Avenue Subway and maximize its benefits to the neighborhood. As the community struggled to recover from the economic downturn, RPA's East Harlem Community Link Initiative evolved into a multifaceted strategy highlighting the potential of transit-oriented development to link low-income urban communities to major regional infrastructure investments.

Shortly after 9/11, the Community Link Initiative picked up speed. RPA worked with its local partner, Community Board, to conduct a visioning outreach to understand the goals of local residents and businesses. This culminated in the 2002 Community Design workshop, bringing together sixty community participants with planners and architects to achieve consensus on the link's character and density to benefit the community and local businesses.

The community has remained involved in the design and phasing of the Second Avenue Subway so that it yields the most benefits for East Harlem. November 2005 saw passage of the Transportation Bond Act, allocating $450 million toward the first phase of the new subway line.

An emphasis on public space investments gave rise to the Second Avenue Corridor Working Group. This ad hoc group of community and agency representatives has identified streetscape enhancements as assets that are readily visible, valuable to all residents, and enjoy widespread support. A Streetscape Framework was published in September 2005 for securing government support and funding.

Future actions will focus on private development that is likely to come, first with rezoning and later with subway construction. This will involve projects near subway stations that are to include affordable housing and career opportunities for local residents. This effort should also involve the growing precedents for inclusionary zoning in New York City and continuing refinement of the community benefits agreements model.

The Importance of a Regional Framework

RPA's main contribution has been to provide a regional perspective to help shape both broad policies and specific projects. The RPA's 1996 comprehensive plan—*A Region at Risk: The Third Regional Plan for the New York–New Jersey–Connecticut Metropolitan Area*—laid out a framework for tying the rebuilding of Lower Manhattan to a vision of sustainable regional development, premised on the need to balance economic competitiveness, social equity, and a healthy environment.[2]

After 9/11, many of the initial transportation strategies were drawn from ideas in RPA's plan or projects that had emerged since its release. This regional framework also informed debates over Downtown's role in the region's economy. Finally, policies that were promoted to connect communities in the five boroughs, New Jersey, Upstate New York, Long Island, and Connecticut proved relevant to strategies to rebuild Lower Manhattan.

Prior to RPA's plan, the Second Avenue Subway had stalled to the point of being written off by politicians and citizens alike. RPA's plan made the case for the subway expansion as strengthening the economy and making new connections between Manhattan and the outer boroughs, linking communities like East Harlem to the north and neighborhoods in Brooklyn and Queens to the east.

RPA's 1999 MetroLink proposal was instrumental in putting the Second Avenue Subway back on the political map. The MTA Board allocated $1.05 billion in its 2000–2004 Capital Program to complete the planning and design, and to initiate construction of a full-length Second Avenue Subway. The upside potential of the new transit service—drastically reduced travel times to regional employment markets, new development and business opportunities, and rising community incomes and property values—offers a once-in-a-century opportunity to transform East Harlem into a vibrant and prosperous district.

The spring 2004 release of the Second Avenue Subway Final Environmental Impact Statement represented a milestone in moving the project toward implementation.

Next Steps: Regional Equity at Multiple Scales

The dynamics shaping redevelopment in both Lower Manhattan and East Harlem demonstrate that issues of regional equity must be addressed simultaneously at several geographic scales. At the community level, the conflicts of class, race, and interest groups must be understood in the context of changing social and economic trends within the region. The pressure for upscale housing development in East Harlem comes not only from the growing attractiveness of this neighborhood but also from the shrinking number of new development sites in both city and suburban locations. These same forces intensify the battles over housing preservation in Chinatown and the Lower East Side and the increasing demand for housing in Lower Manhattan, where conversion of outdated office space into luxury condominiums is changing the character of the Financial District. Strategies that attempt to capture some of the value that is being created, such as inclusionary zoning or community benefit agreements, stand a greater chance of success for securing gains for existing residents, including affordable housing, than ones that resist any new development that might contribute to gentrification.

Moving outward to the metropolitan scale, RPA is addressing the challenge of citing the next generation of growth while addressing the problems created by fifty years of sprawling development. Even in a mature region like the tristate area, as many as three million new residents may have to be absorbed during the next twenty-five years. Many suburban towns are closing their doors to growth, pushing development even farther out and exacerbating the housing crisis in the inner suburbs. In urban areas, redevelopment is clearly better for the poor than continued decay and job loss, but the problems of displaced residents and bypassed communities can be even more acute.

Regional Plan Association's *Vision to Action*, a regional visualization project, addresses these issues using new technology and public participation techniques to create a consensus for equitable and sustainable regional development. Starting in Long Island and working throughout the region, RPA will use GIS technology to demonstrate alternative development scenarios to local communities, so they can anticipate and prepare for future growth. Similar efforts have been implemented in places as disparate as Southern California, Chicago, and Utah, and can provide a model for long-term metropolitan development.

America 2050: National Strategies for Regional Growth

Hurricanes Katrina and Rita demonstrated the local impacts of regional planning decisions such as levee investments and regional disaster preparedness. The hurricanes also exposed the racial and economic inequities in the Gulf Coast, and the shared environment, geography, and economy of the region, which stretches from Houston to the Florida Panhandle. The aftermath and recovery have shown that the local communities, and even the states, are not yet working together to articulate priorities to the rest of the nation for the rebuilding process. The experience in Lower Manhattan, and the role that RPA's regional framework played in helping to define priorities, suggests that a regional approach could benefit the Gulf Coast in clarifying its rebuilding objectives, to obtain more long-term regional funding, and to make the region more resilient to future disasters.

The Gulf Coast is one of the extended "megaregions" that RPA is looking at as the geographical scale for large-scale investments in transportation infrastructure, environmental protection, economic development, and disaster preparedness. In New York, the problems of rising housing costs, the job gap for urban communities, and the continued outward march of urbanization are also felt up and down the Northeast, from Virginia to Maine. The Northeast Megaregion, encompassing the Washington-Boston Corridor, includes strong market cities like Boston, New York, and Washington, D.C.; weak market cities like Baltimore and Philadelphia; and a number of smaller, bypassed cities like Bridgeport, Connecticut, and Camden, New Jersey.

RPA's America 2050 initiative, developed in collaboration with other national and regional organizations, supports a national conversation on growth trends in America over the next fifty years to create a national strategy for prosperity, equity, and sustainability for future growth in America. This agenda emerged from a roundtable discussion on national planning, held at the Pocantico Conference Center of the Rockefeller Brothers Fund in September 2004.

The population of the United States is expected to increase by 120 million people by 2050. Seventy percent of the population growth and 80 percent of employment growth is expected to take place in eight to ten emerging megaregions across the nation—extended networks of metropolitan areas like the Northeast, where the economy, environment, and cultures are linked.

Working with national partners, RPA will advance the America 2050 project by quantifying and analyzing growth trends, coordinating strategies in the emerging megaregions across the nation to create synergies between fast-growing and bypassed regional centers and rural regions, and by advocating infrastructure investments and policies in each of the megaregions that can prepare these places for increased growth. The strategy is akin to East Harlem and Lower Manhattan writ large.

Notes

1. According to New York City Department of City Planning, August 2004, and the U.S. Census 2000.

2. The analysis and recommendations were organized around five campaigns: A Greensward Campaign that identified eleven regional reserves as the "green infrastructure" for defining future growth patterns, a Centers Campaign to promote growth in the five boroughs and eleven regional downtowns, a Mobility Campaign built around a new regional express rail network to integrate and expand the mass transit system, a Workforce Campaign to broaden prosperity to bypassed communities and populations, and a Governance Campaign to create stronger regional institutions and collaboration.

21

Faith-based Organizing for Metro Equity in Detroit (Detroit, Michigan)

Victoria Kovari

Once Upon a Meeting

The story of MOSES's Fix It First campaign began in October 2000 in a basement room in St. John's Parish Life Center in Plymouth, Michigan. Organizers from the faith-based organization MOSES (Metropolitan Organizing Strategy Enabling Strength) and from the Gamaliel Foundation gathered with several clergy and the three guests who would shake up our organizing forever: john powell from the University of Minnesota Institute of Race and Poverty; Myron Orfield, a lawyer and state senator from Minnesota; and David Rusk, a national urban policy expert from Washington, D.C. These experts had been brought to Michigan by the Gamaliel Foundation as "strategic partners" to help local organizations analyze and strategize about important emerging regional issues.

Most of the people in the room on that chilly October day did not have a clue about how to translate the concept of "metropolitan equity" into an actual organizing campaign. They barely understood what the term meant. Some could not even get past the word "metropolitan." We were deeply rooted in the City of Detroit and its problems. To look to the larger region seemed to be a betrayal.

After all, this is *Detroit*—at or near the bottom of every quality-of-life scale, and plagued by urban decay that seems to just never go away. Detroit's schoolchildren rank lower than those in rural Mississippi, Detroit's neighborhoods are home to more than ninety thousand abandoned properties, and Detroit's fiscal crisis is among the worst of any urban area.

Between 1950 and 2000, Detroit lost half its population—nearly one million people. Significantly, the half that stayed in the city was the older and poorer half. In the past twenty years, the flight of the black middle class to the suburbs has changed the landscape dramatically because the center of black economic power has been moving to the suburbs as well. Detroit's music topped the charts in the 1960s and 1970s; now the only chart that Detroit tops is the segregation index, ranking first as the most racially segregated region in the country.

At the center of the economic and racial segregation in Detroit is the automobile. Cars are the lifeline in metro Detroit, but in the world's car capital, nearly 383,000 people are without one. The average annual cost of owning a car in metro Detroit is more than $6,000. With one of the worst public transit systems in the country, it is no wonder that as of 2005, Detroit is the poorest large city in the United States.

The question that both organizers and leaders were forced to grapple with was how does a poor city like Detroit pay to repair its infrastructure when it cannot even keep up with emergency services like police, fire, and EMS? The answer was it can't. This meant that a regional perspective would be required. Yet, as MOSES sought regional solutions to some of Detroit's problems, it faced considerable criticism—both from those inside the city who saw regionalism as a power grab by the suburbs, and from suburbanites who saw regionalism as a grab by the city for their precious and more plentiful tax dollars.

The Gamaliel Foundation's national leadership training teaches that even the most well-intentioned people cannot solve their community's problems if they are alone and do not have power. Looking at the City of Detroit and its leaders, we see that the same principle applies: no matter how well-intentioned its leaders were, and no matter how much federal or state aid it received, Detroit could not solve its problems in isolation. To revitalize Detroit and other urban areas of the state, the flow of resources within southeast Michigan would need to be redirected.

In Michigan, policies that control the flow of resources are often determined at the state level. Inevitably, then, changing these policies would require learning how to work at the state level.

A Political Watershed

In October 2000 at the gathering of MOSES leaders, Rusk, powell, and Orfield put forward a scenario about a political watershed that was about to happen in the state of Michigan, about which most in the room were totally unaware. They hammered home the message that if

MOSES was to have any impact on revitalizing Detroit and its older suburbs, it would be necessary to understand the political decisions being made at the state level, paying particular attention to the upcoming 2002 state elections.

In Michigan in 2002, term limits had set the stage for a nearly complete turnover in state government. Michigan was to elect a new governor, a new attorney general, a new secretary of state, two-thirds of the state senate, and half of the state representatives. This offered an unprecedented opportunity to put the MOSES agenda before a whole new slate of politicians.

The vehicle chosen to advance the MOSES agenda was a mass public meeting. MOSES's then-director, Bill O'Brien, decided that if the organization was ever going to undertake to organize the biggest mass meeting in its history, 2002 was the time to do it. Plans were launched for a meeting of five thousand, to which the candidates for governor would be invited. This would be the occasion to put forward the organization's first central organizing issue related to urban sprawl. However, the *particular* issue to be put forward had not yet been decided upon.

After a series of trainings beginning in 1998, the MOSES Metro Equity Task Force was formed. In December 2000, a group of leaders from MOSES attended the Gamaliel Foundation's National Leadership Assembly, an annual gathering of leaders and organizers from around the United States. One lesson became very clear: if you want to become a more powerful player in your region and begin to redirect the flow of resources in a more equitable way, then you better be ready to take on any regional institution that controls some of those resources.

This was the first time most people had heard the term *MPO* (metropolitan planning organization). Who knew that a single organization controlled a billion dollars of federal money flowing into southeast Michigan every year, and that this money was literally and figuratively driving urban sprawl? The federal money was for transportation—the biggest item in the federal budget after defense spending, and the process of spending it was filled with more pork than a piece of sausage.

MOSES leaders came back to Detroit and began to engage in a "power analysis," a comprehensive process of research, reading, and dialogue to determine who controlled what, and which pieces of the process or the pie could be changed. With the help of some students from the University of Michigan, the leaders discovered how transportation dollars flowed in and out of metro Detroit. They began attending meetings of the MPO and the Southeast Michigan Council of Governments (SEMCOG). They interviewed dozens of people inside SEMCOG and outside: suburban mayors, state bureaucrats, legislators, and business people. At the end of each session, the leaders asked, "Who else should we talk to?"

While the power analysis was under way, a deep sense of intimidation began to surface regarding all the experts and information. Some leaders began to question their own capacity for taking on such a campaign. After all, they hardly knew anything about transportation. The more they learned, the more they realized how much they didn't know, and the more paralyzed they became.

At this point, action—any action—was the best antidote to combat their fear. Having a victory, however small, would help them get over this hurdle.

Giving Birth to "Fix It First"

MOSES began to do a series of actions and gained several wins. Among the victories were getting a seat on the SEMCOG Transportation Advisory Committee, defeating an attempt in the state legislature to cut the budget for public transit by 50 percent; and getting the Big Three auto companies to issue a joint statement in support of regional mass transit. Several public meetings of 300 to 1,000 people were held, at which some of these victories were celebrated.

At one of these meetings, leaders announced the formation of a Metro Equity Allies table. One of the allies at the table was a statewide environmental group that had come up with the phrase "Fix It First" several years earlier. This phrase resonated with our own leaders because it evoked images of crumbling bridges, roads, sewer pipes—images that were all too familiar whenever we drove around the state of Michigan. The issue of metropolitan equity could be very complex and abstract, but "Fix It First" seemed to capture the essence of metro equity and is in a language most people understand. Most important, it offered the potential to put MOSES on the radar of the 2002 candidates for governor.

In the summer of 2002, MOSES's leaders and organizers visited dozens of congregations with a PowerPoint presentation to detail issues of urban sprawl. Additionally, the presentation addressed the need not only to repair and maintain existing roads, sewers, and schools before building new ones but also to invest in regional mass transit. The issues of mass transit and Fix It First seemed to resonate with nearly every congregation across the city and suburbs, cutting across the lines of race and class.

In the course of those presentations, MOSES evolved a three-part Fix It First platform:

1. Devote 90 percent of the money that the state spends on roads, sewers, and schools to fixing the existing infrastructure.
2. Appoint a smart growth commission to make recommendations on comprehensive land use reform in Michigan, and appoint a MOSES/ MY VOICE representative to that commission.
3. Commit to developing fiscal incentives that encourage regional cooperation across the state.

MOSES had finally decided on the issues to be featured at their blockbuster meeting!

¡Sí Se Puede! Yes, We Can!

On a beautiful Sunday afternoon in September 2002, five thousand people packed the newly built Greater Grace Temple on Detroit's northwest side. Dozens of state and local officials were in attendance, along with members of 120 congregations from across metro Detroit and 500 students from various high schools and universities. The two candidates for governor of Michigan came to the front of the church to sign on to the MOSES Fix It First platform. Democratic candidate Jennifer Granholm, who was ultimately elected governor, added: "I will make this a cornerstone of my campaign and one of the major focuses of my administration."

MOSES leaders and organizers were thrilled. More than one thousand members of the largely Spanish-speaking congregation began to break out into rounds of "*¡Sí se puede!*" Yes, we can! And we have.

In the two years following the 2002 public meeting, MOSES continued to win significant policy victories. Two months after taking the oath of office in February 2003, Governor Jennifer Granholm appointed a statewide commission called the Michigan Land Leadership Use Council. In May 2003, during a bitter budget process, the governor came out with a mandate to defer more than one billion dollars in funding for thirty-four new road projects across the state. She eventually prevailed on seventeen of these projects, including the largest ones, slated for metro

Detroit. In September 2004, at a meeting of four thousand people, the governor reaffirmed her commitment to Fix It First.

Detroit and other urban areas across the state can sustain their revitalization efforts only when state lawmakers commit to rebuilding the aging infrastructure of Michigan's central cities and older suburbs.

Laying a Foundation for Lasting Change

Campaigns organized around regional equity strategies are powerful tools for faith-based organizations to influence state policies—especially when coupled with a political analysis and an issues-based electoral strategy. This kind of organizing has enormous implications for faith-based and community organizations working together to shape our future.

22

Values, Vision, and Message: The Spirit of Metro Equity

Introduction

Gamaliel Foundation has organized interfaith coalitions of suburban and inner-city congregations and groups across the United States. These faith-based networks are working to heal the race and class inequities that divide our metropolitan regions.

Here, through three voices from the Gamaliel Foundation, we describe the conceptual underpinnings of faith-based regional equity organizing. First, Gamaliel Foundation's founder and executive director, Greg Galluzzo, discusses the underlying causes and the human impacts of concentrated poverty and urban sprawl. Next, longtime community organizer Mike Kruglik offers reflections on organizing for metropolitan equity. Finally, Rev. Cheryl Rivera speaks to the values and vision underlying a faith-based approach to social change. A sidebar offers strategy tips for organizing a faith-based metro equity campaign.

COMMUNITY ORGANIZING THROUGH FAITH-BASED NETWORKS

Greg Galluzzo

The Gamaliel Foundation is a network of more than fifty faith-based organizations in twenty states. We are also active in five provinces of South Africa.

We created the Gamaliel Foundation to help ensure that the promise of community organizing would be realized. As time went by, we came to realize that something was fundamentally wrong with our analysis of the problems that arise in the work. In many cases we had a good orga-nizer in a community for many years, yet the community continued to deteriorate. Even though victories were won—to improve schools, to get a health clinic, to get some houses built—something else was happen-ing that was, at the time, mysterious to us.

Sprawl and Concentrated Poverty: The Human Toll

We learned a great deal about the underlying causes of our problem when we were introduced to a group of policy experts, including john powell, Myron Orfield, and David Rusk. We learned that much of the problem resulted from high-income people moving out of a community into more outlying areas, and leaving behind a population trapped in a deteriorating situation. This is part of what is known as urban sprawl, and it often results in enormous concentrations of poor people.

In places like Detroit, East Chicago, East St. Louis, and Buffalo, prop-erty values sink as people move out. This means that anyone who wants to protect any equity in one's home has to move away from the commu-nity before the equity dissolves.

Many people think that what drives urban sprawl is that people yearn for the "American Dream": their own home with its garage and its lawn to mow. In fact, much of what drives it is racism and greed. Racism en-codes itself in unspoken messages: move away from where "those peo-ple" are, and move to where "they" aren't, and your life will be better.

Urban sprawl generates profits for land speculators, banks, insurance companies, construction companies, and unions. But the towns they de-velop or expand reap the financial benefits at the expense of older areas.

Opportunities for Wealth-building: Separate and Unequal

Vividly illustrating this process, john powell tells this story: His father bought a house in Detroit; around the same time, his father's friend bought a house in the suburbs. Both men fought in the Korean War and benefited from the Veterans Administration loans program. However, embedded in the V.A. loan program legislation was the clause, "This money cannot be used to integrate a neighborhood." This meant that the white man could not move into the black neighborhood, and the black man could not move into the white neighborhood.

Fifteen years later, the black man's house is worth only $25,000, the same as what he paid for it. The white man's house has ballooned in value from about that same price to $200,000. Both homeowners were highly responsible people who took good care of their property. Yet one man, because of national policies formalized in legislation, was in effect cheated out of $175,000 in equity.

It is not a person's income that determines economic well-being—it is a person's equity, powell likes to point out. With $175,000 in equity, you can do a lot of things. You can start a business. You can send your kids to college. But if you have no equity, you can't do any of those things. If you lose your job or miss work for a while, you might even lose the house you have.

Driven by developer greed, sprawl happens first in the city, then the first ring of suburbs, and then the second ring of suburbs, as people move farther and farther out. A "burn and churn" process takes place: first create a neighborhood and make it appear to be the promised land, then make sure a few African-Americans and Latinos move in, then begin to send the signals: "You'd better move on to the next neighborhood." This process played out all around us as we neighborhood organizers tried to organize in the inner city.

Confronting the Underlying Causes of Concentrated Poverty and Sprawl

After fifteen years, I began to feel that our community organizing work was like cleaning the engine room of the *Titanic*. We were down in the hold of the ship with the working-class people, doing our job well: keeping the fire stoked, getting the grease off the floor, polishing the knobs. Yet, the whole ship was headed for an iceberg, and everyone was going to sink. In the face of a larger set of problems that loomed all around us, our urban organizing work seemed just as narrow and futile as attempts to save that ship.

Sprawl has left a landscape of social devastation in its wake. It has been proven over and over that when people are trapped in poverty, it is rare that any of them manage to escape the damage to their psyche, their sense of possibility for upward mobility, or their perception of political strength. The devastation being wreaked on areas like Detroit is a moral tragedy.

FAITH-BASED ORGANIZING AND CORE VALUES

Mike Kruglik

A faith-based organization is a grassroots organization whose members come mainly from religious institutions, across lines of race and class, relating to one another with mutual respect, human recognition, and love in the broadest sense of the word. They share a moral commitment to creating opportunity. Working together helps us withstand the opposition of the institutions and the people that resist change.

From the relationships forged by deeper connections with one another, we create a larger community that reflects our core values. A key faith value is that all people deserve to be treated with respect and fairness, as befits a child of God. Our religious values command us to love God and love our neighbor.

Wealth, Poverty, and Loving Thy Neighbor

As poverty and inequality continue to increase in America, how can we, as people of faith, love our neighbors in practical terms? The view of our organization, the Gamaliel Foundation, is "The lines that divide people by wealth and race are the real problem and the real tragedy."

The tragedy of Hurricane Katrina revealed how deeply these fault lines run in American society, fault lines that can throw the poor and people of color into vulnerable, disastrous, or even fatal circumstances. The lines fracture every region of America, not just New Orleans. If we do not act now to attack the problem of segregation by wealth and race, there will be many more tragedies in our future.

The Gamaliel Foundation has faith-based grassroots leaders in twenty-two states who are facing up to this problem, at a time when few others even talk about segregation and integration. Many people say, "Segregation? Oh, Dr. King took care of that forty years ago." In fact, there is more segregation now than there was thirty years ago. From the 1954 *Brown v. Board of Education* decision until about 1975, schools experienced less segregation and more integration. However, during the period from 1975 to 2005, schools became more and more segregated.

Educational Opportunity and Metro Equity

Metro equity is about battling segregation and winning integration of opportunity. It works to change the geography of segregation into a geography of integration. This means creating communities that are whole communities, where all kinds of people reside, where jobs are plentiful, and where children can attend good schools.

Concentrations of poverty develop when people are segregated from good schools or from good jobs, when cultural assets such as libraries and performing arts centers do not exist in their communities, or when people are trapped in housing that is depreciating in value relative to other communities. Conversely, kids who attend good schools in neighborhoods that are economically integrated have role models for success, and are going to fare much better in their lives.

Studies conducted in the 1960s by James Coleman, professor of education at Harvard University, found that the most important factor for children's learning was being in a culturally and economically integrated school. Many studies since then have underscored the same point. But if economic inequities keep families away from good schools and prevent parents from providing the type of family support that their children need, or if home is a place of routine anger and violence, then the road to success is much steeper.

Education, Housing, and Transportation as Pathways to Opportunity

In sum, metro equity is simply about facing up to segregation and creating policies that connect people and integrate them into situations of opportunity. In particular, metro equity organizations have found three issues to be the most integral to dismantling segregation and connecting marginalized people to economic opportunity: education, housing, and transportation. If even a fraction of the resources that are being pumped into traditional community economic development in urban neighborhoods was redirected into strategies for achieving equity gains in these three key areas, the impact would be enormous.

"LET MY PEOPLE GO"

Rev. Cheryl Rivera

I am a licensed ordained Baptist pastor and a full-time Gamaliel community faith-based organizer. Currently I am organizing in predominantly African-American communities in metropolitan Chicago.

I am passionate about a regional metro equity analysis because it provides a way to think about the larger community in the way that Christ thought about community. Faith-based organizing calls people out of their places of comfort. We are calling them out of the pews. We are calling them to get into the streets, to get into the boardrooms, to get into the dialogue.

I disagree with those who see metro equity as a white folks' agenda. African-Americans have much to gain from being involved in metro equity, working in coalitions to impact the public policies and decisions that affect our lives. I believe that regional equity is, in fact, central to the black agenda. The problems within African-American communities are much too great for the African-American community to solve alone. The resources are simply not there, because so many African-American communities have been devastated by the loss of jobs, population, and tax base.

Regional Equity: The New Civil Rights Movement

In faith-based organizing, our goal is to move people out of the four walls of worship and into a place where real worship begins: a place of caring about and working toward the greater good and the greater humanity. We must draw upon all of our resources to solve what is perhaps our biggest social problem: a small percentage of people own almost everything. The majority of us fight over crumbs, while the few make self-serving decisions behind closed doors—not only about what happens with the local supply of bread but also what happens to the land where the wheat is grown.

Many see regional equity as the new civil rights movement. The earlier civil rights movement was a multiracial coalition, led largely by African-Americans, with significant involvement by the faith community. I see the church as again being uniquely positioned to affect social justice issues. In virtually every community, one can find a place of worship—a church, a mosque, a temple, a synagogue—where people are concerned about the well-being of other people and of the human family.

The church and the community can provide the kind of support and the kind of womb that can be a first place of comfort, or even a first place of agitation, when necessary. I try to give young people an important message: you don't have to settle. Stop settling! We can all aspire to greatness, and we must aspire to greatness.

Our goal is to make sure that the prophetic voice is not silenced. My faith tells me that to whom much is given, much is required.

My faith also tells me that a kingdom, a city, or a nation divided against itself cannot stand. If a region is divided against itself based upon race or class or ethnicity or language, that region will fall, because it fails to use the best of its resources.

We live in a society that is more pluralistic and multiracial than it has ever been. We *must* figure out how to work together in coalitions: black, white, brown, red, and yellow; city and suburb; Christian, Muslim, and Jew; Catholic and Protestant.

Our Greater Aspirations

Faith-based organizing is about transforming people who can transform places of worship, who can transform neighborhoods and communities and regions and states and nations, who can transform the world.

I get disturbed when people of faith say, "I don't want to get involved in politics." I say, "Don't go to church, then, because politics is everywhere."

The ultimate voice, and the ultimate message, of faith-based organizing is "let my people go." All of them: black, white, red, and yellow. Let them go—and let them share in the best that America has to offer.

Tips for Organizing a Regional Equity Campaign

Adapted from the Template for Regional Campaigns prepared by Metropolitan Congregations United (MCU) and downloadable at www.gamaliel .org/MetroEquity/MELibrary.htm.

1. Organize a group of leaders (veteran and new, city and suburban) as a task force to take on a metro equity issue.
2. Train leaders on metro equity analysis using experts, strategic partners, or experienced staff.

3. Do a power analysis of your region, including one-on-ones with key players: mayors, county executives, MPO directors, committee heads in the House and the Senate (or key aides).

4. Learn about your region's history. Often history yields insight into the uniqueness that defines your city and region, the most important issues, and political relationships and rivalries.

5. Understand the political environment in and outside your region (present and past), through reading, reflection, and discussion.

6. Decide on an issue, regional target, and action.

7. Begin to develop a plan that identifies targets, demands, and actions to take.

8. Assess winnability and possible or likely political outcomes. What possible reforms might have the biggest impact on revitalizing older communities (urban/suburban core)?

9. Identify one or more local public champions for your issue (e.g., a suburban leader, a city leader).

10. Identify other regional institutions as possible targets or allies (county government, RTA, water board, airport authority, chamber of commerce, corporate government relations officers, etc.).

11. Hold a public meeting to declare your metro equity strategy and get commitments from officials.

12. Build relationships with powerful regional players that have resources: money, publicity and support on issues, votes; etc.

13. Create a position statement that states the problem, proposes a specific solution, and specifies the reform(s) you are seeking (these become your demands on public officials).

14. Schedule meetings with congregations to get their endorsement and ask them to send a delegation to a public meeting or event.

15. Stage a public event to raise the profile of your issue and identify your organization with the issue (e.g., hold a press conference or rally, or take over a hearing, summit, or legislative caucus).

16. Use public events to clarify elected officials' interest and level of commitment to your agenda. Follow up afterward to hold them accountable.

17. Identify achievable short- to medium-term policies and campaigns to create genuine urban core and inner suburban rebuilding. Present a credible case that you are serious about rebuilding the urban core and depleted inner suburbs, and you have a winnable plan and the resources to implement it.

Saying Yes: Framing Regional Collaborations to Win

A major challenge for progressive social movements is to visualize what winning looks like. In *Surplus Powerlessness: The Psychodynamics of Everyday Life and the Psychology of Individual and Social Transformation*, Rabbi Michael Lerner argues that people often undermine their own capacity for success by their inability to imagine winning. In Lerner's view, this makes us less powerful than we need to be, with "surplus powerlessness" infecting both our personal relationships and our work world. This undermines our willingness to pursue large-scale visions of personal or social change.

Once a community organization has clarified its goals, considered its assets and liabilities, become grounded, and examined new horizons, it is in position to choose a proactive agenda for social change. The organization is ready to establish priorities and focus energy on getting the job done, securing new allies, delivering results, building long-term commitments, and creating new institutional arrangements.

Successful mobilizations accomplishing these tasks require careful attention to the way issues are framed. Framing means focusing attention on some bounded phenomenon by imparting meaning to specific elements within the frame and setting them apart from what is outside the frame. Through framing, the organizing group identifies a problem and its causes, and develops a target for its actions. The organizing group also develops a solution for the problem, provides a rationale to motivate participation by its members, and positions the movement to engage a larger segment of the population.

The following chapters' stories advocate for metropolitan regional equity work with a shared proactive agenda for sustainability, economic opportunity, and racial justice. Creative collaborations for sustainability in three different contexts are described. In the San Francisco Bay Area, the collaboration between Urban Habitat and the Center for Justice, Tolerance, and Community (CJTC) demonstrates how a new generation of grassroots leaders and established policy centers can work and learn together while making positive things happen in communities. In the Camden/southern New Jersey region, organizers work to build new strategies within an impoverished city, while seeking to change the rules of development in a statewide context. In the greater Los Angeles area, farmers are bringing fresh produce to schools, teaching children nutrition, and helping them to experience their connection to the land and to the cycle of life.

This is the power of saying yes.

San Francisco Bay Area

The greater San Francisco Bay Area is a remarkably beautiful and diverse region made up of ten counties. It extends north to Napa, south to Santa Cruz, east past Antioch, and west to the Pacific Ocean. Combined, its major cities (San Francisco, San Jose, and Oakland) create one of the five largest metropolitan areas in the United States.

The San Francisco Bay both connects and divides people. Prior to bridges and modern transportation, the Ohlone and Miwok peoples lived in self-sustaining villages in the fingers of land surrounding the bay. In the context of history and prehistory, the bridging of peoples across the bay is a very new experience.

Long associated with both a culture and political climate that have favored progressive causes, the greater Bay Area has in fact experienced divisive struggles, such as racial isolation. Gaps in public transportation have exacerbated social and economic divisions. For example, while the waterfront community of Richmond has 40 percent unemployment, Marin County (just across the Richmond-San Rafael Bridge) is one of the richest counties in the United States.

Yet, the San Francisco Bay Area is also a locus for unique collaborative efforts. In particular, it has proven to be fruitful ground for nonprofit social justice groups to come together with other sectors such as labor, business, and academia. One such example is documented in the story "Bridging the Bay," coauthored by Manuel Pastor and Rachel Rosner of the CJTC at the University of California, Santa Cruz, and Juliet Ellis and Elizabeth Tan of Urban Habitat (UH) in Oakland, California.

Since the founding of the Social Equity Caucus in 1997, Urban Habitat and the University of California have worked together to build coalitions that foster regional equity. They helped to shape the Great Communities Collaborative, which brings together residents and local organizations to participate in community planning processes across the San Francisco Bay Area. The ultimate vision of these collaborative planning processes is to create a region of vibrant neighborhoods with affordable housing, shops, jobs, and services within convenient walking distance and near public transit hubs. To this end, Urban Habitat, the University of California, and others have also collaborated on a revision of the General Plan in the City of Richmond.

Camden, New Jersey

The unfolding story of Camden, New Jersey, since World War II reveals the "slow moving hurricanes" that have devastated our older cities. In the 1970s and 1980s, the historic Mount Laurel decisions sought to guarantee fair housing and access to opportunity for all residents of New Jersey. Despite these intentions, in the opening years of the twenty-first century, residents of the city of Camden were trapped in poverty, locked out of jobs and housing in the surrounding wealthy metropolitan region of southern New Jersey. Camden was in administrative receivership and was rated the "most dangerous city in America" in 2005 (Associated Press 2005). Opening a three-part examination, Howard Gillette, Jr., Rutgers history professor and author of *Camden After the Fall*, outlines the need for a collaborative effort that strategically links the city and its surrounding suburbs to one another.

Gillette's piece is followed by contributions from Jeremy Nowak, president of the Reinvestment Fund, and urban policy expert David Rusk. Nowak suggests a comprehensive investment strategy for rebuilding the ravaged landscape of this poor city. Rusk, in his concluding essay on the "outside game" of mobilizing surrounding suburban regions, gives a blow-by-blow account of what may well be the most decisive victory of the regional equity movement in the opening decade of the twenty-first century—the adoption of housing and school policies to achieve an economically and racially integrated society. One of the lessons from the Camden experience is the value of bringing all stakeholders together, exerting the leadership needed to change the rules of the game.

Today, much of the landscape of the United States points to opportunities—while at the same time locking many people out of these very opportunities. At the turn of the twentieth century, W. E. B. DuBois introduced the concept of double consciousness, referring to "this sense of always looking at one's self through the eyes of others, of measuring one's soul by the tape of a world that looks on in amused contempt and pity" (DuBois 1903). Illuminating the adversarial relationships of race and class that drain our cities and abandon people, the contributors to this section simultaneously point to the possibility of framing regional collaborations to win a more inclusive social order that does not leave people behind.

Southern California

The heartbreaking loss of family-owned farms, as well as most people's lack of awareness about where their food comes from, point to our deep disconnection from the land that sustains us. Parallels can be seen between our loss of direct relationship to the land and declines in our health and well-being, as well as the health and well-being of our children.

In the twenty-first century, food that has been processed and chemically treated predominates in supermarkets. Consumers, while flooded with choices, are caught in distribution systems that focus less on proper nutrition than on the bottom line. Natural food stores and organic products have crossed over into the mainstream to a degree. However, many rural communities struggle with economic and social challenges as family farms give way to industrial farming—a sweeping shift from the original circumstances upon which these rural communities were founded and developed. Those who continue as family farmers are working to preserve a long-standing way of life, with value that extends beyond "per-acre yields" and "economies of scale."

The interdependence of our urban core and the outlying rural areas is essential for healthy regional development. In fact, local food systems are integral to what defines a healthy community. To close this section, Robert Gottlieb, Mark Vallianatos, and Anupama Joshi unveil a strategy that teaches children about the value of whole foods, as well as about the role each of them can play in stewardship of the land. With an education that includes ecological literacy, today's young people—and tomorrow's decision makers—will be in a position to make more healthy choices, as well as to advocate for sustainable environmental practices that are at the heart of a community's sustainability and well-being.

23

Bridging the Bay: University-Community Collaborations (San Francisco Bay Area)

Manuel Pastor, Rachel Rosner, Juliet Ellis, and Elizabeth Tan

Progressives often find themselves racing from one campaign struggle to the next, usually in reaction to initiatives crafted by right-wing forces. Meanwhile, conservative leaders and organizers—using a variety of well-funded strategies that tie together constituencies, think tanks, and political advisors—have successfully pursued long-term changes in both framing issues and moving policy.

Forward-looking thinkers and community leaders have recognized the need to develop a long-term view coupled with cross-sector partnerships of untraditional allies. In particular, the cultivation of strategic ties between progressive university researchers and community groups is an unusual and promising step for the regional equity movement. In the San Francisco Bay Area, a number of collaborative efforts between the Oakland-based nonprofit group Urban Habitat and the Center for Justice, Tolerance, and Community (CJTC) at the University of California, Santa Cruz, have worked to overcome the chasms that too often separate universities from community groups.

Since its founding in 1989, Urban Habitat (UH) has brought together environmentalists and social justice advocates, and has played a key role in the environmental justice movement. The Center for Justice, Tolerance, and Community, founded in 2000, has drawn on the experience of its researchers collaborating with community groups to inform agendas in regional equity and other areas of research.

Urban Habitat

Urban Habitat's mission is to build power in low-income communities and communities of color by combining education, advocacy, research, and coalition building to advance environmental, economic, and social justice in the Bay Area. UH focuses primarily on issues of transportation and housing, environmental health and justice, and equitable development. Since 1998, UH's capacity-building efforts have been centered in its Leadership Institute, which applies a race and class analysis and popular education techniques to provide customized training to a broad range of partners including community organizations, labor, and government officials. To date, more than seven hundred leaders spanning the entire Bay Area have participated in Urban Habitat's Leadership Institute. Through its national semiannual publication, *Race, Poverty, & the Environment*, and increasingly, its Web site, www.urbanhabitat.org, UH provides in-depth analysis and viable, real-world solutions for addressing community development and environmental justice issues.

One of Urban Habitat's most powerful tools for increasing community leadership in the region is the Social Equity Caucus (SEC), a coalition of more than seventy-five organizations committed to building a regional equity agenda in the San Francisco Bay Area. SEC participants represent economic, social, and environmental justice community-based groups, as well as labor, public agencies, faith, and youth organizations. Connecting local issues and priorities to a broader regional equity agenda, UH works in strategic partnerships to build power within the Bay Area's low-income communities and communities of color.

Over the years, Urban Habitat's community-based efforts for equity have helped to broaden the agenda of the environmental justice movement from its primary focus on toxic pollution to issues such as military-base conversion, transportation policy, brownfields development, immigrant rights, inner-city disinvestments, suburban sprawl, and local and regional planning and development. Instead of organizing around a single issue within a particular community, UH takes a regional look at the Bay Area's political, economic, cultural, and social forces and their impacts on low-income communities and communities of color. By emphasizing the commonality of problems among communities, including between inner-city areas and older suburbs, the regional approach makes it possible to create new broad-based coalitions for urban reinvestment.[1]

For example, UH has been a leading member of the Richmond Equitable Development Initiative (REDI), a coalition that was formed in November 2003 to ensure that the current and future development of Richmond benefits the city's low-income communities and communities of color. REDI is currently engaging community partners, elected officials, labor, and public health advocates to advance concrete equitable development policies within the city's revised General Plan, a blueprint that will guide development priorities over the next twenty-five years. UH is also a founding member of the Great Communities Collaborative (GCC), a coalition that is developing a coordinated set of strategies that will address the housing crisis, offer alternatives to sprawl, and support equitable development. The collaborative's overarching goal is to make transit-oriented development and infill the primary paradigm for regional growth. GCC's primary objective is to ensure that half of all new homes built by 2030 are affordable to families of all incomes and in walkable communities within a half-mile from transit.

The Center for Justice, Tolerance, and Community

The Center for Justice, Tolerance, and Community at the University of California, Santa Cruz, is an interdisciplinary research center that links themes of social justice and community building. The CJTC emphasizes rigorous, policy-relevant research that addresses the concerns of community-based groups representing low-income communities of color, and collaborates with them to organize locally and regionally on issues of economic and environmental justice.

Regional equity issues have been central to the CJTC since its inception. Indeed, in the wake of the 1992 civil unrest in Los Angeles, many community groups recognized that economic development needed to be more effective in including and benefiting the poor. At the same time, they realized that they could not solve this problem through traditional local community development, especially in light of the dismal track record of such efforts and the fact that policy makers were focused instead on regional revitalization of the slumping L.A. economy (Pastor et al. 2000, 12).

Some community leaders argued that organizers needed to think long-term, to overcome economic isolation and reconnect with the region. This view inspired the book *Regions That Work: How Cities and Suburbs Can Grow Together* (Pastor et al. 2000), which is comprised of research that became fundamental to CJTC's focus. Regional equity quickly became a hallmark of the center's work. New projects were launched to examine regional labor markets to help community groups discern how best to leverage workforce development efforts. CJTC also partnered with three neighborhood-based groups in the Bay Area to develop strategies for regional interaction, helping to form a regional collaborative network in Southern California. Along the way, the CJTC created a core group of regional researchers based at UCSC and other universities. Much of this work was done in collaboration with community partners. In the process, CJTC built mutual respect and productive working relationships, establishing a credible track record as an effective partner.

Initial Steps: The Social Equity Caucus

The first collaboration between CJTC and Urban Habitat occurred during the founding of the Social Equity Caucus in 1997. Although it is an independent group, SEC grew and took shape with support from UH, which was already engaged in efforts to strengthen multicultural leadership in the Bay Area.

In 2002, UH reached out to CJTC to request assistance in developing a strategic plan for the SEC to clarify its vision, goals, and strategies. This collaborative strategic planning process led to the identification of the following set of shared values and guiding principles: 1) all communities have the right to thrive and flourish, 2) no single community or individual is disposable, 3) public institutions and elected officials must be accountable to the region's most impacted communities, 4) policies must address the needs of our communities' most vulnerable members, 5) low-income communities and communities of color must participate effectively in the decision-making processes that impact their lives, 6) diversity is strength and power, and 7) a clear, alternative vision for justice is needed to guide our work.

Bridging the Bay

After the strategic plan was in place, UH proposed a regional summit entitled Bridging the Bay, the goal of which was "to bring together people who have been active in the SEC as well as to expand participation in order to facilitate the development of a regional equity agenda." To help plan and facilitate the event, UH reached out to the Center for Justice, Tolerance, and Community. UH and the CJTC had by that time established a strong track record of working together, a shared set of values rooted in social justice, and a common vision for regional equity.

In April 2003, ninety-five members of the Social Equity Caucus came together for the two-day Bridging the Bay summit to help the SEC move from ideas to action. In attendance were representatives of community-based organizations, environmental and social justice groups, labor, civil rights groups, environmentalists, regional advocacy groups, funders, scholars, public agencies, and elected officials. Participants learned about community-based regionalism and concrete opportunities to make an impact in areas such as tax policies and revenue sharing, transportation, and welfare reform.

One of the attendees asked a fundamental question that helped highlight the central task of the conference: "Can we move beyond any conflicts and just start *somewhere?*" Organizers from AGENDA/SCOPE of Los Angeles, a leader in regional equity organizing, conducted a power analysis to create a visual picture of opportunities for building and leveraging power in the Bay Area. Subsequent sessions highlighted specific examples of how regional coalitions had moved policy and organizing agendas at the local, regional, and state level. Presenters discussed strategies such as community benefits agreements, regional transportation agencies, and state policies on housing and employment. Based on all this input, SEC members identified possible priorities for specific campaigns.

Tangible Results

A few months later, the SEC was in the thick of the fight to defeat Proposition 54, the so-called Racial Privacy Initiative designed to eliminate any consideration of race in the collection of state data in California. While this might not at first glance seem to be a traditional "regional equity" issue, the focus on Proposition 54 was appropriate for several reasons. First and foremost, the proposition posed a fundamental threat to all the communities of color in the region. Second, a coordinated regional strategy was clearly needed, as the state coalition working against Proposition 54 was not funding or prioritizing education and voter engagement efforts in the Bay Area. Finally, SEC members were drawn to the Proposition 54 campaign because it was winnable, it offered the opportunity for concrete experience and action, and a victory with it would lead to further collaborations.

This decision to pursue a joint campaign was an enormous leap in the SEC's development. Now the SEC would have to increase its capacity to manage the complex tasks needed to support a regional campaign. As coordinator of the SEC, UH understood that any SEC campaign would have to be staffed and resourced, and that working with organizations from multiple sectors throughout the region would present new challenges. To defeat Proposition 54, the SEC created customized fact sheets on the proposition for a variety of audiences such as labor, business, faith-based communities, immigrants, and nonwhite voters. It also conducted community educational meetings throughout the region and implemented media strategies including radio shows, public service announcements, and opinion pieces in local newspapers.

After collaborating with partners at the local, regional, and state level to defeat Proposition 54, the SEC began to focus on developing a regional transportation justice agenda for the Bay Area. They formed the Transportation Justice Working Group (TJWG), which helped mobilize a multisector coalition that convinced state legislators and the governor to stop a $20 million annual cut to Alameda County Transit. The TJWG also successfully protected the reduced fare bus pass for youth.

One of the TJWG's most important victories was the establishment of the Lifeline Program of the Regional Transportation Plan (RTP), a program dedicated to meeting the transportation needs of the region's low-income communities. The RTP allocated $200 million over twenty-five years to support the priorities of Lifeline. In May 2007, the TJWG won additional funds for Lifeline by targeting new money generated by the passage of Proposition 1B. As a result of this campaign, one-third of Proposition 1 B funds ($143 million) were allocated to Lifeline.

In 2008, Urban Habitat and Manuel Pastor cohosted the First Annual "State of the Region" summit. This historic event brought together hundreds of community allies and leaders in government, labor, business, and philanthropy to identify strategic opportunities to advance an agenda for regional equity in the Bay Area for the years ahead.

Linking Regional Equity to a Broader Global Justice Agenda: The SEC and the World Social Forum

The SEC sends delegations to national and international conferences to expose them to alternative strategies and models for advancing social and environmental justice. An SEC delegation attended the World Social Forum (WSF) to learn how global economic and political forces are impacting the Bay Area. Reports from many U.S. activists who attended the forum indicated that they were simultaneously inspired and overwhelmed. Participants needed better preparation to maximize the benefits of the experience.

To address this need, CJTC and UH collaborated to create a comprehensive orientation process for the SEC delegation to the 2005 World Social Forum in Porto Alegre, Brazil. A series of orientation meetings were held, culminating in a Bay Area summit to learn from groups that have linked their work successfully to a global analysis, such as the Miami Workers Center and the International Forum on Globalization.[2] During this daylong event, SEC delegates gained a better understanding of the impacts of global economic trends, such as those affecting oil, retail, and trade on low-income communities in the Bay Area. UH and CJTC also collaborated to develop a report examining globalization-related efforts around the world, looking at the impacts on local struggles for social justice—such as the threat of the "Wal-Martization" of the Bay Area retail economy.

After working with CJTC and others to prepare for the WSF, the SEC delegation traveled to Porto Alegre. While there, SEC delegates shared stories and strategies with activists and organizers from many different U.S.-based organizations, as well as attendees from Brazil, Nigeria, Haiti, Germany, Venezuela, and Bolivia. The SEC delegates reported that the orientation organized by the UH and the CJTC gave them the background they needed to optimize their learning at the international forum.

Mutual Benefits to Collaboration

Despite institutional differences in power, style, and goals, partnerships between a university and a community organization can be an important strategy for advancing regional equity. Looking back on their history of collaboration, UH and CJTC see several key lessons that might be of use to others and believe that there are significant mutual benefits to collaboration. For example, UH understands that alone it cannot build power in the Bay Area's low-income communities of color. Therefore, UH has worked hard to build strategic relationships outside the nonprofit sector, with labor, business, government, and academia. Similarly, one of the central elements of CJTC's mission is conducting research in the arena of social justice. One of the best ways to do that is in concert with community organizations that can actually move the work from theory to action.

There are benefits specific to each side of the university-community relationship. For UH, a key benefit is *greater access to the human and financial resources of an academic institution* (Suarez-Balcazar, Harper, and Lewis 2005, 95). CJTC, for example, assisted in securing funds to cover part of the SEC delegation to the World Social Forum in Brazil. In addition, the Center for Community Innovation (CCI) at the University of California, Berkeley, created an inventory of all the vacant and underused properties in Richmond and provided researchers from UC Berkeley's graduate programs to support UH and the Richmond Equitable Development Initiative. A university partnership can also connect the community group to other research efforts taking place throughout the country. Because decision makers are often in search of quantitative analysis and "evidence" to support a community's recommendation, a supporting university study or report can add credibility and impetus to such recommendations.

Another key benefit is *assistance in the development of policy campaigns*. The SEC recently partnered with Dr. Steven Pitts from the UC Berkeley Labor Center to research and create a shared definition of a "quality job" and to advance concrete economic justice policies. Dr. Pitts has been involved with the SEC for many years and has provided the SEC with an important link to the labor movement. He argues that although there is an unemployment crisis in the black community (Pitts 2005), the community also faces an equally daunting crisis of low-quality jobs (jobs that pay poorly, with few benefits, and few opportunities for advancement). The SEC conducted an informal survey of the work being done to improve the quality of jobs for people of color in the Bay Area, and UH and the SEC created a working group to further explore ways to promote economic justice in the region. Recently, UH and members of the SEC, including Dr. Pitts, contributed to a report published by East Bay Alliance for a Sustainable Economy and the Oakland Network for Responsible Development titled, "Putting Oakland to Work: A Comprehensive Strategy to Create Real Jobs for Residents."

Finally, a *university partner can provide outside facilitation* to help advance an agenda or process. An academic is often seen as a desirably "neutral" player because he or she is not personally associated with a particular community organization, or is seen as bringing special expertise to a program. This was the case for UH when Manuel Pastor cofacilitated the Bridging the Bay summit and helped guide the SEC's strategic planning process. The role of consulting expert can be very valuable, although it should be played with great modesty and with recognition that it is the group and not the facilitator who is driving the process.

Of course, benefits go both ways: *community collaborations often lead to better research*. This is partly because the community's "radar" about what is important can suggest topics of study that are at the cutting edge, such as environmental inequities, gentrification, and related issues. Indeed, the entire regional equity agenda of the CJTC was largely inspired by community voices. Collaboration can also help researchers develop new paradigms and methods. For example, CJTC was asked by an interfaith federation working in the Monterey Bay to develop a regional analysis that could serve as a vehicle for identifying key issues for a social justice agenda. The result was a new regional "audit" framework that has since been used in other areas of California and has been featured in a key academic journal.

Similarly, CJTC was asked by a long-standing collaboration partner, Communities for a Better Environment (CBE), to help understand the dynamics of the siting of environmental hazards. CBE wanted researchers to examine the argument that the contemporary pattern of environmental inequity was not the result of hazards being placed in minority communities, but simply emerged when minorities moved into contaminated neighborhoods. Painstaking research demonstrated that the cause was not "minority move-in," but rather was, indeed, discriminatory siting. This resulted in better arguments for community leaders and also yielded a landmark, frequently cited, academic article.[3]

Community collaborations also provide *an arena for testing ideas and approaches*. Moving from regional equity theory to neighborhood-level practice can be a tremendous learning experience for researchers and community leaders alike. An important guideline for designing research and action projects is that the agendas and goals should come primarily from the communities themselves.

It is also the case that *community collaborations help researchers contribute to moving agendas forward*. Many university-based researchers are motivated by pure science or intellectual curiosity, which are perfectly valid goals in the context of academia. However, many university personnel also have political and social agendas that, while helped by solid research, can be acted on only by engaged constituencies. Collaborative work ultimately brings more meaning to academic research by making it more likely that the research will be used in a practical way. The Bridging the Bay collaboration—and the subsequent SEC work on transportation justice—are clear examples of this. Another example is the way in which CJTC's work on environmental inequality has been used to push changes in air quality rules in Southern California (Liberty Hill Foundation 2005).

Finally, *community collaborations can help overcome the isolation of the university*. Universities can be remote and detached from what is going on outside academia, which means that university resources are lost to community actors, and the detached environment can be alienating for students and faculty. Working with the SEC has helped CJTC avoid this disconnect by widening its set of community relationships, particularly in the Bay Area. The partnership has also helped CJTC maintain credibility in the community, which is useful when researchers hope to work alongside communities that may feel "overstudied" and not open to yet another research project.

Lessons Learned and Recommendations

As described throughout this chapter, moving a regional equity agenda involves building intersectoral alliances that can cross the usual boundaries of race, class, and geography in the interest of a common regional future. One important boundary crossing is between community leaders and university researchers. We draw several key lessons from both our UH-CJTC collaboration and from our respective collaborations with other community and university partners. The first is that *collaboration is a long-term process*. Treating a university-community collaboration as a onetime affair is unlikely to succeed in terms of achieving goals, because there isn't enough time to *build the necessary foundation of trust and effective communication* to sustain the relationship through challenging times. Collaborators need to go beyond being project partners and instead become allies.

Ideally, investment in personal relationships will occur before work begins on a specific project, with the partnership emerging from shared interests and goals. Collaboration between universities and community groups is also being encouraged by funders (Washington 2004, 18–29). We would caution funders, too, that when making a long-term commitment, they need to understand that it takes time to develop the personal relationships needed for a successful university-community partnership.

When a community organization partners with a university, it is important to understand both the balance of power and each partner's needs. In our view, *the community organization should play the lead role in identifying the goals and desired outcomes of the collaborative project*, especially if the project involves other community groups who may not have a direct relationship with the university partner. For Bridging the Bay to be a success, with meaningful community engagement, UH had to play a lead role in defining the purpose and goals of the collaboration because UH had a clear sense of the history and the current needs of the SEC. UH also had a much greater investment in the event, compared with CJTC, because UH had to deliver an exceptional experience to its SEC partners. UH had nurtured these relationships over many years and had made the commitment to work with these organizations over the long term.

While each organization needs to understand its role in a shared project, the planning should be done collaboratively. *Partners need to be sensitive to each other's capacity, interests, demands, and work styles.* For example, universities may be more concerned with publishable products, while CBEs may be interested in more popular documents. In fact, there is great room to do both. Community groups should work closely with the university to ensure that the work has real-life applications and is communicated to a broad target audience including decision makers, community members, labor unions, and funders. The information should be communicated in an accessible format that speaks to each constituency. When possible, *the community group itself should gain an increased capacity for research as a result of the collaborative experience.* Indeed, a key goal is to build the internal research and facilitation capacity of the community partner, rather than encouraging dependence upon the university researchers.

There is also a need to *balance research agendas and research results.* Community partners should play a lead role in setting the research agenda, as community leaders often raise important and timely questions that may not have occurred to researchers in a more isolated research world (Prakash 2004). Researchers need to be open to being critiqued about results that do not resonate with community common sense. At the same time, academic researchers cannot guarantee that their results will turn out any particular way. They must not design their research to favor one side, such as work that a communications department might produce for a policy campaign.

Finally, the question of *who actually is the "owner" of the research* should be agreed upon in advance. While this is usually not problematic, it can become so if it is not addressed before the end of the project. Coauthored articles can be powerful tools, as can reports written solely by the university partner that address the questions of the community partners.

Academics must be aware of the *imbalances of power and privilege* that come from being part of elite institutions. This will be amplified if there are class and race differences among colleagues. At the same time, there is a misperception that all university departments or centers are well endowed. This is often not the case, especially for social justice researchers who are most likely to be interested in collaboration with community groups. Still, these perceptions can undermine thoughtful collaboration. More often than not, the university partner does have more resources to dedicate to a project, making it easier for the university to sustain its part of the work over time, while the community group must be funded to participate in all aspects of the collaboration to ensure a consistent and effective level of engagement.

It is important to *create mutual learning experiences*. Too often, university partners are used to communicating in a didactic way—that is, presenting facts and lecturing about how they might be interpreted. It is important to avoid the expectation of "teaching" and instead focus on creating learning experiences. Bridging the Bay, with its emphasis on interaction and the development of strategy, was a good example of such a learning experience.

Conclusion

Collaboration, when done well, is synergistic and strategic. Partners must value each other's expertise through their actions—by listening and communicating with respect and an open mind. At its best, a university will see its community collaborator as an intellectual partner in the research, helping to ground truth and inspire quality work. The joint research on globalization by UH and CJTC is a prime example of the kind of excellent results that can be used practically, in this case by the SEC campaigns and WSF delegates.

Breaking through assumptions—such as that universities are concerned only with academic standards, and that community groups are too busy for research—can lead to exciting partnerships. Working together is much like linking neighborhoods in a region: intuitively we know it is needed, but it does not always seem like a natural match.

Yet, much like regionalism itself, working through the challenges of collaboration with foresight and perseverance can produce both information and action of the highest quality.

Notes

1. See the early analysis in the Bay Area commissioned by Urban Habitat and developed by M. Orfield, "What If We Shared? Findings from Myron Orfield's San Francisco Bay Area Metropolitics: A Regional Agenda for Community and Stability." A report by the Urban Habitat Program (May 1998).

2. See "Bringing Globalization Home: Portraits of Popular Education at the Global-Local Junction," available at http://www.cjtc.ucsc.edu/pub_reports.html. See also the Internet clearing house for popular education techniques for global-local organizing, available at www.globallocalpoped.org.

3. For a general review of the CJTC collaboration with Communities for a Better Environment, see R. Morello-Frosch, M. Pastor, J. Sadd, C. Porras, and M. Prichard, "Citizens, Science, and Data Judo: Leveraging Secondary Data Analysis to Build a Community-Academic Collaborative for Environmental Justice in Southern California," in *Methods for Conducting Community-based Participatory Research for Health*, eds. A. J. Schulz, B. A. Israel, and P. Lantz (San Francisco: Jossey-Bass, 2003). For specifics on the move-in study, see M. Pastor, J. Sadd, and J. Hipp, "Which Came First? Toxic Facilities, Minority Move-in, and Environmental Justice," *Journal of Urban Affairs* 23, no. 1 (2001).

24

Poor City, Rich Region: Confronting Poverty in Camden (Camden, New Jersey)

FROM DEPENDENCY TO SUSTAINABILITY

Howard Gillette, Jr.

According to first impressions, Camden, New Jersey, might be an unlikely candidate for demonstrating the utility of a regional equity program. At fewer than eighty thousand persons clustered in less than ten square miles, the city has a diminutive quality that would appear to rank its importance well below that of Detroit, Gary, Newark, or other postindustrial sites. It is precisely Camden's experience with big-city problems within a relatively confined space, however, that makes its story illustrative. Nearly universally impoverished, it offers a case study of how an entire political unit, not just some of its constituent elements, operates within a climate of crushing austerity, thus bringing to the fore the central role of public power in dealing with urban poverty. At the same time, its strategic location directly between a similarly beleaguered big city, Philadelphia, and a largely prosperous suburban hinterland raises tough questions about the future of regional solutions to localized poverty. In short, while Camden offers a small stage, all the major issues associated with the drive to achieve regional equity appear in recognizable and comprehensible fashion.

Once one of the most productive cities in the nation—its plants turning out products from soup to nuts to battleships—Camden experienced a precipitous decline in the post–World War II era. Despite the conversion of wartime industries and the pull of the suburbs, Camden maintained its industrial base for a while, listing as many manufacturing jobs in 1960 as it did in 1948. While the city's demographics shifted during the period, as first African-Americans and subsequently Puerto Ricans entered the city, seeking new employment and housing opportunities, the city remained predominately blue-collar and white through the late 1960s, when the closure of New York Ship marked a significant decline in industrial employment. Riots that shook the city for three nights in 1971 accelerated the rate of departure among whites, and by 1981, when the city elected its first African-American mayor, the city had shifted to a black majority and the employment base had been devastated. For newcomers to Camden, the city had become a hollow prize.

Declining real estate values made for affordable housing opportunities but stripped the city of the necessary means to balance its budget without raising taxes. Unwelcome activities—including a state prison located on prime waterfront property, a municipal waste plant, and a trash-to-steam incinerator serving Camden County—helped the city balance its budget but marked its environment as distinctly inferior to surrounding suburbs. Camden thus came to represent the worst of all worlds: a high concentration of poverty requiring special services, and levels of capital investment too low to pay for them.

The *Mount Laurel* decisions of 1975 and 1983 David Rusk describes in his subchapter were intended to overturn patterns of exclusionary zoning that had kept the suburbs around Camden largely white and relatively affluent while the city continued to languish. As compromised by the authority granted under the Fair Housing Act of 1985 allowing towns to sell off half their affordable housing obligation, the *Mount Laurel* decisions failed in their determination to disperse the urban poor. As a new century commenced, Camden provided 45 percent of the county's affordable housing opportunities even though it represented only 14 percent of area population. With the continued effect of poverty concentrated in the city, Camden failed either to meet its budgetary obligations or to stem the associated social costs of high levels of crime and social pathology.

The city tried a number of ideas for invigorating the economy. Seeing many partial reforms fail, the state intervened in 2002, vesting its own appointed chief operating officer with broad powers over city governance for a period of five years. Instead of simply putting the city into receivership, it authorized a multidimensional municipal recovery act, the heart of which was a $175 million fund intended to leverage new investment for the city. Of those funds, more than half went to downtown institutions, primarily educational and medical facilities, seeking to strengthen their roles as engines of the local economy. Waterfront development, which had witnessed the conversion of abandoned industrial property into a regional visitor destination for sports and entertainment, received $25 million to convert the state aquarium to private use, thereby securing in the deal a promise to attract additional private investment for shops and housing to complete two decades of development there. The central test of the effort, however, was the revitalization of the city's neighborhoods, none of which had escaped the burdens of disinvestment.

Under the terms of the legislation, recovery was to be guided by a state-commissioned revitalization plan, which was approved by a new Economic Recovery Board in 2003. Influenced by arguments Jeremy Nowak had advanced to make Camden, in his words, a "place of choice and not just a place of only resort," the plan emphasized improved opportunities both for work and decent housing within the city. In seeking to attract more middle-class residents back to Camden, however, the plan encouraged the redevelopment of whole neighborhoods, a point that immediately provoked local fears and resentment. Had a regional impact council provided under the legislation been established and committed to coordinating Camden's recovery in a metropolitan context as intended, such emphasis still might have assured improved opportunities for Camden's poor to relocate to surrounding suburbs. Instead, Camden chief operating officer Randy Primas told members at the Economic Recovery Board's initial meeting that any residents displaced during renewal would be relocated within city limits. That was not a commitment to assuring Camden residents the full range of regional economic opportunities as they had evolved over the period of Camden's decline.

The Ford Foundation's effort to make up for the recovery legislation's deficiencies involved supporting both Camden-area and statewide regional organizing efforts as well as additional investments in training and capacity building in Camden's public and nonprofit sectors. Taken together, the level of investment from the state and the philanthropic sectors was exceptional, making Camden a prime test of a regional equity strategy for the twenty-first century.

It is still too early to measure the effectiveness of these combined efforts. It is clear nonetheless that to change the rules of the game, as David Rusk terms the way in which monetary resources are allocated among jurisdictions, takes extraordinary leadership. As Jeremy Nowak notes in his subchapter, all area constituents—residents, local institutions, developers, and city government—need to join together in a common plan. To date, such efforts have been uneven, with the COO more intent on chasing development dollars than assuring that those dollars lift the fortunes of local residents. The New Jersey Regional Coalition has secured some important commitments to make the Mount Laurel doctrine work by eliminating Regional Contribution Agreements, but such a prohibition has yet to become law outside of Camden.

The city and its residents still have a long way to go. The elements for a just and equitable resolution have been laid out. It will take more than a village, however, to make regional equity a reality in the greater Camden area. It will take the concerted and coordinated effort of many stakeholders—nonprofit organizations as well as the private sector, secular as well as religious organizations, and leaders in both the city and the suburbs—committed to resolving Camden's fiscal difficulties through a metropolitan framework.

THE "INSIDE GAME": A REINVESTMENT STRATEGY

Jeremy Nowak

In its fifty-year population decline, from a peak of 125,000 people in the 1950s to only 79,904 (a fall of 36 percent), Camden, New Jersey, has suffered serious financial losses through sharply lower real estate values, high vacancy rates, and diminished building permit activity. In addition, Camden exhibits low levels of labor force participation, limited adult educational attainment, and high levels of adult and child poverty. From any place-based or household indicator, Camden is among the most distressed cities in the United States.

Of all U.S. municipalities, Camden has the highest level of population at 50 percent of the poverty level or less. Camden also had the lowest level of population with incomes above 200 percent of poverty. The geocoding of income data provides an extraordinary picture of income distribution belts moving outward, away from Camden. This concentration of poverty remains despite the two *Mount Laurel* New Jersey Supreme Court decisions aimed at expanding suburban affordable housing opportunity.

The concentration of poor residents is compounded by race. Camden is almost entirely African-American and Hispanic. The last majority white neighborhood changed composition during the past decade. The geocoding of racial composition for Camden and the surrounding South

Jersey counties demonstrates racial segregation dynamics that reinforce the income distribution data and heighten Camden's isolation.

While population loss, a weak real estate market, and poverty indicators present a significant challenge for redevelopment, the challenge is compounded by the fact that Camden is located in a relatively slow-growing region. With the exception of the downtown Philadelphia business cluster, all major growth in the region is suburban, although—unlike many metropolitan regions—suburban growth has not coalesced into major edge city developments.

The Philadelphia region does have economic growth assets. It has the advantages of an East Coast location, transportation access, and the infrastructure to access this market. It has amenities of place such as architectural form, a strong residential downtown, and first-rate cultural institutions. And it has several competitive business clusters, including professional services, financial services, and the life sciences/pharmaceutical industry. These clusters are supported by significant numbers of educational and research institutions. Camden sits in proximity to these growth clusters, but is largely disconnected from their value.

The challenge for the region is to uncover and build upon its assets, invest in them, and compete globally. This requires civic and private-sector leadership that can set a long-term growth agenda. It involves public investment policies that transcend parochial administrative interests. For places like Camden and much of Philadelphia, it also means being able to manage the tension between growth and change, on the one hand, and equity expectations and economic inclusion, on the other hand.

Rebuilding Camden is about reconnection and revaluation. *Reconnection* means linking Camden families and communities to regional economic opportunities. *Revaluation* means creating enough value—in terms of residential amenities, business location opportunity, and labor quality—that regional consumers, investors, businesses, and residents view it as a place of choice and not just as a place of only resort.

Constraints and Opportunity

The distress trends in the city of Camden continued throughout the 1990s despite its proximity to downtown Philadelphia, public investments along the river, access to the regional airport, proximity to south

Jersey economic growth, and suburban pressure to curtail development and sprawl. The downward spiral of Camden continued for a variety of reasons including 1) a tax structure that is not competitive; 2) a public sector that lacks resources and talent; 3) affordable housing market alternatives in the inner ring; 4) low levels of labor market competitiveness; 5) the high cost of land assembly; and 6) public concerns about crime rates, political corruption, and the extensiveness of blight. Despite the long-term trends, a variety of factors have come together during the past several years to create a window of opportunity.

A State-driven Camden Recovery Plan with Teeth and Resources

The Camden Recovery Act allocates $175 million of state aid for a variety of institutional, infrastructure, and neighborhood development strategies. It also effectively places authority for governance in the hands of a state receivership executive, while providing funds to rebuild the public sector.

Renewed Investor and Developer Interest in Cities

Antisprawl sentiment in the region has placed limits on green fields development and has generated increased interest by developers and investors in the core areas of Philadelphia and the waterfronts of older regional cities along both the Delaware and Schuylkill rivers—including Camden.

The Impact of Public Investment

The concentration of waterfront investments by the State of New Jersey has created a regional entertainment asset that is slowly changing the perception of Camden. Sports facilities, an aquarium, and an entertainment center have made Camden a destination place for suburban families and young people throughout the region.

Expansion of Downtown Philadelphia Real Estate Demand

Downtown Philadelphia, the hottest real estate market in the region, is slowly expanding north, west, and south. Camden represents its eastward expansion. The recent commitment to create more direct transportation linkages across the river may speed this expansion.

Institutional Expansions and Capacity

Spurred largely by county and state public investment strategies, Camden is undergoing a significant process of new institutional expansion by the county community college, the state university (Rutgers), Rowan University, the University of Medicine and Dentistry, and several hospitals.

Abbott Decision

This court decision relating to the need to rebuild schools in older towns and boroughs will result in the investment of more than half a billion dollars for new school construction within Camden over the next few decades.

People, Place, Public Capacity, and Civic Engagement

To maximize the window of opportunity for Camden and its low-income residents, a four-pronged strategy is required: a place-based real estate strategy, a people-based social service and workforce strategy, a public capacity-building initiative, and ongoing civic engagement or community building. All four strategies also complement the regional policy and opportunity approach.

Place-based Strategy

This strategy is designed to address the physical needs within the City of Camden, to shore up its areas of strength, and to uncover an important and sometimes hidden asset: land. This strategy is location-specific; the approach to revitalization reflects the existing conditions of the market.

There are five aspects to a place-based strategy in Camden: 1) neighborhood asset preservation, 2) waterfront/market rate development, 3) land banking, 4) light industrial and business park development, and 5) institutional expansion.

Place-based strategies encompass the complex issue of environmental degradation and environmental resource recovery. Key to almost all of the place-based development strategies is the restoration of the environmental integrity of Camden: remediating brownfield hazards, making investments along rivers, expanding greenway access, and working toward a future where Camden economic development is not based on receiving the environmentally hazardous businesses that wealthier municipalities are thereby able to deflect.

People-based Strategy

People-based strategies are geared at providing direct assistance to families and households in order to give them the social relationships and personal capacity to improve their lives and take advantage of existing and future opportunities. People-based strategies in Camden can be divided into three categories: 1) basic public education opportunities, 2) workforce development strategies, and 3) social services.

The public education system in Camden is broken, despite the rather substantial per-pupil allocation of resources compared with many other low-income districts throughout the country. Workforce development services must be provided to Camden's residents so that they can be better positioned to take advantage of the region's economy. Social service needs in Camden are considerable. The county prosecutor's office reports that 40 percent of their indictable adult cases originate in Camden, though Camden accounts for only 15 percent of the county's total population. While there is a significant amount of recidivism within the system and not all of the defendants are necessarily Camden city residents, the implication of this trend on the city is staggering.

Public Sector Capacity

One of the most important underlying assumptions of the Camden Recovery Act is the need to rebuild basic public sector capacity. In the words of the Multi-Year Recovery Plan, "Camden's government lacks the capacity to plan for redevelopment, allocate resources, assemble property, implement development projects, coordinate infrastructure approvals, monitor progress, apply for available funding, track existing funds, account for performance, or assess problems." Unless Camden addresses this lack of capacity, it will be able to attract only new residents that don't require government services or have the capacity to privatize many public benefits.

The appointment of a receiver government is an important market signal to investors. It ensures a modicum of predictability to get projects off the ground. But it does not paint a real roadmap to how the interim receivership system will have adequate capacity or how it will leave the city (in its absence) with more administrative talent and modern management systems.

Civic Engagement

No strategy to revitalize Camden can have either short- or long-term success in the absence of a civic infrastructure that sets long-term agendas, provides review and credibility for public and private plans and activities, and works to solve tough problems that cross neighborhood boundaries, institutional loyalties, and racial and ethnic identities. Moreover, the civic infrastructure must be more than a Camden-only club. It must encompass relationships that allow for outside leadership from the region to participate and add value, while building more extensive networks and trust.

The dramatic events of the past several years have catapulted a variety of organizing and civic efforts into the limelight, including church-based organizing efforts, business partnerships, and a new civic roundtable that brings numerous organizations and interests into a neutral assembly. The effectiveness of real estate, family, and public-sector investments will increase to the extent to which there is a civic framework that can maintain local momentum and credibility. This must be an essential element of any investment strategy.

Creating a Political and Civic Culture of Change

Years of decline not only create enormous cost and social barriers to revitalization but also a social climate that makes change difficult. Rebuilding Camden requires a political and civic culture that can take advantage of the opportunities that exist.

Fifty years of decline makes most citizens suspicious of the ability of government and the intentions of the private sector. It also means that the public sector has accommodated itself to a no-growth attitude where the pie gets smaller and the demand from the poor for assistance gets greater. Fifty years of decline translates into widespread institutional decline and a sense that the best among us have gotten out. It means that there is no habit of winning or of getting things done.

Change for Camden requires a transaction-based collaboration that the city has rarely been capable of performing during the past five decades. In its absence, all the great plans and targeted dollars will accomplish little.

THE "OUTSIDE GAME": CAN FAITH MOVE MOUNTAINLESS NEW JERSEY?

David Rusk

"We must improve our Fair Housing Act," proclaimed Assemblyman Joseph J. Roberts, Jr., at a 2004 Princeton University Symposium, "a law so flawed in its application that the use of the term 'fair' in the title is a mockery in itself!"

The majority leader of the New Jersey General Assembly continued, "Where is the 'fairness' in allowing wealthy suburbs to avoid half of their fair share obligation by entering into 'Regional Contribution Agreements'—RCAs—with economically deprived cities?" he asked.

Several local residents shifted uneasily in their seats, knowing that Princeton Township had paid impoverished Trenton $460,000 to take a constitutional obligation to build twenty-three units of affordable housing off the hands of the wealthiest community in Mercer County.

"New Jersey must abolish these odious and exploitative RCAs once and for all," he declared. "To advance as a state, we must move forward together—all of us—*and not pay others to stay behind.*"

While Assemblyman Roberts spoke out of a deep personal conviction that April 2004 evening, he had also been fortified by the arguments and political support of the New Jersey Regional Coalition, a powerful faith-based coalition of congregations committed to achieving social justice and genuine opportunity in the most economically segregated state in the country.

The Birth of the New Jersey Regional Coalition

Begun in 2002, the New Jersey Regional Coalition had sharply different goals than the Bush administration's Faith-Based Initiatives. The White House called upon churches to expand their traditional charitable mission. The New Jersey Regional Coalition envisioned its churches practicing a different tradition—that of social justice.

The New Jersey Regional Coalition was affiliated with the Chicago-based Gamaliel Foundation, a faith-based organizing institute. The Gamaliel Foundation itself represented a radical break from the traditional community organizing focus on neighborhood-level action and City Hall. During the 1990s, Gamaliel Foundation leadership believed that the decline of inner cities could not be reversed without addressing development trends outside their boundaries. Neighborhood problems demanded regional solutions.

Myron Orfield, john powell, and I were key contributors to shaping the Gamaliel Foundation's new regional perspective. Orfield, a former Minnesota state senator, literally mapped out why "metropolitics"—building coalitions of central cities and at-risk suburbs—was essential.

Former ACLU national law director powell argued that antisprawl policies and regionwide "opportunity-based housing" are "the new civil
rights issues," and I, a former mayor of Albuquerque and New Mexico
state legislator, insisted that the traditional "inside game" must be coun-
terbalanced by the "outside game"—changing the "rules of the game"
guiding "what gets built where for whose benefit."

Our writings had caught the attention of Marty Johnson, founder and
president of Isles, Trenton's most successful community development
corporation (CDC). Founded in 1981, Isles had been carrying out a
wide range of programs to foster more self-reliant families in healthy sustainable communities.

In 1996 and 1997, Johnson took a sabbatical from Isles to teach at the
Woodrow Wilson School of Princeton University. While there, he began
to develop a national project to better measure the impact of community-
building work—the Success Measures Project. Upon his return to Isles's
management, the organization began applying different measures to
community health. These new measures posed a challenge, forcing Isles
to understand the limits to their impact on communities. Despite many
successes—new homeowners, more high school graduates and fewer
dropouts, cleaner urban neighborhoods— the indicators of overall community health were suggesting a different story—continued flight, concentrated poverty, and other problems. Isles's programs were succeeding
and failing at the same time.

Johnson invited me to come to Trenton and Princeton in April 2000 to
present his regional analysis for Trenton-Mercer County, marking the
first time that any CDC as practitioners of the "inside game" had ever
sought my "outside game" analysis. Johnson also invited Orfield and
powell to speak at Isles-sponsored community forums and, in June
2001, at my suggestion brought a busload of community leaders for a
tour of mixed-income housing developments in Montgomery County,
Maryland.

Johnson began pulling together leading New Jersey organizations to
discuss Isles's findings, test others' assumptions about regional demographic trends, and to explore the potential of forging new alliances
across the region. With newly acquired funds from several foundations,
he launched a new organization; this group commissioned Orfield to research a report, "New Jersey Metropatterns."

At this point, two worlds—a local CDC and a national faith-based
organizing movement—came together. Long based in the Midwest, the

Gamaliel Foundation had dispatched Paul Scully, a veteran organizer, to extend the foundation's reach to the East Coast. One of Scully's first initiatives was to organize the Jubilee Interfaith Organization in North Jersey. (After a year's sabbatical as a local union's organizing director, Scully was about ready to put on his regional equity–organizing hat again.)

Upon Orfield's and my recommendation (both of us had worked with Scully in the East Chicago region), Johnson hired Scully for his project. Quickly, the two decided to move from research and advocacy to grassroots organizing statewide. Johnson, Scully, powell, Orfield, and I drew upon our relationships in New Jersey. We reached out to Barbara Lawrence, head of New Jersey Future; Peter O'Connor, founder of Fair Share Housing; and Gamaliel leaders.

To this core group (Isles, New Jersey Future, Fair Share Housing, and Jubilee) were added the Regional Planning Partnership (later PlanSmartNJ), Coalition for Affordable Housing and the Environment, New Jersey Community Loan Fund, and New Jersey Institute for Social Justice. With cochairs Marty Johnson and Roland Anglin, head of the New Jersey Public Policy Research Institute, and funding from the Ford Foundation, the New Jersey Regional Coalition (NJRC) was formally organized. With Isles as its initial organizational "home" and fiscal agent, NJRC shared offices) in Cherry Hill near Camden with Fair Share Housing Center (FSHC), a nonprofit legal and policy center, was founded in 1975 to challenge residential segregation and exclusionary zoning in New Jersey.

From its inception, NJRC represented a rare blending of experience and talents. Scully added a decade's experience as a top-notch community organizer with the same organization in which a young Barack Obama was trained. Johnson acknowledged that beyond the contacts, credibility, and resources from twenty years of successful "inside game" programs at Isles, Isles had in his words, "been measuring the wrong stuff...."

NJRC made an impressive public debut in June 2003, hosting the People's Summit on Regional Equity for more than five hundred people. All three strategic partners were featured speakers. Orfield presented multicolored maps charting regional disparities from his recently completed report, "New Jersey Metropatterns." I emphasized that "housing policy is school policy": where children live largely shapes their educational opportunities not in terms of school budgets but who their classmates are. Powell detailed how for almost thirty years, state policy loopholes, like RCAs, and suburban intransigence had frustrated the New Jersey Supreme Court's *Mount Laurel* decisions. The court's goal had been to open up New Jersey's suburbs (with their growing job markets and low-poverty schools) to poor minorities trapped in city barrios and ghettos (where both jobs and high-quality schools are vanishing).

Knowing that building grassroots power would require more than the ability to attract a large audience to a onetime event, NJRC launched organizing efforts in all three sections of the state.

Organizing South Jersey

The organizing luncheon for the South Jersey Regional Equity Organizing Project in October 2003 was typical of the three events. The luncheon was held in the new parish hall of Christ the King Catholic Church in Haddonfield.[1] The parish hall was packed with more than 120 persons but was particularly noteworthy for the tremendous diversity of its attendees. There were 32 pastors and 26 other religious leaders representing 43 different congregations.

The geographic diversity of the religious participants was further expanded by participation of numerous civic organizations, such as three NAACP chapters, the Puerto Rican Action Committee, and the Pennsauken Stable Integration Governing Board. In all, thirty-four municipalities in the Burlington-Camden-Gloucester-Salem region were represented, as well as the cities of Vineland, Millville, and Bridgeton (Cumberland County), and Lakewood Township (Ocean County).[2]

The meeting was chaired by Monsignor Dominick Bottino, pastor of St. Stephen's Parish of Pennsauken, Camden's largest first-ring suburb. I delivered the keynote address, which was designed consciously to appeal to the self-interest of "Orange Land" municipalities (at-risk suburbs, in Orfield's typography) and to indicate how they were now being affected by "city problems." The address emphasized that "poor communities pay higher taxes" and that "wealthy communities pay lower taxes."

Such tax disparities directly reflect the uneven distribution of jobs and housing types. Contrary to all economic sense and social fairness, work-force housing is being built in the direct opposite locations to where jobs are growing. In the 1990s, the ten fastest-growing job centers in South Jersey—all outer suburbs—saw more than 42,295 new jobs created but only 1,260 new subsidized homes built. By contrast, the ten municipalities that lost the most jobs—25,264 jobs—saw more than 16,845 low-income homes built or renovated. Camden, for example, lost 6,212 jobs but added or renovated 7,112 subsidized housing units. Almost as striking, first-ring Pennsauken lost 4,199 jobs—the third-highest job loss in South Jersey.

I argued that Regional Contribution Agreements nullified the whole concept of "opportunity-based housing." For example, wealthy Washington Township sold 152 units back to Camden for a payment of $20,000 per unit. During the 1990s, Washington Township had seen a 30 percent increase in jobs when only 7 percent of its elementary school pupils were low-income; by contrast, Camden lost 22 percent of its jobs when 88 percent of its elementary school pupils were low-income.

"RCAs may help create more affordable *shelter* in the cities, but RCAs are literally cementing children into poverty-impacted schools where most are doomed to fail," I summarized in the address. "Affordable housing must be built where the jobs are being created. Anybody who is good enough to work in a community is good enough to live in that community." These comments were met with enthusiastic applause.

To fire the audience up further about faith-based organizing, the program ended with impassioned appeals from the Reverend Dr. Hoffman Brown and the Reverend Karen Brau, both cochairs of BRIDGE, the Gamaliel affiliate in Baltimore. Their contributions showed the value of having a nationwide network of faith-based organizations that a local organizing effort can draw upon for talent, experience, and inspiration. As the initial organizing meeting ended, twenty-eight congregations signed pledge cards to send one to five members to a leadership training workshop in 2004.

The South Jersey meeting set the mold for organizing events in Central and North Jersey where I presented parallel data for each region. During the Central Jersey event, Kevin Walsh, Fair Share Housing Center's attorney, took me to visit Majority Leader Joe Roberts at the state capitol in Trenton. Walsh had helped persuade Roberts to amend the $175 million Camden Recovery Act to ban Camden's accepting more RCAs.

Throughout the winter and spring, NJRC continued to build its grassroots strength in North, Central, and South Jersey. Newly organized congregations in all parts of the state identified their most critical issues: lowering property taxes and reforming state housing policy. Neither was surprising. New Jersey's local governments are more dependent on property taxes than are local governments in any other state except Connect-icut. Economic recession and shrinking tax bases had driven city and many inner suburb property-tax rates to dizzying levels.

The Epic of *Mount Laurel*

The housing issue is more complex—and has a long history in New Jer-sey. In 1971, Peter O'Connor, a young Legal Services attorney, and two other lawyers he worked with agreed to take on the cause of African-American residents of then-rural Mount Laurel Township twenty miles from Camden. They had obtained a federal grant to build new affordable housing, but township leaders who, with visions of high-end housing on the drawing board, denied their zoning application: If you people can't afford to live in *our town*, then you'll just have to leave, the mayor told them.

O'Connor filed suit in state court and, in 1975, won the New Jersey Supreme Court's stunning *Mount Laurel I* decision. Every municipality in New Jersey, the court ruled, has a constitutional obligation to provide a realistic opportunity for the provision of the full range of housing types for all income groups. Moreover, a municipality's affordable housing target would not be limited to serving current residents but would be based on *regional* need. Mount Laurel Township's affordable housing obligation, for example, would be based in part on the needs of low-income residents currently living in Camden.

The court called upon the legislature to implement the court's decision, but to no avail. After eight years, tiring of the legislature's and suburban municipalities' inaction, the supreme court issued instructions for the lower courts in *Mount Laurel II* in 1983. "We may not build houses," the court said, "but we do enforce the Constitution." Among the court's remedies for exclusionary zoning was the so-called builder's remedy. A builder could propose a housing development that included as little as 20 percent affordable housing. If the town council turned the proposal down, the builder could seek a court order directing that the project to be built over local objections.

To blunt the court's determination, the New Jersey legislature finally acted. The Fair Housing Act of 1985 established the Council on Afford-able Housing (COAH) and charged it with setting every municipality's "fair share" obligation. However, responding to then-governor Thomas Kean's veto threat, the legislature instituted the system of Regional

Contribution Agreements by which wealthy boroughs and townships could sell up to half their fair share obligations back to poor cities.[3]

Taking COAH's claims at face value, its achievements have been substantial. As of September 2004, according to COAH's monitoring reports, "the opportunity for approximately 66,600 affordable units has been provided. This includes about 34,900 units that have been built or are under construction, 9,100 units that have realistic zoning in place, 8,800 RCA units and 13,800 units that have been rehabilitated.... More than $175 million has been approved for transfer [through RCAs]," COAH's Web site reports.

The reality on the ground is substantially different. Mistaking a group of NJRC pastors for opponents of fair-share housing, one town manager was unexpectedly candid. "Litigate, litigate, litigate," he advised them. "Then build some senior citizen housing, then RCA the rest."

Throughout these decades, Peter O'Connor persisted. When he left the War on Poverty–funded Legal Services agency, O'Connor established Fair Share Housing Center in Cherry Hill. Despairing of ever getting affordable housing built in Mount Laurel itself, O'Connor then founded Fair Share Housing Development (FSHD), a nonprofit housing provider. Through a front buyer, FSHD purchased a suitable tract of land in Mount Laurel. In 2000, O'Connor broke ground on the first one hundred units of Ethel Lawrence Homes—twenty-nine years after the case was filed.

The physical results were on display when Ford Foundation executives toured the 140 handsome town houses in January 2005. Michelle Baraka, resident services manager, memorably filled in the human dimension. Resident family incomes range from $6,000 to $48,000 for a three-person family. About one-third came directly from Camden itself, and generally formed the lowest income group. Baraka described the adjustments the residents are making. "Now that they don't have to struggle constantly just *to survive*," she explained, "they can begin *to grow*—get a better job, get a better education, become involved in the community."[4]

Kicking the RCA Habit

Housing advocates were split over RCAs. Orfield, powell, and I had all urged outright repeal. Some affordable housing providers, however, were loath to give up RCA money without an alternate funding source. Looking at one of Orfield's maps graphically illustrating how "Blue Land" (wealthy suburbs) was funneling RCAs into "Red Land" (poor cities) and "Orange Land" (at-risk suburbs), Paul Scully observed, "It's like the drug trade. These Blue Land guys are the pushers. The Red Land and Orange Land guys are the users. They know it's not good for them, but once they're hooked, they become addicted. They can't kick the habit."

RCA addicts included Mayor Douglas Palmer of Trenton and Mayor Sharpe James of Newark.[5] Asked to comment on another legislator's characterization of RCAs as "plantation money," Mayor James retorted, "I *love* plantation money. I'll take plantation money to build a city. Anyone who has a 75–25 allocation and wants to give away that 75 percent, call Mayor Sharpe James, City Hall, and I'll be there tomorrow to get it." Indeed, as a senator James supported an amendment in the 2004 legislative session to raise the ceiling on RCAs to 75 percent of a township's fair share allocation.

Pushed by strong, anti-RCA sentiment from its three substate coalitions in North, Central, and South Jersey, founding NJRC board members finally agreed on outright repeal of RCAs. The campaign to repeal RCAs also received a strong boost when the General Baptist Convention of New Jersey, the state's largest group of African-American churches, passed a resolution urging RCA repeal.[6]

Never resolved, however, was the split generated by Fair Share Housing Center's legal challenge to the New Jersey Housing and Mortgage Finance Agency's allocation of federal Low Income Housing Tax Credits (LIHTC).[7]

FSHC challenged the state's allocation of LIHTC in state court, showing that its policies compounded concentrated poverty. As the Gamaliel Foundation's national strategic partners, powell, Orfield, and I assisted FSHC in this suit that, if successful, would have had nationwide implications. However, some erstwhile "inside game" allies, such as the Reinvestment Fund, that receive LIHTC funds opposed the suit.

FSHC had succeeded in blocking the state's fiscal year 2002 LIHTC allocation, and community development corporations and city mayors were screaming all around the state. O'Connor and Walsh were trying to force the state to allocate tax credits to build genuine "opportunity-based housing" where jobs were growing and schools were successful. However, LIHTC users just wanted to build housing even if it would be only more affordable housing "on the affordable housing side of town."

A partial "settlement," however, unblocked the freeze on fiscal year 2002 allocations while the litigation proceeded to trial. The settlement was achieved in classic, New Jersey, hardball-style. At the same time that the suit was brought, Fair Share Housing Development was proceeding with the final forty units of Phase II of Ethel Lawrence Homes. Phase II required a funding package of $6.9 million, including $1.6 million from the state's Department of Community Affairs (DCA). With all other financing in place, FSHD was advised by DCA that its application had been approved, that only minor technicalities were holding up formal closing of the entire package, and that it was encouraged to have the contractor go forward with construction.

Then, in late October 2003, O'Connor received word from DCA that if the LIHTC suit were not withdrawn, the DCA loan for $1.6 million would be denied. After exhausting other possible solutions, FSHC and the plaintiff NAACP branches reluctantly agreed to a partial settlement of the suit so that FSHD could complete the forty-unit addition.

Targeting Tax Reform and Opportunity-based Housing

Meanwhile, NJRC was refining its reform agenda. It commissioned Myron Orfield's Amerigis Group to simulate different property tax reforms and new formulas to replace COAH's inadequate system. In a series of issue summits in June 2004, Orfield presented the results to the three substate regional groups. All adopted a common agenda for statewide action.

Regional Housing Reform
Link low-income housing to opportunity

• **Abolish Regional Contribution Agreements**
Eliminate the loophole in the Fair Housing Act that permits wealthy communities to buy out of their share of affordable housing by paying poorer communities to take in more poor people.

• **Base affordable housing obligations on a fair "growth share"**
Link low-income housing development with a town's growth in homes and new jobs. A fair growth share should be 20% of all new housing and 1 house for 5 new jobs created.

• **Housing in growth areas should be affordable to the most poor New Jerseyans**
Currently state regulations do not require that "affordable housing" be affordable to families earning less than 40% of median income. Workers should be able to live near the high job-growth areas.

Property Tax Reform
Lower tax rates, increase services, and reduce disparities

• **Reduce reliance on local property taxes for schools**
Shift more responsibility to the state. This will bring down property taxes and will reduce inequities between communities.

• **Explore tax-base revenue sharing**
Require that a portion of all new growth is shared throughout the region. This will discourage wasteful competition, slow sprawl, increase investment in older areas, share regional resources, and reduce tax disparities.

Each organization's commitment to this common reform agenda was reaffirmed at their Rolling Thunder rallies in October 2004.[8] On Thursday, October 21, South Jersey Regional Equity Organizing Project (SJREOP) staged its rally at the Cherry Hill Hilton. Upward of 550 people were present—a very geographically and racially diverse crowd.

That diversity was reflected on the podium with leaders from different denominations (Baptist, Catholic, and Methodist) and from different municipalities (Sicklerville, Pennsauken, Atlantic City, Blackwood, Paulsboro, Cherry Hill, and Haddonfield). Rev. Vivian Rodeffer, United Methodist district superintendent, offered the opening prayer and the Most Rev. Joseph Galante, Bishop of the Roman Catholic Diocese of Camden, offered closing reflections.[9]

The rally presented issue demands and secured public pledges of support from the four state legislators present and two members of the Governor's Constitutional Convention Task Force.[10] The demands covered

• housing reform (repeal RCAs, support a "five and five" growth share formula, provide an alternative funding source to RCAs for housing rehabilitation in distressed communities, and assure that COAH's regulations "have a true impact on de-concentrating poverty");
• property tax reform (set up regional tax-base sharing by pooling and redistributing 40 percent of the growth in property value on an equitable basis and increase the state's share of local education costs);
• regional planning (appoint the Regional Impact Council for the Camden Recovery Plan and provide a seat for the SJREOP); and
• state constitutional convention on property tax reform (limit the convention to property tax reform, allow regional disparities to be addressed through tax-base sharing, increase state support for school funding, and assure that property tax reforms support open space, equal education opportunity, racial integration, and urban revitalization goals).

The four legislators pledged their support, even though several had refused to commit to RCA repeal prior to the rally. At evening's end, rally leaders also read out the names of South Jersey legislators who had been invited but had not come to the rally.[11]

Killing RCAs Town-by-Town

While gearing up for a massive public education campaign in support of repealing RCA authorization, NJRC's congregational core teams pressured their local governments to abandon their use. For example, the Town of Haddonfield was in the process of adopting its affordable housing plan (after they were ordered to do so in response to litigation brought by Fair Share Housing Center). Two-thirds of the town's plan was admirable (an inclusionary zoning ordinance and commitment of town funds to an affordable housing fund); one-third of the town's plan was not (using RCAs as a third option). Father Rob Sinatra, a young Catholic priest and NJRC leader, "communed" with the mayor, urging her to drop the third option.[12] After he did so, Haddonfield agreed in litigation with FSHC to not do RCAs. South Jersey's wealthiest town would build all its "fair share" allocation within the town.

The pervasive presence of RCAs was evident when the Ford Foundation officials toured the Camden area in January 2005. The group stopped at George Fine Elementary School where former Pennsauken mayor Rick Taylor is the principal. Mayor Taylor recounted how he had been approached that week by an "RCA broker" representing Medford Township, a wealthy fifth-ring suburb. Medford proposed to sell Pennsauken up to half of its COAH allocation of affordable housing units through an RCA at $25,000 per unit—a total of $3 million.

Given Pennsauken's tight fiscal situation, some council members may be tempted by the money, the mayor observed, but he personally was strongly opposed. "RCAs just concentrate more affordable housing in poorer communities that already have far more than their fair share," the mayor explained to the foundation officials. "The longer-term impact would increase Pennsauken's social and fiscal stress."

Two weeks later, NJRC reported the end of the story. The five-member town council had met in executive session. Medford's RCA broker had lobbied the Pennsauken town council hard. "How are your constituents going to vote in the next election when they hear that you turned down $3 million for the town?" he threatened. Over Taylor and the new mayor's opposition, three council members were ready to accept the RCA money.

Then the city attorney—who is also Pennsauken's democratic chairman —spoke up. He had received a call that morning from Joe Roberts's office, he said. Assemblyman Roberts opposes RCAs, he explained, and Roberts urged Pennsauken to reject Medford's offer.

A committeeman changed his vote to oppose the RCAs. He did so, he explained, because of the growing opposition to RCAs among Pennsauken residents and churches. The organizing done by St. Stephen's and Pennsauken's Stable Integration Governing Board ensured a large crowd would oppose the proposed transfer of housing obligation. The committeeman also said he would feel uncomfortable asking for a favor three months later if he didn't support Majority Leader Roberts on this issue. The town council voted to reject Medford's bargain.

Epilogue

Months passed as now-speaker of the Assembly Joe Roberts looked to NJRC and its allies to build support for repealing RCAs and clear away opposition around the state. Every bishop called for the law's repeal. Major daily newspapers editorialized in favor of repeal.[13] Other towns, such as Montclair and Maplewood, followed Pennsauken's lead in passing resolutions opposing RCAs.

Anti-RCA rallies grew larger. In November 2006, fifteen hundred NJRC members and allies from all over the state packed the venerable St. Nicholas of Tolentine Roman Catholic Church, demanding RCA repeal.

A clear hurdle to overcome was the public position of the urban mayors. As two of the three biggest "users" of RCAs, Newark and Trenton were clearly hooked. After longtime Newark mayor Sharpe James retired in 2006 rather than face a second challenge from young reformer Corey Booker, Trenton's mayor, Doug Palmer, became the most vocal urban mayor supporting RCAs. His support gave the wealthy white suburban RCA "pushers" some vital moral high ground to hide behind. Turning Mayor Palmer around was essential.

NJRC strategized around the mayor's self-interest. NJRC leaders from Trenton, especially Marty Johnson and the mayor's pastor, Rev. Daryl Armstrong, were long-term allies of the mayor, and worked for months to find common ground.[14] After protracted discussions, Palmer agreed to end his advocacy of RCAs, claiming his only interest was to replace future RCA monies with an alternate revenue source to continue housing rehabilitation and new construction within Trenton.[15]

Meanwhile, with Kevin Walsh as principal litigator, Fair Share Housing Center, NJRC's primary ally, was racking up a string of impressive court victories. Most decisive was getting the court to throw out COAH's Third Round rules for Fair Housing Act compliance. "Incomprehensible," the court ruled, giving COAH six months to come up with new regulations and freezing approval of all pending RCAs, effectively shutting down the system after wealthy suburbs had forked over $210 million to sell off 10,256 units from their constitutional obligation to build affordable housing.

The controversy surrounding state policies' being rejected decisively by the courts in three suits brought by FSHC plus grassroots pressure from NJRC helped convince New Jersey governor Jon Corzine to change state housing czars, replacing a longtime antagonist with a much more favorable commissioner of the Department of Community Affairs. The change in leadership was evident when COAH unveiled its revised Third Round rules in December 2007. The mandated set-aside for affordable housing was boosted from 10 percent to 20 percent of all new housing construction, and an additional new affordable unit was required to be provided for every sixteen new jobs rather than every twenty-five new jobs previously. Overall, Third Round targets had been doubled over the previous proposal. However, with the state law still on the books, the RCAs would still be countenanced (though COAH boosted the going price from $35,000 to $70,000). Litigation continued; despite substantial progress, the proposed regulations contained other major flaws.

By fall 2007, Speaker Roberts unveiled his comprehensive housing reform bill with the cosponsorship of now-majority leader Bonnie Watson Coleman. Their proposals were a measure of how powerful and influential the NJRC, the FSHC, and their allies had become. The Assembly leaders' twelve-point legislation proposed

- to eliminate Regional Contribution Agreements—the primary goal of NJRC's four-year campaign;
- to require a 20 percent set-aside for workforce housing in all state-aided developments—an NJRC/FSHC policy;
- to establish a new school-funding formula that provides special state funds for low-income children anywhere—an NJRC proposed revision for state school aid;

• to allow federal Low Income Housing Tax Credits to be used in mixed-income, market-rate developments in low poverty/high opportunity towns—an NJRC-supported/FSHC-litigated policy;
• to set aside 25 percent of affordable units for extremely low-income families (less than 30% area median income)—an NJRC/FSHC policy;
• to require towns to spend municipal housing trust fund dollars on affordable housing within their borders—a FSHC policy;
• to mandate that municipalities provide density bonuses to developers constructing inclusionary developments—an NJRC-championed policy;
• to require COAH to *document* existing affordable housing units it claims when allocating a town's fair share goal—an FSHC-litigated issue (successfully; state court threw out current state allocation formula, calling it "incomprehensible"); and
• four other provisions not specifically part of the NJRC/FSHC agenda (create state affordable housing trust fund; require one-for-one replacement of affordable housing lost through redevelopment; create a state Comprehensive Housing Plan; and require COAH to publish affordable housing statistics annually).

The Assembly Committee on Housing and Local Government held its first-ever hearings on RCA repeal in December 2007. At committee staff behest, NJRC Housing Task Force chair Paul Bellan-Boyer coordinated supporters' presentations. The hearings emphasized the strength and breadth of anti-RCA forces. As Bellan-Boyer summarized afterward,

We helped arrange the coordinated testimony of three mayors, four nationally recognized policy experts,[16] three powerful statewide religious leaders, four advocacy organizations, four community groups, two union leaders, one housing developer, and a leader of the historic civil rights struggle, C. T. Vivian.

It is apparent to me that RCAs cannot stand much longer. We have changed the debate over this policy. We have exposed them as a moral and policy disaster. And we have changed the balance of power on this issue, through our organizing and persistent effort over the past three years. The opposition is in disarray. We have the support of powerful allies. And, working with Speaker Joe Roberts and the Assembly Majority leadership, we have brought this issue to the front of the state's agenda.

The fight is not over. The forces that think they profit from this segregation will continue to resist reform. Some of the legislators will change with the new session in January. There will be squabbles over how to fund RCA replacement money. The Senate has yet to take any action on this reform.

Yet our own testimony, given by Pastor David Thornton and Rohn Hein, reminded us and the rest of New Jersey that evil cannot long stand when light is turned upon it. I claim no gift of prophecy when I say that victory will be ours, by the grace of God and through your continued faithful support and hard work.

Bellan-Boyer's optimism was instantly rewarded on one key front. Urged on by Governor Corzine, Speaker Roberts, and senate president William Codey, the legislature took up a total revision of the state school aid formula in the lame duck session. The basic reform had been formulated by NJRC with the support of Myron Orfield and his Ameregis analysts. Its principles had been emphasized at NJRC's two major statewide rallies (gaining public commitments from legislative leadership). NJRC had laid out the school funding proposal at ten forums in legislative districts (attracting 100 to 300 constituents each time). At the legislative leadership's request, Orfield and Rohn Hein, NJRC's Tax Reform Task Force chair, had presented the reform formula to a key legislative committee and to the administration's top education officials.

In recent years, by court order, the lion's share of state education aid had flowed previously to thirty-one high-poverty *Abbott* districts. Now, with an infusion of $450 million in additional state funds, state aid was proposed to flow to all school districts in proportion to their number of low-income students and their relative tax capacity. The consequences would be revolutionary.

• The thirty-one *Abbott* districts would be "held harmless" financially for three years (despite per-pupil expenditures that averaged one-third higher than the rest of the state's school districts, with such high concentrations of low-income pupils, academic improvement in *Abbott* schools was insignificant);
• hard-pressed inner-suburban school districts, with steadily rising numbers of low-income students and shrinking tax bases, would receive significant state funds, both boosting school resources and allowing modest tax relief; and
• wealthy, outer-suburban towns would be deprived of their "fiscal zoning" defense against meeting their affordable housing obligations ("We can't afford to educate low-income children")—now, the money would come with the child.

The battle over school aid was a cliffhanger. The governor, assembly speaker, and senate president drove the legislative process hard. NJRC was the only community group unambiguous in its support. A panel of NJRC leaders—Hein, Bellan-Boyer, Ben Coates, and Dianne Brake (of PlanSmartNJ)—acted as the lead public witnesses. Orfield again teleconferenced for his testimony. With the lame duck session required to end by Tuesday, January 8, at noon, the senate approved the Assembly-passed bill by a one-vote margin around midnight Monday.

When asked how much of NJRC's proposed formula ultimately made it into law, Orfield replied with a broad grin, "About 85 percent." The school funding reform, as I reported to the Ford Foundation, was a huge victory for NJRC, the most significant victory yet achieved in regional opportunity campaigns anywhere.

From a successful blitzkrieg for school funding reform, NJRC reverted to trench warfare over the full housing reform package. Opposition had been reduced to the New Jersey League of Municipalities—supported by its wealthy suburban members—and legislators from RCA "sending" districts, but it was still a potent combination.

Thirty-three years after the court's decision, twenty-three years after the legislature created the giant RCA loophole, and five years after NJRC began its campaign to repeal RCAs, the State of New Jersey had finally been brought to adopt housing and school policies to achieve the vision of the historic *Mount Laurel* suit—an economically and racially integrated society.

Notes

1. Several months before, Scully had helped organize a gathering of about fifty Catholic priests and lay leaders at St. Stephen's to recognize locally Peter O'Connor for receiving a national award at a testimonial dinner in Washington, D.C., where the national Catholic Campaign for Human Decency honored O'Connor for his three-decade-long campaign for the *Mount Laurel* doctrine. Scully and I had attended the CCHD testimonial.

2. Of the ninety-five participants from the Camden area, though the largest single municipal group (33) came from Camden, the central city ("Red Land"), only nine came from wealthy suburbs ("Blue Land"), and fifty-three came from declining or at-risk suburbs ("Orange Land").

3. Governor Kean, who had earlier called the supreme court's *Mount Laurel* decisions "communistic," would become cochair of the 9/11 Commission two decades later.

4. O'Connor believes that there is a limitless demand for "opportunity-based housing" (affordable housing in safe neighborhoods, located in job-rich areas with low-poverty, high-performance schools). There were 868 applicants for the first 100 ELH units opened; more than 1,800 persons (again, about one-third from Camden) waited in line to pick up applications for Phase II (only 40 units).

5. By the last years of the economic boom of the 1990s, poverty rates in most central cities had declined modestly; even in Camden, the poverty rate dipped slightly from 36.6 percent in 1989 to 35.5 percent in 1999. In Trenton and Newark, two RCA-addicted cities, however, poverty rates *increased*—from 18.1

percent to 21.1 percent in Trenton and from 26.3 percent to 28.4 percent in Newark.

6. The resolution was the result of Rev. Daryl Armstrong's leadership, who chaired the General Baptist Convention's Social Action Commission. Rev. Armstrong was also the chair of the Central Jersey Regional Equity Coalition and pastor of Shiloh Baptist, a powerful church in Trenton. NJRC cochair Roland Anglin helped organize a gathering of African-American leaders with john powell in September 2003 during the NAACP state convention. Rev. Armstrong was not only there but he also brought to the dialogue with powell Rev. William McKinley, the president of the General Baptist Convention.

7. The federal Housing and Urban Development Department estimates that LIHTC is involved in financing 90 percent of all affordable housing construction nationally.

8. The Gamaliel Foundation affiliates sponsored twenty-six Rolling Thunder rallies that fall, drawing more than 60,000 participants.

9. When shown Myron Orfield's map charting how wealthy suburbs are sending their fair share housing obligation back to poverty-ridden cities through RCAs, Bishop Galante commented angrily, "That's just like the Civil War draft." During the Civil War, wealthy draftees regularly paid poor men bonus money to take their place in the ranks.

10. Attendance by legislators at the South Jersey rally was undoubtedly boosted by events at Jubilee Interfaith Organization's North Jersey Rally in Vailsburg on October 14, which drew a thousand members. When all but one legislator deliberately reneged (without advance warning) on previous promises to attend the rally, Jubilee loaded thirty leaders onto a church bus. They drove to the nearby site of a $500-a-plate fund-raiser for the West Orange mayor (who is also its state assemblyman) where the absent legislators were assembled. Marching into the middle of the fund-raiser, the Jubilee leaders upbraided the group for snubbing a thousand of their constituents who were waiting for them just a few miles away. Though incensed, the mayor and other legislators left his own fund-raiser to come to the rally. The word got around quickly, both boosting legislator attendance at the South Jersey rally and leading to state senate president pro tem and incoming acting governor Richard Codey to schedule immediately a meeting requested by NJRC. (He had been ducking them.)

11. It is a standard rule of Gamaliel network organizing technique that public officials must be made to understand that they can neither renege on commitments nor refuse to meet with representatives of the faith-based coalitions with impunity.

12. Last-minute leadership of the town council came from the planning commission chair, who had experienced a recent epiphany on the issue. For several years, she had had a housekeeper come out weekly from Camden and had come to like her. Ms. Haddonfield asked Ms. Camden what she could do to help her. "You know what you could really do to help me?" replied Ms. Camden. "Help me move to Haddonfield so that I can put my son in your good schools."

13. Typical was the position of *The Star-Ledger*, the state's biggest newspaper: "Regional contribution agreements ought to be eliminated.... It is time to make all communities confront their obligation to help with affordable housing needs and get creative in finding solutions" ("A Spur to Affordable Housing," January 11, 2008).

14. At one point, the NJRC had to threaten to bring priests and ministers to the Washington, D.C., banquet where Palmer was being installed as president of the U.S. Conference of Mayors; they would hand out flyers, urging his fellow mayors to encourage Palmer to change his position on RCAs.

15. Shortly thereafter, in July 2007, Mayor Palmer turned back a proposed $3.25 million RCA from Hopewell Township, to which a Trenton hospital was relocating. The mayor denied that the move was a form of reprisal against Hopewell for the loss of the hospital. "This is not a way of sticking it to them. It's reality," he said. "They will need to build affordable housing with 1,500 jobs (moving from the city to the township). Quite frankly, they're going to need it."

16. I was the lead public witness and Orfield teleconferenced in from Minneapolis.

25

Farms to Schools: Promoting Urban Health, Combating Sprawl, and Advancing Community Food Systems (Southern California)

Robert Gottlieb, Mark Vallianatos, and Anupama Joshi

A Radical Departure

On a morning in late April 2003, students at Widney High School in Los Angeles were given a class presentation in nutrition education. Instead of the classic lecture on why students should eat five fruits and vegetables a day, students were presented with a box of fruits and vegetables delivered from a farm in San Diego County at the urban edge of the huge Southern California metropolitan region. The fruits and vegetables had been picked the previous day and dropped off at the classroom by the farmer. A lesson had been prepared by the teacher to include not only an understanding of where the food came from but also the value of fresh-from-the-farm food.

Tasting the food was part of the lesson, providing a way to talk about the benefits of fresh fruits and vegetables. Some students hadn't previously seen some of the greens that were provided. The students made a salad with the greens, then added the carrots and tasted the strawberries. They marveled at the strawberries, sweeter than any they had tasted before. They compared what they made favorably to salad bars they had seen at restaurants and began to discuss the nutritional value of all the foods from the box. Teachers described how students were trying foods they wouldn't normally try, and how the class was "much better than just talking about what 'healthy' is" (Center for Food and Justice 2003).

Two years later and sixty miles to the east, at Jefferson Elementary School in Riverside County, California, in the fast-growing Inland Empire, its predominantly Latino and African-American students were experiencing another radical departure concerning food, but this time in the school cafeteria. The school lunch included a new option—a fresh fruit and vegetable salad bar, with some items picked the previous day from nearby farms. The students lined up in droves, waiting patiently and excitedly, despite the limited lunch period, for their fresh and tasty options.

Rodney Taylor, the Riverside Unified School District's nutrition services director, was also enthusiastic, convinced that his work in school food service now had a new meaning. "Some of the comments we heard from children would warm your heart," Taylor said the next day. "[Students] truly appreciate the opportunity to eat farm fresh fruits and vegetables on a daily basis." The school principal had warned Taylor that some of the children who qualified for a free lunch never ate in the cafeteria, including one girl who, the principal remarked, just "didn't like school cafeteria food." "But there she was yesterday," Taylor said, "eating lunch, a salad bar lunch of course, and very happy with the choices available on the salad bar." For Taylor, the program represented a breakthrough in the way he defined his job and the role of the school lunch program (Taylor 2005).

For the farmers, the Riverside and Los Angeles programs had also expanded their horizons—and viability. One of the Riverside farmers, Doug Powell, had been farming since 1992 on land that had been held by his wife's family since the 1970s. Powell is an organic farmer who specializes in onions, garlic, lettuce, carrots, beets, fava beans, and blackberries, and he also tends some two hundred fruit trees, producing apples, pears, Saturn peaches, and jujube. The Jefferson Elementary School program contributed to his farm's viability. Powell, like Taylor, saw the farm's connection to the school as representing a paradigm shift. The experience of watching the schoolchildren line up, chirping away as they picked out items he had brought to the school that morning, validated everything he loved about farming (Schwartz 2005).

Halfway across the county, a similar program in the Midwest, operational for more than three years, was also successfully changing school lunch menus. Working closely with the staff and students at three Madison area schools—Lincoln, Shorewood Hills, and Chavez Elementary—the Wisconsin Homegrown Lunch (WHL) program has developed a range of innovative educational programs that have also enhanced opportunities for local farmers. WHL included classroom-based nutrition education sessions, seedling and planting activities, a farmer in the classroom program, field trips to local area farms, and, in collaboration with food service staff, special locally sourced meals for festivals and picnics.

The program typically begins at a new site with a taste test of seasonal fruit and vegetables. Students rate these items by drawing a happy face next to their favorites. If approved on three taste tests, the products are incorporated into the school lunch menu. Feedback is intriguing and sometimes idiosyncratic. One participant, five-year-old Sophie Chumakova, was among the many youngsters who gave each vegetable a smile. "Spinach is good for you," she was quoted in the local newspaper. "It makes your cough go away" (Cullen 2004).

At a lunch line at Lincoln Elementary School, Iris Tirado, the food services director, introduced a new menu as part of the Friday Harvest Festival—seasoned tortilla wraps with grilled chicken, yogurt cream cheese dill sauce, and locally produced organic spinach and shredded cabbage. Filling out the recycled paper tray were apples from a nearby orchard, vegetarian chili with blue chips, and a cookie with dried cranberries. Students, as well as parents who had come to help out, were giving the lunch positive reviews and leaving little on their plates. Julian, a third grader, said that his lunch was "way better" than the normal offerings, such as the Pizza Hut pizza that had been his favorite. Tirado commented to the paper that she was aware that she could face barriers in institutionalizing the program, including the common impression that local produce would be more expensive. Nevertheless, she commented, "I don't see why spinach from Wisconsin is going to be more expensive than any other spinach. I think it can work" (Cullen 2004).

For the farmers, the WHL program opened up opportunities they might not otherwise have considered. Doug Wubben, a farmer linked with the program, spoke of plans for a cooperative processing facility that could turn spinach, cabbage, and other vegetables into the precut, prewashed produce that institutions such as schools are equipped to handle. "If it happens, it could open the door to the huge market of institutionalized food," Wubben exclaimed (Sensenbrenner 2003).

The students, school food service staff, and farmers in the Los Angeles, Riverside, and Madison areas are on the cutting-edge of a new movement called "farm to school." In low-income schools like Jefferson, Widney, and Lincoln, the health and nutrition benefits are obvious, especially amid concerns about childhood obesity and its related health impacts. But in addition, this new movement embraces a wide range of other issues and perspectives that are at the heart of a sustainable-metropolitan-communities approach. Farm-to-school connections can improve the viability of small farms and provide a new framework to preserve farmland and combat sprawl at the urban edge. Farm to school strengthens a "food systems" approach emphasizing local and seasonal food over food that is produced in distant areas and highly processed. This "community food systems" goal, in turn, provides the direct link between improving daily life and establishing a sustainable development model for both the urban core and its outer edge.

This chapter describes how farm to school addresses four key goals of a community-food-systems and sustainable-metropolitan-communities approach by

• improving the health and nutrition of school-age children, particularly low-income youth;
• strengthening the capacity of local and regional farmers, particularly those engaged in sustainable farming practices;
• adding to the tool kit of strategies for containing and ultimately reducing sprawl-inducing developments, by helping preserve farmland while improving daily life in urban core communities; and
• establishing an important component of a community-food-systems approach—the development of a viable regional food system no longer entirely dependent on the global system that has come to dominate food growing, processing, distribution, and consumption patterns around the world.

What Is Farm to School?

Farm to school connects local farmers with educational institutions such as K–12 schools, colleges, and preschool programs. It provides a new framework for addressing sprawl issues in conjunction with core inner-city needs. Farm to school fits directly within the sustainable-metropolitan-communities framework as a concrete, hands-on program with broad institutional and policy-related implications.

Schools seek to implement the crosscutting, holistic farm-to-school concept at multiple levels. They buy and feature on their menus farm-fresh foods such as fruits and vegetables, eggs, honey, meat, beans, and other items. They incorporate a nutrition-based curriculum that includes students becoming directly engaged with local farmers and their farms. They develop school gardens as a learning experience, a form of much-needed physical activity, and a strategy for changing students' perspectives and willingness to try fresh fruits and vegetables. They also provide students with experiential learning opportunities through composting and recycling programs that focus on what gets eaten, what gets left, and where it goes. Farmers selling to schools also have access to a new market and participate in programs that educate children about local food and agriculture, while creating a new type of "farm to consumer" or "farm to table" connection.

Serving meals to as many as 28.4 million children each day, the National School Lunch Program is a tremendous marketing opportunity for small farmers who rely mostly on direct marketing through farm stands and farmers' markets. While farmers benefit from increased sales to nearby school districts, the students and staff of these schools enjoy the freshest and most nutritious and appealing meals possible, through the inclusion of locally grown and seasonal ingredients. Many food service directors incorporating farm-to-school programs report an increase in meal participation (and therefore revenues) for the school district. In a Congressional briefing on the benefits of farm-to-school programs, Doug Davis, a food service director from Burlington, Vermont, said that he has "increased the number of fruits and vegetables going home in the bellies of students by buying locally and working with a local nutrition education organization, Vermont Food Education Every Day. The nutrition education events in the classroom help inform and engage students about healthy choices so they are more likely to try the foods when they get to the cafeteria." Davis also observed that school budgets are approved locally and contribute to the local economy, and that "buying from local farmers can have a positive effect on the school's reputation in the community" (School Nutrition Association 2005).

Each of the thousand-plus farm-to-school programs in the United States is tailored to suit such criteria as geographic location, growing season, school food environment, and available resources. In California, where fruits and vegetables can be grown year-round, school cafeterias commonly feature farmers' market or farm-fresh salad bars once or twice per week. On those days, students can choose a salad with other items including milk, bread, and protein sources such as tuna as an alternative to the traditional hot lunch. While seasonality may appear to be a barrier, schools in the Midwest and Northeast have found some very creative ways to develop farm-to-school programs that reflect the flavors and features of their regions.[1]

One successful strategy has been to emphasize farm purchases in the fall and spring, when an abundant variety of crops is available. New York State established "NY Harvest for NY Kids," an annual weeklong event that takes place in October and is backed by the state legislature. During this week, field trips to farmers' markets or nearby farms give students the opportunity to churn butter, make applesauce, milk cows, and participate in other farm-related activities. In New England, the farm-to-school program has focused exclusively on apples, buying apple cider and "small-sized apples" (perfect for kids) from local orchards. Several states, such as New Mexico, Kentucky, North Carolina, Michigan, Mississippi, Florida, Georgia, New Jersey, Washington, Illinois, and New York, have used the Department of Defense Fresh program that facilitates the purchase and transport of state-grown products for schools' use in breakfasts, lunches, and snacks (National Farm to School Web site 2007).

In addition to the obvious benefits of improving the taste and quality of foods offered in school cafeterias and enhancing local farmers' incomes, the farm-to-school approach provides a platform for educating children about agriculture and the environment, and the relationship of food to health. The programs provide hands-on or "learning by doing" educational lessons regarding where food comes from, how it is grown, and the value and joy of eating fresh fruits and vegetables.

"Cooking with Kids," a Santa Fe, New Mexico–based program, has developed an experiential food and nutrition education curriculum that models interdisciplinary teaching and learning. Students participate in at least fifteen hours each academic year of classroom cooking and tasting classes—incorporating math, science, social studies, language arts, music, and art—using locally grown varieties of tomatoes, apples, citrus, other fruits, and greens. Classroom recipes are adapted for food service and offered about twice a month as school lunches. Green and White Fettuccine with Tomato Basil Sauce became so popular that it is now on the regular school lunch menu in all Santa Fe elementary schools.

In the first ten months of operation, the vermicomposting and school garden project at a Laytonville, California, middle school saved $6,000 in Dumpster fees by reducing the amount of paper and food waste collected in commercial Dumpsters and destined for the landfill. This successful program involving more than four hundred students continues to provide a model for students as environmental stewards (Center for Ecoliteracy 2004).

Finally, farm-to-school programs can lower barriers that industrial-scale agriculture has placed between farmers and consumers. This is of crucial significance for the place of agriculture within a sustainable-metropolitan-communities and regional-planning framework. When local farmers, school staff, and PTA members collaborate to launch a farm-to-school program, or when students visit a farm, this establishes a broader-based constituency for understanding the importance of local and regional farming within the region as a whole.

Farm to School: A Growing Movement

Since its origins in the late 1990s in Florida and California, farm-to-school programs have been mushrooming all over the country. A group of farmers formed the New North Florida Marketing Cooperative (NNFMC) and started supplying to thirteen schools in Gadsden County, Florida. Within a few years, NNFMC had expanded its operations to fifteen school districts across Florida, Alabama, and Georgia, supplying mainly collard greens, field peas, muscadine grapes, and turnips. In California, support from parents and the community helped facilitate the first "Farmers' Market Salad Bar" in the Santa Monica–Malibu school district. What started as a pilot program at one school expanded to every school in the district by its third year of operation. The Santa Monica program received a number of awards (such as "best lunch menu" by the California School Boards Association and a "City Livability Award" from the United States Conference of Mayors), emerging as the flagship of the new farm-to-school movement (Mascarenhas and Gottlieb 2000; U.S. Conference of Mayors 2003). In Berkeley, California, renowned chef Alice Waters sowed seeds for the Edible Schoolyard program by offering a school garden, kitchen, and classroom program for students at the Martin Luther King Jr. Middle School. These pioneering programs have paved the way for school districts embracing the farm-to-school model.

In the past decade, farm-to-school programs have progressed from a pilot or trial stage to fully institutionalized operations recognized and supported by food service, school administration, parents, community organizations, and local and state agencies. Farmers have responded to the increased demand and have streamlined operations to serve the school food service market more effectively. For example, the NNFMC purchased basic processing equipment to wash, chop, and bag collards that were ready to cook or ready to freeze, so that school districts could receive the best possible product throughout the year in precut and prewashed form.

Large food service distribution companies are also jumping on the "buy local" bandwagon. Sysco Corporation recently launched a "Born in New Mexico" campaign to set up a statewide network to buy fresh produce from local producers for distribution to food service operations in New Mexico and surrounding states (Robinson-Avila 2005).

Just the sheer number of farm-to-school programs that have been initiated in the past decade is proof enough of a burgeoning movement. From only a handful in the 1990s, more than one thousand farm-to-school programs in thirty-four states are now operational in the country, with a March 2007 National Farm to Cafeteria Conference bringing together more than eight hundred farmers and participants from schools, colleges, universities, hospitals, prisons, and other institutions to share strategies and learn from one another's programs.

When Rodney Taylor, the Riverside, California, food service director, first heard the term *farm to school* in 1996, he was food service director of the Santa Monica–Malibu school district. He was not only dubious but was also convinced that the initiative was a kind of crackpot idea of some parents "who just had too much time on their hands," as he has characterized his attitude at the time. Taylor reluctantly agreed to test the farm-to-school program for one week during the summer session at one low-income school in the district, launching it on a day when the students would have to choose between pizza and the farm-to-school salad bar. "It blew me away," Taylor recalled, the day the salad bar opened. More than three-quarters of the students were choosing salads! "The kids lined up," Taylor recalled, "excited, waiting for their turn, and then wolfing down their salads. I was astounded, and that moment made me recognize that I could have a different kind of mission as a food service director, to offer a different kind of choice for the kids and to make the schools healthier places" (Taylor 2005). Taylor's paradigm shift, an experience now shared by hundreds of food service directors, farmers, parents, teachers, and, perhaps most important, students, indicates how the links between healthy schools, healthy farms, and healthy communities can be established. The next step is figuring out how such a paradigm shift can become common practice.

From Pilot to Institutionalization: A Farm-to-School Policy Framework

Farm to school resonates as an obvious and much welcomed approach with clear benefits such as improving the quality of school food, expanding opportunities for local and regional farmers and potentially countering the tendencies toward sprawl, as well as promoting healthier eating habits among school-age children, including, prominently, low-income children who have become increasingly prone to obesity and other diet-related health problems. However, farm to school, even as it has grown and captured the imagination of players like Rodney Taylor, requires new policies, and, ultimately, a new policy framework so that the approach does not become limited as a niche program. Formal policies are critical if programs are to become fully institutionalized, with their benefits a solid part of a broader sustainable communities framework. The next section of this chapter looks at the policy context, including the impacts and barriers for developing a program and the issues that need to be addressed.

Impacts and Barriers

The first generation of the farm-to-school movement has been shown to have significant impacts on children's knowledge, attitudes, and dietary behaviors and on income potential for farmers. Preliminary information from the programs clearly shows that children will eat healthy food if it is offered and promoted in the cafeteria, and if it is connected with educational experiences at school. The Ventura, California, school district found that when fresh farm products were introduced into the cafeteria and supported by nutrition education in the classroom, students chose healthy meals 75 percent of the time, compared with 46 percent before the educational component was introduced (Christensen 2003). In the Santa Monica–Malibu, California, program, the number of students choosing a salad bar lunch jumped by over 500 percent when fruits and vegetables from the farmers' market replaced standard produce, reinforcing the fact that even young elementary school students will choose and appreciate taste, freshness, and flavor (Mascarenhas and Gottlieb 2000).

Most farm-to-school programs require start-up funds in the range of $2,000 to $3,500 per school site to buy equipment (such as child-size salad bars or extra refrigerator space), and some ongoing costs to pay for additional labor (most often one part-time position). Foods purchased from organic or small farmers are frequently not as costly as many people expect, especially when menu planning incorporates items that are in season locally.

Once operational, many farm-to-school programs can pay for them-selves. Salad bar participation rates and overall participation rates in the school lunch program have increased each year since the launch of the Santa Monica–Malibu program. Growth in school meal participation provides additional revenue for school food services, since the USDA reimburses each meal. For example, Rodney Taylor estimates that an increase in student participation of approximately 8 percent would be sufficient to cover the additional labor costs of the Riverside farm-to-school salad bar program (Taylor 2005). Similarly, increases in participation by adults (i.e., teachers, staff, or parents) who pay a higher price than the students and choose school cafeteria food when a farm-to-school component is available, have increased revenues and provided a signal to the students about the high quality of the salad bar.

Data on farmer incomes has been more difficult to compile, but the limited information available is promising. During the 2003–2004 school year, the Davis Joint Unified School District in northern California spent 38 percent of its total produce procurement budget on direct purchases from local farmers. In just three months, the Ventura Unified School District purchased $40,000 of locally grown fruits and vegetables. The Mississippi farm-to-school program reports $416,650 worth of products bought from local farmers for schools, an increase of 45 percent from the previous year. Organizers of the North Carolina program report a 65 percent jump in school orders for locally grown strawberries (Feenstra and Ohmart 2004; Delta Farm Press 2005).

High participation rates in farm-to-school programs provide one indication of possible increased fruit and vegetable consumption. A UCLA School of Public Health study, using a twenty-four-hour dietary recall method with students from three Los Angeles Unified School District schools that piloted a Farmers' Market Salad Bar, showed an increase in student consumption of fruits and vegetables by more than one serving per day, attributable to the salad bar school lunch choices. The same study revealed that students were not only making healthier choices in the school cafeteria but were also continuing the healthy eating pattern in their homes (Slusser et al. 2007).

Despite these positive results and the importance of the objectives associated with farm to school, significant barriers have been identified to full-scale institutionalization of the program. Some of the programmatic barriers follow.

School Food Service Operations Are Under Tremendous Pressure to Produce Standardized Meals at a Low Price
Price, predictability, and convenience greatly influence food service purchasing decisions. Participation in the National School Lunch Program allows a school food service operation to purchase commodity meats, cheeses, eggs, and processed fruits and vegetables at prices below market levels, thus disadvantaging direct purchases from local farmers. Kitchen facilities are frequently inadequate for cleaning, processing, and storing fresh products, and staff often requires training in the techniques of handling and storing fresh produce. Many kitchen facilities at individual school sites have been eliminated. Food service divisions operate as independent businesses and do not receive any money from the general funds, even for starting innovative programs such as farm to school. Encouraging success has been achieved in working with the food service staff to find creative ways to address these issues—such as offering products that are ready to use, or developing relationships to ensure that local products can be sourced at wholesale prices (through existing distributors, from grower cooperatives, or through the Department of Defense Fresh program). Start-up grant funds have also been used to pay for additional equipment and staffing needs related to farm-to-school programs.

The Logistics of Ordering, Billing, and Delivery Has Been an Obstacle in Nearly Every Pilot Program Developed across the Nation
It has become clear to farm-to-school advocates that procurement and distribution systems must be developed that meet the needs of both school food services and small-scale local farmers. School food services commonly receive precut, prepackaged produce from distributors who provide nearly any product in nearly any form at any time of year. Although taste and nutritional value suffer in storage, shipping, and processing, the product delivered to the kitchen door is predictable, convenient, easy to use, and inexpensive. Small-scale farmers, who have had limited access to institutional markets, are unaccustomed to the invoicing, packing, and delivery needs of school districts. School food service directors have also identified problems arising from the added time and paperwork required for working with multiple farmers (often the case with farm-to-school programs). Assisting farmers to modify practices to meet the needs of school food service and assisting school food service to understand the constraints under which farmers work benefits both

constituencies. Encouraging distributors to purchase from local farmers has also been very effective in some areas.

Aside from these kinds of programmatic issues, farm to school can meet its key objectives only if it is supported through policy instruments and institutional support, whether at the school district, local government, state government, or federal level. Such a commitment is likely to become available when the objectives of health promotion, farmland preservation and urban edge land use strategies, and the development of a community-food-systems and sustainable-metropolitan-communities approach come to be directly associated with support for farm to school.

The Obesity Factor: Changing Diets

The rise of farm to school has paralleled the increased attention by the media, the public, researchers, and policy makers to the issue of obesity and other diet-related health problems. Extra weight and obesity are considered today to rival smoking as the nation's leading cause of preventable death. Rates of obesity among U.S. adults nearly doubled between 1980 and 1999, rising from 15 percent to 27 percent. Over the past few decades there also has been a dramatic increase in the prevalence of obesity among children and adolescents. In the 1960s and 1970s, around 5 percent of young people aged 6 to 19 were obese—by 2000, that figure had tripled to 15 percent. This increase was largest among Mexican-American and African-American adolescents (Ogden et al. 2002).

The epidemic of obesity among children and adolescents is an important risk factor for the onset of type 2 diabetes. The U.S. Centers for Disease Control and Prevention has estimated that upward of one in three Americans born in 2000 will contract diabetes unless the public begins to eat more healthily and exercise more.

Fast foods and sodas, as well as the large portion sizes of those foods and drinks, have become both the symbol and substance of a negative dietary shift in many communities over the years, associated with profound changes in food production, processing, storage, and distribution. These dietary changes have been directly associated with the onset of both type 2 diabetes and obesity.

Health practitioners have increasingly focused on health-promoting dietary changes and physical activity as among the most effective strategies to prevent obesity and associated type 2 diabetes. However, research has shown that such dietary choices are influenced by factors such as cost, accessibility, and availability of foods. One doesn't eat five fruits and vegetables a day if access to fruits and vegetables is limited and the available produce isn't fresh. American children eat one-third of their meals outside the home, where foods are higher in calories, fat, saturated fat, and sodium, and lower in other nutrients (Cavadini, Siega-Riz, and Popkin 2000; Lin, Guthrie, and Frazao 1999).

In low-income communities, positive dietary changes have been facilitated by local food outlets that provide a wider variety of healthy, affordable foods. In a 2002 study of the association between the local food environment and residents' reports of recommended dietary intake, African-Americans' fruit and vegetable intake increased by 32 percent for each additional supermarket in their census tract, often their only source of fresh produce. School nutrition environments also have been shown to exert strong influence on student eating patterns (Contento et al. 1995).

Foods available in schools through the National School Lunch and Breakfast Programs are another major "away from home" food source for school-aged children, but the sources also, increasingly, include food sold in vending machines or through a la carte vendors. These "competitive foods," as they are called, are exempt from regulations requiring school meals to be balanced and nutritious. Unfortunately, the trend of the school food service industry until recently has been to mimic rather than challenge the onslaught of fast food. According to California school district food service directors who responded to a mailed survey, 90 percent of high schools sold these competitive foods and 72 percent permitted advertising of brand-name fast foods and beverages on campus (Public Health Institute 2001).

One of the strongest arguments for the development of farm-to-school programs has been their association with these obesity and diet change issues. While the school lunch program provides a far more balanced and nutritious meal than the competitive foods sold in vending machines or the fast food restaurants that surround many school campuses, until recently the school food service industry has been reluctant to challenge the fast food status quo. A review by Center for Food & Justice researchers of the primary school food service trade publication, *Food-Service Director*, found that between 1997 and 2000, numerous articles described efforts to use "branding" techniques and other fast food–related strategies to help maintain participation levels for school lunch programs. However, where data is available, such as from the programs in Santa Monica, Ventura, and Davis, California, overall school lunch participation rates have increased when farm-to-school options are made part of the lunch offerings.

Fighting Sprawl: Farm to School as a New Approach

Farm-to-school programs have been recognized for their significant role in promoting healthy diets on school sites. There is an additional benefit to the farm-to-school approach, one that extends beyond the participating schools' hometown into the surrounding landscape. By linking farmers with institutional buyers, farm-to-school programs can help prevent farmland loss and slow down urban-edge sprawling development.

According to the American Farmland Trust, the United States lost six million acres of agricultural land to development between 1992 and 1997. This loss occurred at a rate that was 51 percent higher than during the previous ten-year period (American Farmland Trust 2002a). Poor planning and sprawling land use choices are driving much of the loss of farmland. From 1982 to 1997, the amount of land devoted to urban uses grew nearly three times as fast as did the U.S. population.

Farms are vulnerable to sprawl because the agricultural economy fails to provide most farmers with a sustainable livelihood. A number of decades-long trends, from consolidation in the food processing industry to unbalanced federal farm payment programs, have stacked the deck against smaller farms.

Small farmers lack markets where they can receive a reasonable price for their produce. In 2002, nearly 80 percent of farms had annual sales of less than $50,000. More than 60 percent of the nation's farms had sales of less than $10,000 (USDA 2002). Most very small farmers survive on income earned outside of the farm (USDA 1998b). It is no surprise that thousands of farmers reluctantly give up their livelihood and sell their land each year, sometimes to developers.

Nationwide, there were 300,000 fewer farmers in 1997 than in 1979. An additional 100,000 farms were lost between 1997 and 2002. About 94 percent of the nation's farms are small, family-owned farms, defined as having less than $250,000 in annual sales, but this large majority of small farms received only 41 percent of all farm income. The large farms, many of them corporate owned, also receive the majority of government commodity payments (USDA 2002; USDA 2000; USDA 1998a).

The correlation of the conversion of land to urban uses and the conversion of *farmland* to those same urban uses is striking. Riverside County to the east of Los Angeles is one of the fastest-growing regions in California. The county's population grew by 60 percent between 1990 and 2004. Riverside County lost 57,896 acres of agricultural land between 1992 and 2002. Close to 85 percent of this area (48,408 acres) was classified by the state as "important farmland." Nearly ten thousand acres of agricultural lands were converted to urban uses during the 2000–2002 period alone (California Department of Conservation 1992–2002). As a result of the pressures of sprawl and other factors pressuring farm livelihoods, the county lost 18 percent of its farms between 1997 and 2002. Small farms fared the worst; the county's remaining farms are on average 32 percent larger than they were six years ago (USDA 2002).

The population of nearby San Diego County increased 17 percent between 1990 and 2004. The county lost 21,869 acres of agricultural land from 1992 to 2002. Nearly half of this loss (10,667 acres) was of important farmlands. Almost 5,000 acres of agricultural lands were converted to urban uses between 2000 and 2002 (California Department of Conservation 1992–2002). The county lost 28 percent of its farms between 1997 and 2002. Remaining farms were 13 percent larger than before (USDA 2002).

Madison and Dane County, Wisconsin, are less populous than Southern California, but they too are feeling the impacts of sprawl. Dane County's population registered a 23.6 percent increase between 1990 and 2004. In fact, by some measures, Madison is a more sprawling city than San Diego or the Riverside–Los Angeles metropolitan region. A greater percentage of residents in the Wisconsin capital live in the suburban and exurban fringes of their city, where new homes, roads, and developments cut into what were once rural communities (El Nasser and Overberg 2001). Dane County lost 9 percent of its farms and 8 percent of its land in farms between 1997 and 2002 (USDA 2002).

It has become clear to researchers and policy makers alike that unless small farm economies are improved and development is channeled into denser, smarter patterns, sprawl will continue to transform family farms into suburban outposts. In fact, 86 percent of the nation's fruits and vegetables and 63 percent of dairy products come from farms near urban areas, putting some of the country's most vital farmlands directly in sprawl's path (American Farmland Trust 2002a).

In the past several decades, as sprawl and farmland loss trends became magnified, a number of strategies and policy initiatives were developed to slow down or reverse these trends. These included land trust purchases, habitat protection measures, urban growth boundaries, purchase or transfer of development rights, property tax relief, creation of agricultural districts, and development of strategies to compensate farmers for the loss of their right to develop their properties (American Farmland Trust 2002b). Although much of the antisprawl literature includes language concerning the need to enhance farm income at the urban edge, there often has been an insufficient understanding of the value of urban edge farming, both as a vocation and for its contribution to an alternative community-oriented *food-system* approach. Too often, antisprawl activists fail to understand the significance of the value of the farming, and consequently see urban edge farmers not as allies but as part of the problem. This is due in part to the hostility of farmers to any land use controls or tax measures that appear to limit farmers' future choices about what to do with their land.

Placing urban edge farming in a food-system context focuses attention on the issue of farm income. Urban edge farmers, as well as small farm operators in general, have increasingly relied on income from direct sales to consumers or institutions. This trend is reflected in the recent growth of urban farmers' markets whose sales have passed the $1 billion range and are now located in all fifty states. Farm to school is an extension of that direct marketing approach and could represent a more substantial source of income for urban edge farmers.

A Community-Food-Systems and Sustainable-Metropolitan-Communities Approach

The rise of farm to school helps to challenge the dominant trends of an increasingly globalized food production and distribution system, the increase in fast food consumption and its health and diet impacts, and the squeeze on small local and regional farmers. The community-food-systems approach is a synthesis of ideas surrounding health; environmental, social, and economic justice; and sustainable agriculture. It also references a new type of social movement that seeks to empower its participants including farmers, farmworkers, community residents, students, parents, teachers, and school food staff. As the Community Alliance with Family Farmers has put it, this is a movement of producers and eaters, producing a new food ethic while shortening the distance between the food produced and food consumed (Redmond 2002). Farm to school is the newest and perhaps most promising of the new ideas and alliances of the community-food-systems approach, indicating that change in school food operations is capable of extending its impact far beyond the school cafeteria itself.

Similarly, the sustainable-metropolitan-communities approach links the equity, environmental, and economic issues associated with suburban and exurban development with core urban or inner-city needs. The sustainable-metropolitan-communities framework prominently highlights issues of transportation, land use, and community economic development. Food systems issues need to be incorporated as central to the argument about the suburban, sprawl, and inner city links. The farm-to-school program makes those links explicit through a more just and sustainable set of relationships, from farm to consumer and from farm to urban institution. The effect of highlighting a new set of relationships around food is to enhance the opportunities for creating healthy communities, healthy schools, healthy farms, and, ultimately, healthy regions, significantly strengthening the sustainable-metropolitan-communities approach.

Note

1. See National Farm-to-School Web site for description of regional Farm-to-School programs throughout the United States. Available at http://www.farmtoschool.org.

III

Regional Equity and the Future of Sustainable Metropolitan Communities

Introduction to Part III

The breakthrough stories in part II describe an emerging movement that is strengthening metropolitan communities through defining common interests. Coalitions bring together people from wide-ranging backgrounds to advocate for practices that are more equitable as well as more ecologically viable. Informed by the past five decades of metropolitan development in the United States, this movement can improve the quality of life for communities that face both internal pressures and the realities of global economic forces.

The previous part's stories shine light on the innovators and early adaptors of the movement. If the work of these pioneers follows the pattern of other social innovations, the field will continue to evolve around innovative methods and tools that demonstrate a potential to become standard practice. Such a transition will draw upon examples such as the ones outlined in this book.

If successful, regional equity advocates will face challenges similar to those experienced in all social movements. The dominant metropolitan paradigm is deeply entrenched. This book's concluding chapters look at the ways to build and strengthen the movement for regional equity— through large gatherings, new institutions, public and private partnerships, strategic communications, documenting and measuring successes, addressing the continuing impact of segregation, and facing the challenges of global urban development. In each case, a commitment not to leave communities behind is implicit.

Building the Capacity of the Regional Equity Movement

In May 2005, the Advancing Regional Equity Summit in Philadelphia brought together 1,300 people to share current thinking and best practices for building more livable communities. The conference was organized by Angela Glover Blackwell, executive director of PolicyLink (www.policylink.org), and Ben Starrett, executive director of the Funders' Network for Smart Growth and Livable Communities.

This inspiring national convention brought attention to the work being done nationwide on issues such as smart growth, reduction of sprawl, social and environmental justice, affordable housing, transportation equity, and living-wage jobs. Including a diversity of voices and avoiding divisive, single-issue struggles are essential aspects of nurturing the collaborative leadership that can spark innovation and bring about lasting change.

In 2008 almost 2,000 local, regional, state, and national leaders gathered in New Orleans for a third summit, sharing ideas and resources and furthering the quest for equitable development, smart growth, and social justice.

Summarizing important lessons learned from these summits, the authors gaze toward future horizons, reflecting on the challenges and opportunities that a growing regional equity movement may face.

26

Building the Capacity of the Regional Equity Movement

Angela Glover Blackwell and L. Benjamin Starrett

A National Convening for Regional Equity

In May 2005, nearly 1,300 people converged on Philadelphia to plot a path for the future of change. Representing a broad array of racial, ethnic, religious, and socioeconomic groups, people came, attracted to the vision of Advancing Regional Equity: The Second National Summit on Equitable Development, Social Justice, and Smart Growth.[1]

Building upon the first national summit on regional equity, which in 2002 drew 650 participants to Los Angeles, the second summit confirmed the beginnings of a movement. In Philadelphia, a broad range of groups—faith-based institutions, labor unions, academics, organizing networks, environmental and community-based organizations, government, the nonprofit community, and the private sector—rallied around an inspiring goal: to create vibrant, mixed-income communities throughout America and a society in which everyone, including low-income families and communities of color, can participate and prosper.

In plenary sessions and workshops, advocates energetically discussed how to develop policies and strategies intentionally focused on those most isolated from regional opportunities, and those low-income residents who have weathered hard times and want to benefit from the forces of revitalization. Strategies and ideas focused on making homes and apartments affordable to people of all income levels and ensuring access to communities rich in good jobs, transportation choices, good schools, supermarkets, banks, parks, and cultural institutions. Much enthusiasm was generated around building public will and furthering the movement for regional equity.

A Challenge for Movement Building: Developing a Shared Agenda

Galvanizing organizations and individuals into a cohesive movement that advances regional equity is not without challenges. Differences in and disagreements over priorities, issue areas, and approaches to change can stall progress.

Fortunately, equitable development has emerged as an approach around which many issues coalesce and intersect—bringing together housing activists, environmentalists, transit advocates, community health experts, and social justice groups—in crafting a collaborative agenda to redevelop and revitalize America's cities, suburbs, and rural areas. Equitable development is guided by four principles:

Integrate Strategies for Supporting People and Targeting the Places Where They Live and Work

Community development efforts should focus on linking the needs of people—such as good housing, good schools, a healthy environment, accessible and affordable transportation, and opportunities for jobs and job training—with infrastructure and other amenities supportive of strong communities. Without a "people and place" focus that is inclusive of low-income communities of color, gentrification and reinvestment risk ultimately displacing longtime neighborhood residents. In contrast, being intentional about the needs of people and place can result in vibrant, mixed-income neighborhoods that are communities of opportunity, places where people want to live.

Reduce Disparities That Exist Between the Region and Its Parts

Metropolitan areas are more likely to thrive when attention is paid both to regional growth and central-city poverty. Wherever people live throughout the region, all should have access to the services and opportunities essential for healthy, livable communities. Win-win policy solutions to combat poverty can simultaneously improve conditions in low-income communities of color and build healthy metropolitan regions.

Promote Triple Bottom-Line Investments

Public and private investments in low-income communities are keys to revitalization. However, to reduce poverty, promote advancement, and encourage sustainability, these investments must produce a triple bottom line: financial returns for investors *and* economic and social benefits for residents in the form of jobs, services, entrepreneurial opportunities, and access to desirable homes and apartments that people can afford— including ownership options and a healthy environment for all.

Ensure Meaningful Community Voice, Participation, and Leadership

Broad, well-supported participation of community residents and organizations in planning and development helps to ensure that the community will benefit from development and revitalization. Community residents and organizations need access to the tools, knowledge, and resources that guarantee meaningful participation and engagement in the new civics of regional leadership.

Building on the Track Record of the Regional Equity Movement

Throughout the country, individuals and communities have been raising their collective voices to confront socioeconomic disparities and inspire meaningful policy change.[2] In Massachusetts, for example, a coalition of twenty organizations has met regularly for four years as Action for Regional Equity (Action!). Action! includes representatives from environmental, affordable housing, transportation, and community development organizations across the state of Massachusetts. With the goal of achieving "a better Commonwealth through better public policy," Action!'s policy priorities include increasing rental assistance for working-class families, securing financing for a statewide affordable housing trust fund, bringing an equity framework to transit-oriented development, instituting a data collection system to bring accountability to regional fair housing, and increasing citizen engagement in policy advocacy (Marsh 2003). In 2005, a policy victory was achieved through a successful campaign to mandate data collection and an analysis of the state's investments in affordable housing.

Similarly, in Washington, D.C., the Campaign for Mandatory Inclusionary Zoning won a policy requiring developers to set aside units for people making below 80 percent of area median income, as well as units for those making below 50 percent of area median income (Fox and Rose 2003). In neighborhoods plagued by gentrification, the three-year campaign brought together more than fifty organizations united in their quest for housing that is affordable to families in every income category. The participation of lower-income residents in the process helped ensure that D.C.'s policy reached levels of affordability not normally achieved in most inclusionary zoning plans.

Stories of success have been critical to the movement and are fueling change. As the movement goes forward, however, advocates have also taken the time to reflect on lessons learned.

The Continuing Challenges of Race and Class

As Hurricane Katrina's devastation of the Gulf Coast region made painfully clear, race and class issues persist in the Gulf Coast and throughout the nation and are supported by policy choices made over decades that have exacerbated conditions and isolated poor people and people of color from civic participation, opportunity, and prosperity.

Public policy—and even private and business policy—determines where people live; the quality of the schools their children attend; the availability of fresh, quality food for purchase; and the ability to engage in physical activity that supports health and well-being.

Public policy failed New Orleans, and failed its people of color. The collapse of the city's levees and floodwalls in the wake of Katrina, and the heartrending scenes of predominantly African-American, poor, or elderly New Orleans residents stranded and neglected at the Superdome, the convention center, and on sweltering highway overpasses served as a reminder of the continuing racial and class dimensions of economic and social inequity in America. In one fell swoop, the word *infrastructure* had been pushed into the nation's consciousness. Yet, few people not versed in the intricacies of planning and engineering understand how aging bridges, tunnels, and transportation systems are straining municipal budgets and impacting the daily lives of residents in every part of the country.

In failing to protect people and property from flooding, government proved incapable of honoring a covenant with its citizens—a pact rooted in the belief that government is responsible for keeping communities safe and making it possible for residents of all races and incomes to participate and prosper. Acknowledging the role of race and poverty in policy decisions is central to achieving the goals of regional equity. Further, to build public will and engage a broad constituency of Americans in the policy debate, the regional equity movement must answer two critical questions: Why should I care? What's in it for me? Again, the cornerstone of this discussion comes down to race, a topic that remains uneasy for most Americans. Still, the movement must push through the discomfort and unease if it wants to reach its full potential.

Even when movement organizers have addressed the challenges of race and class, conflicting priorities still can force a stalemate. Assorted constituencies struggling to address differing specific needs may be wary of connecting with organizations whose priorities lie elsewhere. Negotiating and reconciling competing priorities is often a challenge for movement leaders but cannot be overlooked.

For example, environmental organizations have raised consciousness about issues of water, waste, and pollution, advocating for crucial legislation to protect the environment and the earth's inhabitants. Yet, since the first Earth Day in 1970, the movement has been criticized for its lack of attention to equity. "The Soul of Environmentalism," a key report released in 2005, highlighted the environmental movement's failure to connect its admirable goals to social justice issues (Gelobter et al. 2005). The report underscores the reality that to be successful, the regional equity movement needs to confront the ways in which issue, priorities, race, and class separate potential partners.

Finding Common Ground

A movement for regional equity must also address three critical questions: Who benefits? Who pays? Who decides? These questions speak to the belief that everyone should have the opportunity to participate and prosper in society, especially those most disadvantaged. Answers, however, are difficult and emerge from a common agenda, carefully crafted from the many different perspectives and issues of concern to collaborators.

Too often people working for social change lose sight of their commonalities, focusing more on areas of disagreement than agreement. We cannot make the mistake of fighting over crumbs while supporters of the status quo divide the whole loaf. Building a sustainable regional equity movement requires that people move beyond differences to the places where mutual points of agreement can be found, explored, and expanded. Ongoing efforts to bridge perceived divides among issues such as environmentalism, affordable housing, and health are essential if we are to create an integrated, holistic regional equity movement.

To be successful, the movement must place equity concerns at the heart of its agenda. For example, proponents of smart decisions about growth may emphasize building more compactly to conserve open space, reduce automobile dependence and energy consumption, and integrate land uses into more mixed-use, pedestrian-accessible neighborhoods. However, employing well-intentioned smart growth principles and strategies without incorporating equity goals can result in housing that is unaffordable to many people of color and low-income individuals and families, thus effectively reducing their access to the opportunities afforded by vibrant, economically robust communities. Without attention to fairness for existing residents—many of whom have lived in disinvested neighborhoods for several generations—largely white or middle- and upper-class newcomers moving into gentrifying neighborhoods will be the only ones reaping the benefits of smart growth.

Communications as a Movement-Building Tool

The basic rules of communications—clarity of message and knowledge of audience—inform all movement-building activities. Raising awareness and building public will for equity and opportunity begin with clear communication. Fostering a diverse and inclusive movement requires sensitivity in the use of language and messages.

Effective communications campaigns must reach beyond established constituencies to engage new communities, opinion leaders, and policy makers with messages about how *everyone* can benefit from policies that advance regional equity.

Movement organizers need to reach out to ethnic and alternative media outlets and cultivate relationships with a racially and ethnically diverse range of congregations, organizations, and community groups. Building an inclusive movement also requires reframing messages. Organizers must look beyond the black-white paradigm that has characterized discussions about race and culture in America to recognize, understand, and embrace the growth and diversity of Latino and Asian constituencies. These include newly arrived immigrants, as well as residents who have been in the United States for generations.

Ensuring the Movement's Future

In March 2008, more than 1,500 of the nation's leading equity advocates met in New Orleans to continue to build momentum for change through Regional Equity '08: The Third National Summit on Equitable

Development, Social Justice, and Smart Growth. A rich menu of panels and workshops were offered, covering the latest policy framings, such as infrastructure and access to healthy foods, lessons learned from New Orleans and the American Gulf Coast, the latest communications technologies to advance equity, the connection between climate change and equity, and the latest case studies and success stories on policy impact. In addition, in a presidential election year, a plenary panel explored why it is important to keep race, poverty, and equity on the political agenda.

In the long run, the movement cannot be sustained without a new generation of leaders who are grounded in community, committed to advocating for policy change, and gifted with leadership skills or the potential for leadership. Foundations and intermediary organizations can help to develop such leadership by supporting programs to build and sustain networks, organize new and existing constituencies, and educate leaders on the policy process. Reaching out to young people and encouraging their participation will be vital to supporting the goal of regional equity.

PolicyLink, the Funders' Network, and hundreds of other organizations are committed to building a diverse and inclusive regional equity movement to achieve a society that enables everyone to participate and prosper. The realization of such an achievement will represent a new horizon for economic, environmental, and social justice in the United States.

Notes

1. For more about the principles of equitable development, see A. Glover Blackwell and R. K. Fox, *Regional Equity and Smart Growth: Opportunities for Advancing Social and Economic Justice in America* (Coral Gables, FL: Funders' Network for Smart Growth and Livable Communities, 2004).

2. For additional examples, see *Signs of Promise: Stories of Philanthropic Leadership in Advancing Regional and Neighborhood Equity* (Coral Gables, FL: Funders' Network for Smart Growth and Livable Communities, 2005).

Reaching Out to New Strategic Partners

To advocates of metropolitan regional equity the question may be posed, "How will we know when we are succeeding?" The chapters in section 2 explore how regional equity is fostered through public and private partnerships, new methods of measuring and interpreting data, networking to share expertise and resources, and innovative communications strategies.

Bart Harvey directs Enterprise Community Partners, a national nonprofit housing organization founded in 1982 by James W. Rouse, a successful real estate developer. Harvey links business and community development leaders in constructive conversation to bring about regional equity goals within the American mainstream. Humans remain fixed in nonsustainable patterns at our peril, Harvey says. He argues that equity can be achieved only through building an inclusive movement that engages the private sector. Enterprise has recently collaborated with the Urban Land Institute (ULI) on the Project on Regional Equity Leadership Forum, organizing a series of live, online regional equity events. With more than 34,000 members, ULI is the preeminent worldwide association of land use and real estate development organizations worldwide.

To know whether or not progress on regional equity is being made, clear benchmarks must be established. The following chapter focuses on the contributions of several analysts, including David Rusk, in applying metrics to the analysis of regional equity outcomes. Metrics, together with GIS mapping techniques, provide analytical tools for addressing existing inequities and persuading business and civic leaders that regional approaches work (Rusk 2003).

To realize the promise of full participation and inclusion in American society, African-Americans and other communities of color must join and help to lead the struggle for metropolitan reform in the United

States. To support engagement from this community, the African American Forum on Race and Regionalism was launched in 2002. Here, the institution's secretariat and president of Global Environmental Resources, Inc. (GERI), Deeohn Ferris, emphasizes the importance of African-Americans becoming agents for change to ensure that redevelopment efforts in their communities reflect the residents' needs. In assessing smart growth's aims alongside the realities often experienced by African-Americans who are impacted by smart growth, the African American Forum on Race and Regionalism works to ensure that a history of displacement is no longer perpetuated.

Crafting a compelling message that informs and motivates is another crucial link in building this movement. Ellen Schneider, president of Active Voice in San Francisco, California; and documentary filmmaker Andrea Torrice team up to describe the powerful uses of media for advocacy. Beginning with Torrice's current work, *The New Metropolis*, the authors detail how a social issues documentary can connect viewers to the experience of spatial segregation and transform their perception of what community can be.

27

Business, Grass Roots, and the Regional Agenda

Bart Harvey

The Business Context for Regional Equity

For regional equity to succeed, grassroots groups need to build relationships with business leaders. Center for Justice, Tolerance, and Community codirector Manuel Pastor, Jr., has noted that we live in a highly polarized society where democratic conversation and goodwill are often overshadowed by antagonism. Community activists may feel their agendas will always be in opposition to business interests and choose not to start potentially productive conversations. This can be detrimental to many regions, especially to poorer residents in dire need of representation and advocacy.

In turn, businesses in the United States need to recognize that regions are the building blocks of the global economy, and if businesses wish to be competitive, they have to be located in a competitive region. For businesses to thrive in an emerging economic order, it is paramount to promote the health of the region where they are located. This is particularly important in terms of workforce readiness. If 30 to 40 percent of a region's population do not have a high school education, this lack of preparedness leads to a shortage in well-trained workers and becomes a drag on the economy.

A well-educated, well-housed, well-paid labor force is in the best interests of both business leaders and community development groups. This is the high road toward regional competitiveness where economic opportunity can be developed for people at the lower end. In *Regions That Work: How Cities and Suburbs Can Grow Together* (Pastor et al. 2000), the editors cite that out of fifty-nine regions in the United States, the regions with less economic polarization tend to be more competitive. Yet, those with misgivings about "competitive" versus "collaborative" often see one to the exclusion of the other. Both can be integral to how a region defines itself. In fact, some business leaders are taking a closer look at how communities actually prefer smart growth measures that limit sprawl as well as increase opportunities for education and job training. They follow in the spirit of Harvard Business School professor Rosabeth Moss Kanter's observation that "the most effective business contributions are made in places where public leadership—from civic activists as well as elected officials—is strong, visionary, respected across interest groups, and change-oriented" (Moss Kanter 2000).

As CEO of Enterprise Foundation, a Maryland-based community development organization, I have been able to help develop a more inclusive framework that links social activists, grassroots organizers, and business leaders in unlikely coalitions. Over the past twenty years, my work with the foundation to foster business and community collaborations has successfully encouraged business leaders to invest more than $7 billion to build affordable housing for low-income communities. This chapter highlights some of the contributions of Enterprise Foundation, emphasizing the importance of working together to address the realities of global economic competition and its impact on a region's well being.

Enterprise Foundation: Win-Win Scenarios for Business and Social Justice

The regional equity movement focuses on how we develop communities smartly and competitively, given the new realities of the global economy, and how traditional employment for Americans is changing. The United States has adopted a policy focused on global markets and free trade without looking carefully at the kinds of investments needed to support communities and create the society we would like to have. In essence, we as a nation have embarked on a wholesale economic strategy without really thinking through how we have to both reach and train people. What new economic policies are going to be critical to our future? And what settlement patterns do we need?

I believe the settlement patterns we need for the future are large and require the wise use of resources, such as transportation infrastructure, as well as action on expenditures for critical human needs, such as health care. For social activists, it is critical now to understand our global markets, our economy and where it is going, and the need to be competitive in a world where wealth is being rearranged. If their priorities are principally adversarial in nature and focused on setting people against business, neither side benefits. In this new economic reality, social justice advocates need to learn when to push and when to pull back.

However, it takes education on all sides. Business leaders need to look at the economic transactions and ask what the benefits are for the community—and what points kill a transaction. So, while social justice advocates should study economics, business leaders should know the needs of the communities they serve. They may have different bottom lines than social justice advocates, but many business leaders are responsive to reasoned arguments that will provide a win-win. Enough of them understand that economic policies cannot be at the expense of the country, the region, or the city. Yes, there will be clashes over differing theories, but it is clear that efficiency at the expense of the population does not work.

There Is Room for Everyone

At the Enterprise Foundation, we are committed to a vision that if you build in the right way, there is room for rich and poor alike. We do not need exclusive movements in this country; we need policies that enhance building sound communities. Key strategies should include integrating workers and jobs in the same place so there is a more efficient and competitive arrangement where everyone benefits. By not investing in our regions so that they become more competitive, we risk killing off parts of our country and are left with the cost of caring for a population that has been totally left behind.

There are some fair policies that the free market never has and never will take care of. But we have to ask what the basis of our society's fairness should be, as well as what the rules of this game are. Assuming we want to go forward with full-scale globalization, what rules are going to keep a connected, democratic, interdependent society together? Without understanding the sets of decisions we need to make, we will suffer the consequences of unsound economic policy.

The community development industry has learned and achieved a great deal over the past twenty years. Now, we must heed macroeconomic forces and take the field to the next level of size, scale, and impact. Consider: Enterprise Foundation is financing affordable and supportive housing in concert with New York mayor Michael Bloomberg's effort to rezone whole areas of the city. Critical to that initiative, Enterprise and its partners are developing a $200 million acquisition fund to put community developers on equal footing with private developers for private site acquisition. Near Johns Hopkins Hospital in East Baltimore, Enterprise has helped land-bank eighty acres in the heart of the city, and has helped create a tax-increment financing system to support the development of mixed-income housing. In Denver, Enterprise has supported mandatory inclusionary zoning, which has been enacted. In Albuquerque, we have supported a downtown redevelopment effort that, if successful, will direct resources into a civic trust to preserve affordable housing and support arts and culture.

At Enterprise Foundation, we work to lift grassroots action to a higher plane and incorporate both economic and regional jurisdictions—which in reality are not *political* but are *economic* in nature. We have learned what can happen when community development joins with economics to create a larger, more powerful framework. Indeed, mergers, joint ventures, and strategic alliances are critical in this time of diminishing resources. As the movement for regional equity evolves, its organizers must get serious about extending the reach of its voice and speak in unison with the environmental movement, transportation advocates, the education sector, and economic development leaders. In time, new and broader alliances will increase the movement's relevance and access to a diverse and enlightened constituency.

28

Measuring Success: Using Metrics in Support of Regional Equity

David Rusk

How do we measure success? As regional equity takes root in the next generation of practice, techniques and tools for measuring progress are critical to building momentum and gaining traction. Basic numerical analyses—whether counting a decreasing number of vacant properties in a neighborhood over a decade or comparing the number of jobs obtained through various community benefits agreements (CBAs) in a year—bring precision and provide hard data to bolster arguments for regional equity policies. More subtle qualitative measures are also being developed. For example, we can now look at housing as not merely affordable but as existing within matrices of opportunities that include transportation to quality jobs, access to green public space, and proximity to healthful food.

As a former mayor of Albuquerque and a pioneer in the application of regional equity metrics for measuring and analyzing human activity and settlement patterns, I advocate using metrics to offer community leaders not only statistical indicators but also a means to interpret data. I am not alone in this advocacy. The Oakland, California–based group Redefining Progress; Manuel Pastor, at the University of California, Santa Cruz; and john powell, at the Kirwan Institute—among many others—are also part of this growing movement to establish community-defined indicators that "expose obstacles to a healthy quality of life, and illuminate economic, environmental, and social trends."[1]

Metrics also offer a way to keep multiple stakeholders committed to a plan of action without requiring congruence of motivation. Comparisons among regions that enable state or nationwide assessments are also possible with metrics. For example, Myron Orfield's analysis of the fiscal capacities of jurisdictions illustrates compelling measurable disjunctions between affluent suburban communities and at-risk suburbs.

Racial segregation continues to be a significant factor that limits access by people of color to good jobs, good schools, wealth-building opportunities through increasing home equity, and many other economic goals. In "Regional Equity Metrics" (2004), I outline specific regional equity goals and their corollary indices, which are based on measurements of both racial and economic segregation. I benchmark the Baltimore and Camden regions against their peers, evaluating them against the best and the worst of "Big Box" and "Little Box" regions, respectively,[2] and arguing that benchmarking provides a framework for assessing what measurable equity goals might realistically be set for the two regions.

Residential segregation indices are commonly measured in three ways: using dissimilarity indices, using isolation indices, and using exposure indices.[3] "Dissimilarity indices measure the degree to which a minority population (e.g. blacks, Hispanics, poor persons) is set apart from the majority population (e.g. 'whites,' non-poor persons)," I explain in the paper. "On a scale of 0 to 100, an index of '0' would indicate an even distribution of a minority group across all neighborhoods (census tracts) of a region; an index of '100' would indicate total racial or economic apartheid. At an index of 100 for blacks, for example, all blacks and only blacks would live in certain neighborhoods and all whites and only whites would live everywhere else."

The use of segregation indicators in constructing a metric is a groundbreaking approach. In addition, a community's voice, or lack thereof, also can be seen as a metric. One example of this can be seen in Thomas W. Sanchez's study that quantifies the composition of metropolitan planning organization boards as a contributor to potential bias in allocating state and federal transportation funds nationwide (Sanchez 2006). And the Gini coefficient, an index that is a measure of inequality of distribution, has been used to quantify disparities in income distribution and has influenced the development of other useful metrics, including the Robin Hood index.[4]

Highly specialized and expensive research is not always required. I have demonstrated this by developing metrics from U.S. Census data collected from all jurisdictions across the United States. At the Brookings Institution, Census 2000 data have been the basis of several important studies, including *Redefining Urban and Suburban America: Evidence from Census 2000* (Katz, Lang, and Berube 2006), a three-volume study outlining the demographic trends defining our metropolitan regions.

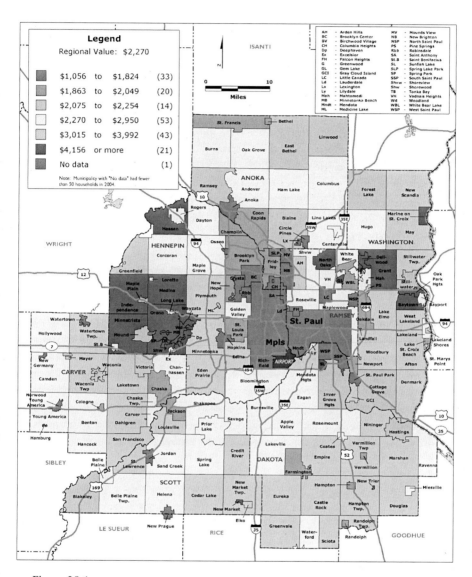

Figure 28.1
Minneapolis–St. Paul region: property tax capacity per household by municipality.

The Power of Images

Regional equity advocates such as Myron Orfield, executive director of the Institute on Race and Poverty at the University of Minnesota, are using GIS mapping tools to increase their impact. A GIS map explains a region's complex social patterns in a way that is easy to grasp. While columns of numbers are mind-numbing, a map illustrating impacts across a whole region helps people to understand key regional social and economic trends that affect their lives. GIS maps can be used to show economic inequalities among communities, or how new growth at regions' edges undermines the inner city. Assisted by such maps, communities have achieved important gains in such areas as fair housing, equitable school financing, reform of transportation spending, and brownfield remediation.

The forces of segregation and inequality are too large to confront in isolation. Activists tackling inner-city poverty often focus on one neighborhood at a time. Yet, the power of racial discrimination and fiscal inequality in a region undermines their efforts. Maps make larger patterns of inequity visible, both to citizens and to decision makers.[5] In the years ahead, metrics will play an increased role in defining outcome goals and will focus efforts to achieve greater regional equity for economically isolated and racially segregated residents. By using presentation strategies illustrated effectively with metrics, advocates will be in a better position to create compelling arguments to help reduce inequalities within regions and address the isolation of the poor from the rest of the society.

Notes

1. See D. Rusk, "Regional Equity Metrics" (2004), a report for the CORE group at the Ford Foundation.

2. "Some state laws provide for relatively few local governments. Maryland's dominant 'Big Box' county governments, for example, have both legal and political ability to adopt county-wide growth management and mixed-income housing rules. However, most metropolitan areas typically have fewer governmental decision-making bodies in a given jurisdiction. They feature a multiplicity of 'Little Box' local governments that resist voluntary compacts on tough, controversial issues. Only changes in state or federal law can set different region-wide requirements." D. Rusk, "Inside/Outside Game: The Emerging Anti-Sprawl Coalition," http://www.ginsler.com/documents/NHC-3.html.

3. Statistically, the isolation and exposure indices are highly correlated with dissimilarity indices. Change the dissimilarity indices and the isolation and exposure indices will automatically follow.

4. The Robin Hood index is a measure of income inequality where the higher the index, the greater the inequality in income distribution.

5. See Myron Orfield's short article in *Yes! A Journal of Positive Futures*, issue 34, Summer 2005, p. 27, on the effectiveness of GIS mapping, illustrated by a map of average household tax capacity from community to community in the Minneapolis–St. Paul region. The issue is full of relevant articles on the theme of "What Makes a Great Place?"

29

Networking for Social Justice: The African American Forum on Race and Regionalism

Deeohn Ferris

During the past decade, a broadening coalition has addressed the confluence of tensions posed by public policy decision making in metropolitan community development. Environmentalists, health professionals, planners, developers, and community advocates are working together to develop a new metropolitan agenda centered on principles of smart growth as an antidote to sprawl. In this context, *smart growth* can be characterized as "sustainability from a conservation perspective." Its focus is on limiting development, advocating for open space, and preserving green space.

Except for a few activists and scholars, few in this expanding coalition are examining the nexus between smart growth and economic, social, and regional equity. Furthermore, despite the negative consequences of sprawl in African-American communities, for the most part the voices and views of African-Americans on sustainability, land use, and regional equity have not been factored into the smart growth public policy debate.

To address this gap, African-American activists and scholars are redefining smart growth in terms relevant to revitalizing black metropolitan communities suffering from sprawl-driven inequities, long-term public- and private-sector disinvestments, and environmental degradation. Regional equity seeks to ensure that a region provides access to opportunities for the broader society, as well as for African-Americans and others whose inclusion, historically, has not been assured. For African-Americans, the central regional equity challenge is to create sustainable metropolitan communities through sound decisions that respect the linkages between community health and prosperity and economics, civil rights, environmental factors, transportation, land use, and development.

Regionalism Agenda: Means and Resources

The African American Forum on Race and Regionalism[1] was established in 2002 as a collaborative project of the Ford Foundation's Sustainable Metropolitan Communities Initiative. The forum works to ensure that the history and contributions of African-Americans are recognized and integrated, that antisprawl solutions are relevant in African-American communities, and that problem-solving approaches protect civil rights and the environment.

The forum brings together an interdisciplinary community of thinkers, writers, researchers, journalists, scholars, advocates, and others who are working to ensure equal protection in key arenas of civil rights and the environment. Constructive dialogue and relationship building leads to strategies, recommendations, projects, and programs. The forum is broadening the interdisciplinary involvement of African-Americans and African-American organizations in policy development and place-based projects. Through assembling a database of experts and organizations and a Web site, the forum is linking people with one another for the sharing of information and expertise. The forum is also working to catalog best practices, write and publish books and articles, and expand philanthropic resources available to organizations.

Regional meetings organized by the forum are facilitating base building in African-American communities by tapping into and broadening this extraordinary African-American brain trust. This is an essential strategic component in organizing and implementing a policy and action agenda that addresses the use of race as a framework to control space, place, and the availability of resources.

Race and Space: Barriers to Opportunity

Framed from an African-American perspective, land use and regional equity could become the cardinal civil rights issues for the millennium. African-Americans constitute a sizeable segment of the population in sprawl-threatened cities and suburbs. Incorporating social, economic, and environmental justice and civil rights advocacy, African-American scholars and activists situate racial inequity at the center of both the causes of sprawl and of regional equity and growth solutions.

Historically, decisions about land and space have shaped the course of people's lives. In the United States, blacks have a several-hundred-years' relationship to *land, race, and space*. For African-Americans, *race and space* have meant ships and chains, plantations, sharecropping, Jim Crow, lynching, "wrong side of the tracks," and urban and suburban ghettos.

Today, spatial arrangement remains one of the most fundamental ways that opportunities are distributed in the United States. Controversial issues of environment, housing, education, taxes, transportation, and jobs are inextricably bound with space, land use, and race. These concerns are characterized by disproportionate environmental impacts, urban disinvestments, white flight, and a cities-versus-suburbs mentality.

Regional inequalities are a major cause of urban inner-city disinvestments. In the absence of a sound regional decision-making approach to shaping public policy on growth and development, conflicts typically arise among cities, suburbs, and exurbs.

Transforming Inequitable Public Policies

For many communities of color, sprawl is the undesirable consequence of state, federal, and local laws, policies, and subsidies that encourage suburban migration from urban areas and inhibit healthy rural development. Many of these same governmental initiatives constrain the benefits of expansion and economic growth in communities of color by promoting segregation and disinvestments in housing, education, transportation, public health, and environmental protection.

Sprawl has relocated jobs, housing, and population—and the associated tax base—to the suburbs. Equally important, racial discrimination compounds the negative impact of sprawl in African-American communities. For example, research data establish that decades of federal spending to construct the interstate highway system supported the rapid expansion of metropolitan areas by building roads that galvanized suburban development at the expense of cities. White flight from urban core areas, which occurred after African-Americans migrated to the North (usually in pursuit of greater economic opportunities), was enabled by these roads and increased automobile use.

In comparison, public transit that serves people with low incomes has remained either inadequate or nonexistent. Additionally, these highway projects often separate, disrupt, and isolate communities of color, while also posing safety hazards, compounding air pollution, and imperiling public health.

The impact of Federal Housing Administration and the Veterans Administration homeownership programs that limited mortgage subsidies to racially homogeneous white neighborhoods also demonstrates the contributions of discrimination to sprawl. By ignoring the importance of home ownership in communities of color, these programs exacerbated housing segregation in the cities and the suburbs.

Some researchers have cited the contribution of federal public housing policies to the concentration of urban African-Americans in poverty and the widening of the income gap. Scholars have uncovered issues related to the links between governmental investments in water and sewer systems, water quality, and sprawl. Data on other sprawl impacts on African-Americans include the contributions of state and local government—such as the exclusionary and discriminatory effects of local zoning on land use in communities of color—that have encouraged environmental degradation, health hazards, blight, and low property values.

Other significant factors include racial steering, lending, and insurance redlining; educational disparities resulting from the expanding chasm between more affluent suburban schools and schools in inner cities; and the ruralization of industry, which has shifted job centers and income away from cities to the suburbs and exurbs. Essentially, sprawl has created incentives for whites to move to the suburbs, while imposing barriers on African-American communities.

A key trend in smart growth that has considerable impacts on African-American communities is the return of affluent whites to the cities. For example, where there is no community involvement in brownfields redevelopment, often these programs catalyze urban revitalization at the expense of community residents and locally owned small businesses. As whites rediscover the inner cities and inner-ring suburbs, market-induced neighborhood displacement and gentrification typically occur. This is epitomized by rapidly accelerating property values and taxes, shrinking affordable and low-income housing stock, and increasing economic development that prices out long-term residents, property owners, and workers.

Achieving Quality of Life for All: African-Americans as Change Agents

Historically, public policies, laws, economic incentives, and investments have not favored African-American communities. Without African-American involvement in development decisions, smart growth threatens to replicate these circumstances over the next several decades. For the African-Americans whose communities are under pressure, community development, regional equity, and smart growth must be situated within the bigger picture of race.

It is unlikely that significant regional equity advances will occur without the African-American community in a central and substantial role. Forging strategic alliances and working in multicultural coalitions are also critical pathways to progress. African-Americans' future is linked with other communities from which we may be segregated, but with whom we nonetheless are intertwined. Increasingly around the country, African-Americans are teaming with other people of color to create political majorities. Successful regional equity advocacy will capitalize on this growing phenomenon.

African-Americans have valuable insights and experiences to share as we continue working to ensure access in our communities to the fundamental economic and social opportunities—such as education, environmental protection, jobs, housing, transportation, and voting rights—that can help make a higher quality of life a reality for everyone.

Note

1. The African American Forum on Race and Regionalism engages in a variety of research endeavors, including a recent report entitled "Regionalism: Growing Together to Expand Opportunity to All," presented to the Presidents' Council of Cleveland in June 2007. Available at http://www.thepresidentscouncil.com/www/news.php. In addition to organizing regional symposia, the forum authored a special series of articles entitled, "Beyond Katrina: An Exclusive Series from the African-American Forum on Race & Regionalism" in *The Next American City* 1, 13: pp. 34–48. Cochairs of the forum are Angela Blackwell of PolicyLink; john powell of the Kirwan Institute, and Robert Bullard, of the Environmental Justice Resource Center at Clark Atlanta University. More information on the forum is available at http://www.aafrr.org/index.html.

30

Sharing the New Story: Regional Equity and Strategic Media

Andrea Torrice and Ellen Schneider

How do we get democracy to work for everybody? How do people share in the goods and the services—the best that America has to offer? The only way that can happen is if we work together to create a new reality and a new kind of community that's based upon equity in the region.
Reverend Cheryl Rivera

These were the closing words spoken by the Reverend Cheryl Rivera in a video segment of a one-hour public television documentary, *The New Metropolis*. This film is dedicated to showing how land use and transportation policies—which have favored suburbanization and sprawl over the past seventy-five years—have created a growing spatial inequity across the country. Viewers of the video excerpt were attending Advancing Regional Equity: The Second National Summit on Equitable Development, Social Justice, and Smart Growth, held in Philadelphia in May 2005. On one level, the video segment told the personal story of an African-American woman who lives near Gary, Indiana, and her work as a pastor with the Interfaith Federation of Northwest Indiana. But the video also did more than that. The juxtaposition of images and characters provided an on-the-ground example of some of the unresolved and urgent civil rights and land use policy issues that many American communities are grappling with today. *The New Metropolis* not only kicked off the conference but it also helped to validate the work of attendees who struggle to respond to those same issues every day.

In this chapter, through the lens of the development of this film, we discuss how the media's role in promoting regional equity can become more targeted and effective in this new century.

Putting a Human Face on Policy Issues

Videos and films are perhaps the most prominent storytelling vehicles in the world. Like plays, operas, and books, they can convey messages and stories about important issues of the moment, framing them in compelling ways that print and speech can never do. They have the potential to inform and shape our decisions and lives, from what we think about a particular issue to what we choose to wear.

Unfortunately, we are oversaturated with these messages, and most of the films and media we see today focus on selling products or providing cheap thrills by raising our heart rates. But despite this trend, a growing body of films and documentaries aim to capture the stories and issues of our times. As both a documentary and a public engagement project, *The New Metropolis* hopes to spur discussion that will lead to better public policies by conveying and humanizing some of the issues that have inspired the regional equity movement.

Toward a Cross-Sector Dialogue

If a film like *The New Metropolis* had been made in the mid-1990s, a strategic communications consultant group like Active Voice probably would have seen the film at a stage when it was quite far along, because the filmmaker would have been spending time gathering interviews, setting up the shoot, and constructing the narrative. The consultant group might have had a screening before it was done, but not necessarily. It is also quite probable that the filmmakers would have been creating the film in some amount of isolation.

Things have changed markedly in the past decade. These days, you don't simply make a documentary. First, you make a film, and then you create a series of public engagement tools and strategies to get the film used by as many different audiences as possible. Filmmakers are invested in making stories that are accessible and resonant, stories that hit home for specific targeted audiences. Some are even beginning conversations with a wide range of stakeholders during the developmental stages to make sure that their product makes a difference.

To that end, these filmmakers have gravitated toward investing in a consultative process with communications strategists. These teams help to facilitate cross-sector conversations, working with targeted groups to find out how they would use these stories, what policy issues are on the horizon that these stories may help address, what advocacy efforts these stories may help to enhance, and what distribution opportunities might lie ahead. If a film goes nationwide on public television, the filmmaker's goal becomes to steer it in such a way that it has an achievable outcome, so that when there is a possibility of a high number of U.S. television households seeing the documentary, there are specific targeted ways for people to talk about the issues that the film addresses.

Better Distribution Tools

It takes a village to make a good documentary and get it distributed to the widest possible audience. Today, we have an array of flexible distribution tools and platforms at our disposal. A work that a decade ago might have been a sixty- or ninety-minute film designed primarily for broadcast, and occasionally for theatrical release, now has myriad new opportunities for getting out there and finding and motivating audiences. Versions of the work may be purposely constructed to achieve different kinds of outcomes—one version aimed specifically for issues of regional equity, or another aimed at policy makers to help them understand the human and community implications of the decisions that they make. There may be one version in Spanish, designed to facilitate bringing Latino communities into contact with advocates focused on equitable land use. A good strategy is to hold focus groups with grassroots activists and policy makers to get feedback *before* completion, identify key issues, and help spread the word about the film.

As a filmmaker, you are not just hoping people tune in. You are hoping that, in a briefing room of a senate subcommittee, your documentary works in conjunction with a senior policy analyst's presentation, giving lawmakers a really deep understanding of what is at stake. You also want it projected at a site-specific, outdoor screening where stakeholders come together to talk about key issues. In the end, the idea is to combine creative application as well as distribution.

Note that application is not only specific to the needs of advocacy media, which is particularly adept at mobilizing people within a movement to get more active. Honing a regional equity message for distinct audiences can create a cross-sector dialog, enabling people to come together, deliberate, and take action that works for everyone. With proper support and innovation, a film can serve as a catalyst for such cross-sector discussions, getting business leaders, developers, and grass-roots leaders to work together in a way that enhances the health of a region and the country as a whole.

Uncovering Global Linkages for Sustainable Metropolitan Communities

The twin global challenges of climate change and poverty arise from the way we live in cities. In the coming decades, rising global temperatures are expected to raise the sea level and change patterns of precipitation, with potentially catastrophic global effects on forests, crop yields, water supplies, human health, animals, and many types of ecosystems.

According to the United Nations's Intergovernmental Panel on Climate Change (IPCC), the warming of the earth is the result of human activities such as agriculture, urbanization, and transportation (Underhill 2007). Scientists predict that global warming will have devastating effects on global metropolitan regions. Expected impacts include flooding, damage to buildings and infrastructure, contamination of drinking water, social dislocation, and the spread of respiratory and infectious diseases.

In light of the climate crisis, society is challenged to retool its economy and re-envision its cities. This process will undoubtedly generate new patterns of growth (IIED 2007; Gore 2006). As part of a global response to this challenge, Van Jones calls for a conscious effort to establish green-collar job opportunities for residents within economically challenged communities. In the course of pursuing innovative strategies for confronting global warming, it is vital to incorporate economic opportunities for the communities of color who have been marginalized within the petroleum economy.

Inequitable patterns of community development predominate in the United States and worldwide. In Africa, Asia, and Latin America, poor people's lack of voice within policy making contributes to a situation in which millions live in shantytowns and lack access to essential services. People who do not have an established place of residence (and the address that goes with it) are invisible in local and national elections. In their chapter, the International Institute for Environment and Development's Celine d'Cruz and David Satterthwaite examine how urban poor people from eleven countries, on three continents, are joining in common cause to improve their neighborhoods and regions. Federations such as Shack/Slum Dwellers International build cooperative relationships among community residents, government officials, and international agencies, achieving significant improvements to the residents' quality of life as they "make a way out of no way."

31

Climate Change and the Quest for Regional Equity

Van Jones

Having been horrified by images of an American city underwater, and stirred by Al Gore's case for urgent action, Americans are finally rising up and demanding action to reverse global warming.

The climate crisis is galloping from the margins of geek science to the epicenter of our politics, culture, and economics: students planning marches against emissions, consumers and investors flocking to carbon-cutting solutions, reporters and editors running environmental stories on the front page above the fold, corporations stampeding one another to showcase their love of clear skies and lush forests—and both blue Democrats and red Republicans suddenly waving green banners.

As the new environmentalists advance, who will they take with them, and who will they leave behind? Will those same activists who will use their growing clout to convince Congress to adopt market-based solutions also seek better aid for black and impoverished survivors of Hurricane Katrina, who still lack housing and the means to rebuild? The rising environmental lobby will fight for subsidies and supports for the booming clean energy and energy conservation markets. But will they insist that these new industries create jobs and wealth-building opportunities for low-income people and people of color?

The sad racial history of environmental activism tends to discourage high hopes among racial justice activists. And yet this new wave has the potential to be much more expansive and inclusive than previous ones.

Environmentalism's First Wave: Conservation

But first, the bad news: No previous wave of U.S. environmentalism ever broke with the racism or elitism of its day. In fact, earlier environmental movements often either ignored racial inequality or exacerbated it.

For example, the first wave of environmentalism was the "conservation" wave. But the original conservationists were not John Muir, Teddy Roosevelt, or David Brower. They were the Native Americans. Before the Europeans arrived, the entire continent was effectively a gigantic nature preserve. A squirrel could climb a tree at the Atlantic Ocean, and climb from branch to branch to branch—until she reached the Mississippi River. So many birds flew south for the winter that their beating wings sounded like thunder. And their numbers could blot out the sun. Native Americans lived in harmonic balance on a continent that was fully populated by humans. In fact, the leading indigenous civilizations achieved world-historic heights of political statesmanship by founding the Iroquois Federation, a model for the U.S. founders.

Unfortunately, those same founders rejected the Indians' example of environmental stewardship. The destruction of nature by the colonizers was so relentless, heedless, and massive that some ultimately balked, and created the famed "conservation movement."

Fortunately, the conservationists also enjoyed some success; their worthy efforts continue to this day. But the mostly white conservation movement owes an incalculable debt to the legacy of the indigenous peoples, a debt that still has not been repaid.

Today's large conservation groups together have countless members, hundreds of millions of dollars, and scores of professional lobbyists. But when Native Americans fight poverty, hostile federal bureaucracies, and the impact of broken treaties, these massive groups are almost always missing in action.

Environmentalism's Second Wave: Regulation

In the 1960s, the second wave of environmentalism got under way, sparked by Rachel Carson's book *Silent Spring*. This wave could be called the "regulation" wave. It challenged the worst excesses of industrial-age pollution and toxins. Among other important successes, this wave produced the Clean Air Act, the Clean Water Act, the EPA, and the first Earth Day (1970).

But this wave, too, was affluent and lily white, and it was largely blind to toxic pollution being concentrated on poor and brown-skinned people. In fact, some people of color began to wonder if white polluters and white environmentalists were unconsciously collaborating in steering the worst polluters and foulest dumps into black, Latino, Asian, and poor neighborhoods.

Finally, people of color began speaking out, and in the 1980s a new movement was born to combat what its leaders called "environmental racism." Its leaders said: "Regulate pollution, yes—but do it with equity. Do it fairly. Don't make black, brown, and poor children bear a disproportionate burden of asthma and cancer."

Two decades later, that so-called environmental justice movement continues to defend the poor and vulnerable. But it functions separately from so-called mainstream (white) environmentalism, which has never fully embraced the cause of environmentalists of color. In other words, since the 1980s, we have had an environmental movement that is segregated by race. Given this history of racial apathy, exclusion, and even hostility, is there any reason to expect much different from the latest upsurge of eco-activism?

Environmentalism's Third Wave: Investment

Well, in fact—there is. For this wave is qualitatively different from the previous ones. The third wave is about investing in solutions for the future: energy conservation, solar power, hybrid technology, biofuels, wind turbines, tidal power, fuel cells, and more.

The green wave's new products, services, and technologies also could mean something important to struggling communities: the possibility of new green-collar jobs, a chance to improve community health, and the potential to build wealth in the green economy.

If the mostly white global warming activists join forces with people of color, the United States can achieve eco-equity at last. Discussions of race, class, and the environment today can go beyond how to atone for past hurts or how to distribute present harms. Today, we can ask: How do we equitably share the benefits of a brighter future?

That kind of question gives a powerful incentive for people of color, labor leaders, and low-income people to come to the environmental table. At the same time, for all their present momentum, the newly ascendant greens cannot meet their objectives without the support of a much broader coalition.

Green-Collar Jobs?

From the perspective of people of color, helping to build a bigger green tent would be worth the effort. The LOHAS (lifestyles of health and sustainability) sector was a $229 billion piece of the U.S. economy in 2006, and it is growing vertically. But unfortunately, the LOHAS sector is probably the most racially segregated part of the U.S. economy in terms of its customers, owners, and employees. Changing that could create better health, more jobs, and increased wealth for communities that need all three. For example, urban youths trained to install solar panels, to keep buildings from leaking energy, to work with eco-chic bamboo, or to fix hybrid engines will find good work.

We need green technology training centers in every public high school, vocational school, and community college. And America needs an Energy Corps, like AmeriCorps and Peace Corps, to train and deploy millions of youth in the vital work of rewiring a nation. Beyond that, people of color must also have the chance to become inventors, investors, owners, entrepreneurs, and employers in the green economy. They should also use their political power to influence the scope, scale, and shape of the green economy.

It makes sense for people of color to work for a green growth agenda, as long as green partisans embrace broad opportunity and shared prosperity as key values.

Eco-Equity Is Smart Politics

To avoid getting outmaneuvered politically, green economy proponents must actively pursue alliances with people of color. And they must include leaders, organizations, and messages that will resonate with the working class. Otherwise, opponents of change will actively recruit everyone this new movement ignores, offends, or excludes.

California provides a cautionary tale: Voters there rejected a 2006 ballot measure to fund clean energy research. A small excise tax on the state's oil extraction would have produced a huge fund, propelling California into the global lead in renewable energy. But the same message about "energy independence" and the bright eco-future that wooed Silicon Valley and Hollywood elites flopped among regular voters.

Big oil spoke directly to pocketbook issues, running ads that warned (falsely) that the tax would send gas prices through the roof. On that basis, an NAACP leader and others joined the opposition, and the measure's support plummeted.

The Hidden Danger Of Eco-Apartheid

But the real danger lies in the long term. The United States is the world's biggest polluter, and to save the earth, Congress will have to do more than pass a "cap and trade" bill. And Americans will have to do more than screw in better lightbulbs.

To pull off this ecological U-turn, we will have to fundamentally restructure the U.S. economy. We will need to "green" whole cities. We will have to build thousands of wind farms, install tens of millions of solar panels, and retrofit millions of buildings. We will have to retire our cars, trucks, and bus fleets, and replace them with plug-in hybrids and electric vehicles, powered by a clean-energy grid.

Reversing global warming will require the work of tens of millions in a World War II level of mobilization. Such a shift will require massive support at the social, cultural, and political levels.

And in an increasingly nonwhite nation, that means enlisting the passionate involvement of millions of so-called minorities—as consumers, inventors, entrepreneurs, investors, buzz marketers, voters, and workers.

All for Green and Green for All

Any successful, long-term strategy to save the planet will require a full and passionate embrace of the principle of eco-equity. And, given that predicted ecological disasters will hit poor people and people of color first and worst, our society has a moral obligation to ensure equal protection from the peril—and equal access to the promise—of our new, ecological age.

Nowhere is the need for a politics of hope and an economy of promise more profound than it is among America's urban and rural poor. Now is the time for the green movement to reach out, to open the door to a grand historic alliance—a political force with the power to bend history in a new direction.

Let the new environmental activists say: "We want to build a green economy strong enough to lift people out of poverty and create new careers for America's children. We want this 'green wave' to lift all boats. This country can save the polar bears and black kids, too."

Let them say: "In the wake of Katrina, we reject the idea of 'free market' evacuation plans that leave families to drown because they lack a functioning car or a credit card. We reject the ideology that says we must let our neighbors 'sink or swim.' Katrina's survivors still need our help, and, in an age of floods, we need a plan to rescue everybody next time."

Let them say: "We want those communities that were locked out of the last century's pollution-based economy to be locked *into* the new, clean, and green economy. We don't have any throwaway species or resources—and we don't have any throwaway children or neighborhoods either. All of creation is precious. And we are all in this together."

A Green Growth Alliance

Those words would make environmental history.

More important, they could begin a total realignment of American politics. The idea of "social uplift environmentalism" could serve as the cornerstone for an unprecedented "Green Growth Alliance." Imagine a coalition that unites the best of labor, business, racial justice activists, environmentalists, intellectuals, students, and more. To give the earth and her peoples a fighting chance, we need a broad, eco-populist alliance—one that includes every class under the sun and every color in the rainbow. By embracing eco-equity as its ultimate goal, the new environmental movement can play a key role in birthing such a force.

A Global Perspective: Community-driven Solutions to Urban Poverty

Celine d'Cruz and David Satterthwaite

The concerns that drive discussions on regional or metropolitan equity within metropolitan areas in the United States are present in all cities. However, the form, scale, and spatial patterns of deprivation, exclusion, and discrimination obviously differ greatly from city to city. So does the extent to which government structures, policies, and practices moderate—or exacerbate—this disparity.[1]

Most cities in Africa, Asia, and Latin America have levels of inequality that exceed the most unequal U.S. cities—inequality in terms not only of income levels and income-earning opportunities but also in housing quality; security of tenure for housing; access to schools and health care; and provision of water, sanitation, drainage, and garbage collection within residential neighborhoods. There are usually very large inequalities in the rule of law and policing, with many predominantly low-income areas having none. In many cities, the government is not elected. In many others with elected governments, citizens may be excluded from the right to vote because they have no legal address. In most cities, such inequity involves discrimination against certain groups because of their social or ethnic identity or place of origin.

The Urban Transformation

The urban landscape has changed radically over the course of the past half-century. Between 1950 and 2000, the urban population in Africa increased ninefold; it is now larger than North America's urban population. In Asia, Latin America, and the Caribbean, the urban population increased nearly sixfold over this same period. Africa, Asia, and Latin America now have most of the world's urban population and most of the large cities. Many cities' populations grew tenfold between 1950 and 2000; some grew more than twentyfold. It is not only the size of the urban population that has grown but also the proportion of people living in urban centers. More than three-quarters of the populations of Latin America and the Caribbean live in urban areas; in Asia and Africa, two-fifths live in urban areas, and the proportion continues to increase in most nations.[2]

This worldwide increase in the proportion of people living in urban areas stems from very large economic, social, political, and demographic changes, including not only population growth but also decolonization, the multiplication of the world's economy, and the shift in economic activities and employment structures from agriculture to industry, services, and information. For most of Africa and Asia, rapid urbanization began around the time of independence, much of it associated with building the institutional base that independent nation-states need, and expanding the education and health systems that had been left so undeveloped under colonial rule. Urbanization was also boosted in many nations by the removal of colonial controls on the right of populations to live in cities.

In recent decades, economic change has been a more important influence than political change on the extent to which national populations urbanize. In general, the wealthier a nation, the more urbanized its population. Also, the more rapidly a nation's economy grows, the more rapidly it urbanizes. The internationalization of world production and trade has influenced urban trends in most nations, and many cities owe their prosperity to this increasingly globalized economy.

The Face of Urban Poverty

In most of Africa and much of Asia and Latin America, cities commonly have between a third and a half of their populations living in makeshift huts and shacks. These are usually built on land that is illegally occupied or that has been subdivided for housing without government approval—and without meeting official standards. A high proportion of these huts and shacks have no provision for piped water or sewage. Rarely is there a government service to collect household waste—and if there is, it is irregular and most inhabitants have to carry their garbage to distant central collection points.

It is common for the inhabitants of these settlements to be denied access to government schools and health care because they have no legal address. This usually means that they are not able to register to vote and cannot use many public services. A high proportion of those living in illegal settlements are at risk of having their homes bulldozed. One of the most powerful indicators of inequality is the mortality rate for infants and children, which is often ten to twenty times higher in poorer districts compared with richer districts within the same city (Hardoy, Mitlin, and Satterthwaite 2001).

Governments often suggest that it was very rapid population growth that caused cities to have such a high proportion of their population living in illegal settlements lacking basic services. The rapid flow of "migrants" into their city is often highlighted as "the problem." Most migration in low- and middle-income nations is the result of people moving in response to better economic opportunities in urban areas, or to the lack of prospects in their home farm or village. Migration may also be driven by flight from disaster or civil conflict, but these are exceptions. In either case, it is cities, small towns, and rural areas with expanding economies that attract most migration.

Thus, a more accurate reason for urban problems is the failure of governments to adapt to rapid urbanization and change their policies accordingly. This failure is not universal, since there are good examples of rapidly growing cities where most people live in legal homes with basic services. For most of Asia and Africa, among the most important reasons for this failure are weak, underdeveloped local government structures and the lack of qualified staff. In addition, in both rich and poor nations, the development of effective, accountable city government is a difficult and politically conflictive process, strongly opposed by many in higher levels of government and by powerful vested interests. In some instances, national governments have resisted more effective city governments as a means of guarding against separatist movements.

Government Structures and Inequality

The failure of city and metropolitan governments in nations in Africa, Asia, and Latin America to cope with the growth in their populations is widespread. But the scale of these failings and the proportion of the population affected by them differ greatly from city to city. For the most successful cities, the proportion of their population living in illegal settlements and lacking basic infrastructure and services can be a low percentage; while for the least successful cities, the proportion can be 40 to 70 percent.

It is not possible to generalize about the measures that have helped to lessen these inequalities, as they vary too much between cities and even within cities over time. Sometimes growing city prosperity contributes to greater equality in most aspects of the city's quality of life, as, for instance, in Porto Alegre in Brazil.[3] In other cities, as in Bangalore in India, it does not. Indeed, most forms of inequality—for example, in health status, income, living conditions, and access to services—have probably increased as a result of Bangalore's rapidly growing prosperity (Benjamin 2000).

Innovative city and municipal governments have sometimes helped reduce many aspects of inequality. This has often been supported by national decentralization reforms that make city governments more accountable to their populations; far more city governments are elected now than thirty years ago. In other instances, city or national government policies have exacerbated inequality—for instance, in the many cities that have done large-scale bulldozing of the informal settlements or tenements where poorer groups are concentrated in order to free up land for infrastructure and profitable real estate developments (du Plessis 2005). Typically, those whose homes are bulldozed are dumped on the city periphery, far from their workplaces and often on land sites with no basic services and no houses.

When considering what has contributed to regional or metropolitan inequity in different nations, the direct contribution of government policies may be less in most low- and middle-income nations. In the United States, the contribution of government inaction and incapacity is greater. Inequalities in income levels, and the lack of government action to increase the supply and reduce the cost of land for housing, exclude much of the low-income population from legal housing.

The extent to which metropolitan inequity is underscored by government structures differs greatly from nation to nation. Many large cities in low- and middle-income nations are made up of agglomerations of municipalities with very large contrasts between the wealthiest and poorest municipalities, and with the poorest municipalities generally having the weakest and worst financed local governments. The extent of the inequalities in services such as piped water, schools, and health care is influenced by how these services are funded; if they are funded by national, provincial, or state agencies, there are more possibilities of more equal funding across wealthier and poorer districts—although such possibilities often do not become realities.

Urbanization and Economic Development

Rapid urban growth and the increasing number of "megacities" with 10 million or more inhabitants in low- and middle-income nations are often viewed as dangers or ecological threats. But there are relatively few very large cities, containing only a small percentage of the world's population; in most instances, rapid urban growth is a positive sign, as it is a response to a national economy that is growing. In each of the world's regions, the largest cities are heavily concentrated in the largest national economies. In addition, by concentrating people and production (and their wastes), cities have many potential environmental advantages over more dispersed settlement patterns for resource use, pollution control, and waste reduction.

The key issue with regard to urbanization is to develop governance structures that can build on urbanization's potential advantages and avoid its potential disadvantages.

The Work of the "Slum" and "Shack" Dwellers' Federations

Perhaps the most significant initiative today in urban areas of Africa and Asia in addressing both absolute and relative poverty is the work of organizations and federations formed and run by the urban poor. In at least eleven nations, these federations are engaged in many community-driven programs to upgrade the neighborhoods in which they live, or to develop new housing that low-income households can afford and to improve infrastructure and services (including water, sanitation, and drainage). They also support their members to develop more stable incomes and work with governments to avoid evictions and minimize the need to relocate poorer groups.

These federations are based on hundreds or thousands of "savings groups" formed and managed by local urban poor groups. Women are particularly attracted to these groups because they provide their members with crisis credit quickly and easily. Their savings also can help fund housing improvements or income generation. These savings groups are the building blocks of what begins as a local process and then develops into citywide and national federations. Each savings group not only manages savings and credit efficiently but also their collective management of money and the trust it builds within each group increases their capacity to work together on other initiatives. The wider federations act both to support the savings and initiatives of each member group and to give each more political muscle as part of a larger entity.

The number and scale of projects within these federations grow as they learn from one another. First, one federation savings group develops a solution—such as a scheme to upgrade their homes or to develop new homes, a community-managed toilet, a partnership with the police for community policing, or a change in land use regulations that cuts the costs of land for housing. Then, other groups within the federation visit and discuss the innovation with those who implemented it. They consider how they might try a similar initiative, adapted to their needs and capacities, and the availability of land and other resources. Also, the different national and citywide federations directly support and learn from one another, as well as supporting the development of comparable federations in other nations.

Not only do these federations make demands on local and city governments (and often higher levels of government) but they also undertake major initiatives themselves to improve their housing, develop new homes, or provide water and sanitation. The federations develop solutions that are almost always less expensive, more appropriate, and higher quality than those provided by government programs. Thus, their demands to government are not so much for government provision— because they know of the inadequacies of such provision in the past— but for support for the solutions that they develop themselves. Savings groups and their wider federations seek partnership with governments to develop more appropriate solutions together, then they draw on government support to increase the scale and effectiveness of these solutions.

All the federations support their savings groups to try out initiatives themselves, for instance, upgrading their existing homes, developing new homes, or building and managing a public toilet block. These experiments allow them to learn how to plan, manage, and finance such initiatives. Most of their initiatives have much lower unit costs than conventional government or international agency initiatives, and draw far more on local resources. The federations can also do what governments find hard to do—for instance, developing detailed maps of their settlements to allow the installation of piped water, sewer, and drainage networks.

Each time that a federation group implements an initiative, other federation groups visit and discuss it. Because it was designed, financed, and implemented by a group like them, they build confidence to try out initiatives themselves. As more groups within the federation develop their own initiatives, the federation can grow to become a citywide and even a national movement.

Many city governments and some national governments and international agencies have supported these community-driven approaches, increasing the scope of what is possible. Such approaches have particular importance for metropolitan equity because some of the federations or networks of savings groups have developed to the point where they are influencing citywide programs. They go beyond supporting community-driven initiatives in the "poorer" areas, actively helping to shape citywide planning, management, and governance that address the needs of poorer people.

Ultimately, no large-scale program is possible without the support of local government. The federations need government agencies to work with them to provide secure land tenure, as most of the homes and settlements in which federation members live are illegal. Many citizen entitlements, including the right to vote and access to schools, depend on having a legal address, so this has to be negotiated. Again, individual community organizations have little negotiating power—but when these organizations aggregate into federations, their power increases. The eleven national federations are also part of a transnational movement, as they work with one another and with urban poor organizations in other nations. They have formed their own international umbrella organization—Shack/Slum Dwellers International (SDI)—to work to change the policies and practices of international agencies to be more supportive of community-driven development. SDI also supports exchange visits between member federations and assists emerging federations in other nations.

Examples of the Federations' Initiatives

• In **India**, the National Slum Dwellers Federation and *Mahila Milan* (savings groups formed by women slum and pavement dwellers), with more than 700,000 members, are working in many cities and smaller urban centers on projects to provide new housing and upgrade existing housing, involving tens of thousands of households.

• In **Thailand**, community organizations formed by the inhabitants of low-income areas are engaged in hundreds of initiatives to improve their homes and neighborhoods—for instance, through upgrading their houses and installing paved roads, paths, drains, and water and sewer pipes. They also negotiate tenure of the land they occupy or reach some other agreement with the landowner and government—sometimes dividing the site they live on so the landowner gets part of the land in return for providing them with tenure on the rest of the site.

• In **Cambodia**, the Solidarity for the Urban Poor federation is working in 200 slums with community-based savings and credit schemes, and working with the government in an ambitious program to upgrade hundreds of slums and develop alternatives to evictions.

• The **South African** Homeless People's Federation represents 1,500 autonomous savings and credit groups, and has an active membership of more than 100,000 families who live in some 700 informal settlements, 100 backyard shack areas, 3 hostels, and 150 rural settlements. Their projects have provided housing and/or land tenure for more than 12,000 households.

• The Shack Dwellers federation of **Namibia** has more than 300 savings groups with 12,350 member households; most live in informal settlements or backyard shacks, although 2,300 member households have acquired land for housing. Their partnership with the city government in Windhoek showed how land subdivision regulations and infrastructure requirements could be changed to reduce the cost of land for housing, making it affordable to low-income households.

Why Are These Federations Significant?

The significance of these federations can be seen in five aspects:

1. The scale of their work is so large that in many nations, their programs are reaching tens of thousands of people—in some, hundreds of thousands or even millions.

2. Their work and their willingness to develop partnerships with governments are changing the approaches of city and national governments and international agencies. Their explicit strategy is not to replace government, but to make government more effective.

3. They are redefining the dynamics of participation. The savings groups are at the center of these federations and all the initiatives they take. While nearly all the federations have support from NGOs, they know that it is the savings groups and the federations that have the lead role.

Women have central roles in all the federations, and all the federations strive to make sure that the poorest households can join.

4. They have the capacity to lower unit costs for building or upgrading homes and neighborhoods and to mobilize local resources so that external support goes further. They recover costs for many initiatives, thus greatly reducing and sometimes even eliminating the need for external funding.

5. They provide support for the poorest groups to move from "clients" to active agents. For governments, working with federations can bring about a fundamental change in how politicians and bureaucrats perceive "poor people" and their organizations. Government staff, and staff from international agencies, often view the "poor" as their "clients" or as "beneficiaries" of their programs. The federations' involvement broadens such perspectives, encouraging a shift from the patron-client relationship to partnership, and requiring that professionals learn how not to dominate the planning and management of initiatives.

Going to Scale

Individual community organizations are unlikely to get governments to change their policies even if they can negotiate some concessions. Federations with hundreds or thousands of community organizations have more chance of success. Changes in government policy and practice are usually required if federation programs are to "go to scale." This has been achieved in many places by a combination of strong community organizations, demonstrations and precedent-setting projects that show governments what federations are capable of, community-managed surveys and enumerations to provide the data needed for citywide programs, and a willingness to develop partnerships with city authorities.

This combination has produced citywide changes in Phnom Penh, Mumbai, Windhoek, Durban, and many cities in Thailand. Some federation programs have attained national significance. For example, the upgrading program of the Cambodian federation received the support of the national government, while in India the community-managed toilet block program stimulated the national government to set up a special funding facility to encourage comparable programs throughout the nation. The work of the Homeless People's Federation in South Africa has influenced national housing policy toward supporting the "people's housing process."

The *Baan Mankong* (secure housing) program in Thailand is perhaps the most ambitious national program. It is unusual in that it is a national government program that supports the initiatives of savings groups and their networks (Boonyabancha 2005). Since 2004, it has sought to provide 300,000 households in 2,000 urban poor communities with improved housing and living conditions and tenure security by the end of 2008. This program gives subsidies for infrastructure improvements and loans to community-based savings groups and their networks for income generation, land, and housing construction or improvement. This support allows each community to design what it considers appropriate, rather than making all plans fit official blueprints.

This Thai program has also shown how support for community-driven initiatives can lead to comprehensive citywide plans driven by urban poor communities and their networks. Citywide plans are important not only because they increase the number of people reached but also because they can change the nature of what is possible—especially in how urban poor groups can become involved in addressing citywide inequities.

The first step in these citywide plans is to build an information base of conditions in all of the areas with poor quality housing. In Thailand and in many other nations, community organizations and their networks of federations have shown how to do detailed slum surveys in ways that fully involve the inhabitants. In addition to providing key information, this process develops links between all the urban poor communities and makes apparent the differences among the many "slums," allowing solutions to be tailored to each group's needs and circumstances. It also provides a way for the groups to help choose which settlements will be upgraded first. If they are not involved in these choices, those that are not selected will feel excluded and often resentful.

The second step involves pilot projects. When designed and implemented by external agencies, pilot projects often fail to develop beyond the pilot phase. But if these are planned within citywide processes involving urban poor organizations, they are seen as centers of experiment and learning that become precedents and catalysts for action elsewhere.

Citywide consultations, data gathering, and pilot projects strengthen the horizontal linkages between urban poor communities so that they engage collectively with city governments in discussing citywide programs. Rather than each urban poor group being able to negotiate with only the politicians or civil servants responsible for their district, these negotiations are at the city level and can address the fundamental problems of the urban poor—such as land tenure, infrastructure, housing, and services—at the city scale.

This is not easily achieved. City governments and professionals find it difficult to see urban poor organizations as key partners. City politicians find it difficult to no longer be the "patron" dispensing "projects" to their constituency. Traditional community leaders may resent their loss of power. Yet, this kind of citywide process can lead to a jump in scale from isolated projects to citywide strategies while building partnerships between urban poor organizations and local governments that support an ongoing process and continuity.

Tools and Methods

All the federations use savings and credit groups, pilot projects, community-driven surveys and maps, and community exchanges. These strengthen the federations (including supporting a continuous learning cycle among its member groups) and help to change the attitudes and approaches of governments and international agencies.[4]

Exchange visits between savings groups and other groups interested in learning more about their work create strong personal bonds among communities, so that they learn to work with one another rather than seeing one another as competitors for government resources. Although exchange visits are primarily to support community organizations, civil servants and politicians are also invited to take part. These visits often have shown the professionals new ways of working. For instance, many professionals have visited Windhoek to see how the city government's changes to plot sizes and infrastructure standards have made plots more affordable for poor households. And Kenyan railway authorities visited Mumbai to see how Indian Railways supported community-managed resettlement for those living along the railway tracks.

All the federations use precedents developed by their members to help change government policies and practices. It is much easier to negotiate with government officials when these officials can see the results of a new house design, a functioning community toilet, or a detailed slum enumeration. When one local government has accepted a change in approach, other officials can be brought there to see how it works.

In ten nations, federations have set up their own urban poor funds, which help finance member groups seeking to acquire land, build homes, and develop better livelihoods. These funds are also where members' savings are deposited and where external funding from governments and international agencies is managed. Such funds provide accountability and transparency for funders. A contribution to the federation fund from a city government can indicate a change in government attitude and the beginning of a partnership.

Changing the Change Process

The tools and methods described here seek to create a more equal relationship between poor communities and government and international agencies in identifying problems and in developing solutions. They also demonstrate to such agencies the capacity of urban poor groups, including the many resources they can contribute to making government initiatives more successful.

The federations avoid any formal political alliance. In the short term, this can bring considerable disadvantages as politicians steer government support toward those in their party and prevent support going to communities that do not back them. But this keeps the federations open to everyone and protects their capacity for independent action. Since any large-scale success depends on support from government, a politically independent stance allows the federations to negotiate and work with whoever is in power locally or nationally. The federations' politics has been called "the politics of patience"—negotiation and long-term pressure, with confrontation used only as a last resort.

The Role of International Agencies

The work of all official aid agencies and development banks (such as the World Bank) is justified by supposedly addressing the needs of "the poor"—the very people who form these federations. Yet, these official agencies and banks have difficulty working with the federations because their structures were designed to work with and through national governments rather than community-driven development. Thus, most external funding for the federations has come from international NGOs. If international agencies wish to support community-driven development, they need to change the way in which their support is provided.

If estimates for the costs of addressing the needs of those living in shacks are based on the costs of conventional government and international agency-funded programs, this will require hundreds of billions of dollars. Most of this would have to come from international agencies. But if estimates are based on federation initiatives, the cost is much less, and local resources from communities and municipal government can cover a much higher proportion.

Changing government and international agency approaches to working with the urban poor is often far more important than generous international funding. This does not mean that international funding is not still needed, but these agencies' roles need to change—to encourage local community-driven initiatives, to support community-government partnerships, and to develop their accountability to urban poor groups and their federations.

Do These Community-driven Processes Have a Downside?

Community-driven approaches have been criticized for absolving national or local governments from their responsibilities. But one of the key features of these federations' work is their demonstration to governments of more effective ways in which the government can act, and of the potential of partnerships between government and community organizations. The federations have demonstrated a capacity to change the approaches of city governments and some national governments. The federations have also been criticized for increasing aid dependence, but actually they do the opposite, as they demonstrate solutions that require far less international funding.

Each of the federations has had some initiatives with disappointing results. This cannot be avoided in large-scale movements such as these, formed by people with the least income and influence, and which encourage their member organizations to try out many new initiatives. There are projects that fail, community organizations that cease to function, loan repayment schedules that are not maintained ... but one of the key roles of the federations is to learn how to cope with these problems and to learn how to avoid them in the future.

These federations also generate opposition. Many slums have powerful slum landlords and other vested interests that oppose representative community organizations. Many politicians dislike the federations because they will not align with their election campaigns. Contractors may dislike the federations because their profitable (and often corrupt) relationships with local governments are threatened. Architects, planners, engineers, and other professionals may find it difficult to work with the federations, because they are used to "being in charge" and making decisions without real consultation and discussion.

Conclusion

It would be naïve to think that more equitable patterns of city and metropolitan development can be achieved easily in Africa, Asia, and Latin America. This is especially the case regarding poorer groups' access to land for housing, given the power of real estate interests in most urban areas. But many federations of "slum" and "shack" dwellers and homeless groups have demonstrated a new approach that has been shown to work in many nations and cities.

Such an approach is rooted in addressing what is perhaps the most fundamental inequity—the lack of voice and decision-making power allowed to "the poor" and their own organizations. Yet, it is their needs on which the entire international aid and development business is based. This lack of voice also includes a lack of accountability to "the poor" by most international agencies that claim to work for their benefit. Also, most such agencies lack the capacity to actually support these representative organizations of the urban poor.

As governments and international agencies recognize the importance of community-driven processes at the neighborhood level, linked together by federations that can work at the city scale, support is needed for the following:

• many community initiatives in each city, and learning cycles that can develop into valuable precedents.
• intra-city, inter-city, and international exchanges for community representatives and, where relevant, staff and politicians from city and national governments.
• community-driven slum surveys and enumerations, to support local action and citywide initiatives.
• citywide plans that involve all urban poor communities and their organizations.

As new programs are developed, most of the attention should be on upgrading conditions within poorer groups' existing settlements, rather than moving them to new locations. This is usually their preferred option because it does not disrupt their livelihoods and social networks. Generally, it is also cheaper. If upgrading housing in their current neighborhood is not possible—for instance, for those living on pavement—land sites for new housing are needed nearby, and those who are to be moved must be involved in choosing the most appropriate sites.

In funding the work of the federations, care is needed in the use of loans because loans always impose financial costs on poor households. It is good practice to help low-income groups avoid the need for loans or minimize the size of the loan they need—for instance, by keeping down unit costs. This implies a different approach from most lending agencies, which judge their success by how many loans they provide and how much they lend. However, when used appropriately, credit can help support improved livelihoods and better housing, while also making limited funds go farther.

Perhaps the most critical determinant of whether a city government or an international agency is effectively addressing metropolitan equity is the quality of their relationship with low-income groups and their organizations. When they listen to and work with organizations and federations formed by slum and shack dwellers and the homeless, the scale of what is possible increases exponentially.

Notes

1. This chapter is drawn from a report by the authors entitled "Building Homes, Changing Official Approaches: The Work of Urban Poor Organizations and Their Federations and Their Contributions to Meeting the Millennium Development Goals in Urban Areas," IIED (2005), download available at http://www.iied .org/pubs/display.php?o=9547IIED&n=7&l=19&c=mdg. The printed version is available from Earthprint at www.earthprint.com. This paper, which is part of a series on poverty reduction in urban areas, should be consulted for more details of the work of the federations and for a full list of sources on which this chapter draws. See also www.sdinet.org. This chapter also draws heavily on the work of several people who work with the federations and its umbrella organization Shack/Slum Dwellers International—especially Sheela Patel, Somsook Boonya-bancha, Joel Bolnick, and Diana Mitlin.

2. Statistics for urban populations in this chapter are drawn from the United Nations Population Division.

3. See R. Menegat, "Participatory Democracy and Sustainable Development: Integrated Urban Environmental Management in Porto Alegre, Brazil," *Environment and Urbanization* 14, 2 (2002): pp. 181–206.

4. This section draws heavily on Sheela Patel, "Tools and Methods for Empowerment Developed by Slum Dwellers Federations in India," *Participatory Learning and Action 50*, London: IIED (2004). PDF can be downloaded at http://www .iied.org/NR/agbioliv/pla_notes/pla_backissues/50.html or http://www.planotes.org.

Beyond Segregation: Toward a Shared Vision of Our Regions

In the mid-1990s, Myron Orfield, executive director of the Institute on Race and Poverty at the University of Minnesota, introduced a breakthrough program for combating the decline of America's suburbs. Since the publication of Orfield's *Metropolitics: A Regional Agenda for Community and Stability*, his essays and books have become widely recognized as seminal documents in the field (Orfield 1997; Orfield 2003a; 2003b). Orfield uses geographic information systems mapping and case studies to demonstrate how social separation and sprawling development patterns harm residents of wealthy—as well as poor—suburban communities.

A persistent challenge within the movement for sustainable metropolitan communities is the ongoing spatial segregation that marginalizes so many African-Americans and other people of color. Policies are perpetuated that fail basic moral requirements of fairness. Orfield's vision calls for America's metropolitan regions to move beyond what he terms "spatial apartheid"—that is, the concentrated poverty of many urban centers that contributes to the racially separate societies that characterize modern American life. This culture of separation hurts everyone, with the damage passed along from generation to generation. Providing more equitable access to opportunities will move us closer to becoming a civilization that is founded upon the core values of equality and integrity.

33

Beyond Segregation: Toward a Shared Vision of Our Regions

Myron Orfield

Outside of the isolated world of fair housing research and advocacy, there is no widely acknowledged, discussed, or accepted belief in the existence or effect of the concentration of poverty and segregation in America. Hardly anyone talks about segregation anymore. Yet, more and more empirical research shows clearly that the concentration of segregation and poverty is caused by racial discrimination in the housing market. Researches have demonstrated how poverty and segregation destroy the lives of young people. Segregation deprives schoolchildren of opportunity, intensifies risks to their health, and destroys the fabric of urban life and the fiscal basis of our cities (Massey and Denton 1998). Each year, each month, each day this becomes empirically clearer—unmistakably clear, undeniably clear. Yet, it is unspoken. It is not part of the public conversation.

Our society talks about the effects of segregation, failing cities and schools, but almost no one talks about segregation as the root cause of our problems. We assume that "problem" neighborhoods exist because of problems with the people in the neighborhoods, rather than contextualizing urban problems within the pattern of poverty and segregation in metropolitan development. Perhaps this is because we fear what talking about racial segregation might mean—how angry the world might become, or how quickly friends and benefactors might become enemies. We blame urban problems on the urban poor and assume that "their" problems can be fixed piecemeal, with better individual choices or more programs and more money.

A Growing Divide

This individualist political worldview does not account for the metropolitan forces that concentrate poverty. This lack of understanding is reinforced by the growing separation of the races. We live in a society that is more and more different because it is separate. Wealthy whites, who usually live far away from poor inner-city neighborhoods, mainly see race through the bloodthirsty and exploitative crime-weather-and sports production we call "the news."

If there is evil, it lives in the forces of segregation. The evil of segregation destroys people's lives and causes them to hate themselves. It destroys the mightiest manifestation of popular democracy in human history: America's melting pots; strong, old industrial cities; multiracial, multiethnic, multiclass democracies. It causes the haters to hate even more than they did yesterday, or to hate without knowing they are hating. It erodes the possibility of hope to build progressive coalitions that can help us achieve humanity's better purposes. It prevents us from acting to stem an economic, moral, and environmental poison that could— but need not—destroy humanity, a poison that each day prevents humans from solving, rather than, wallowing in their problems.

We stand today in a new century. In the decades following the emergence of the civil rights movement, the inequalities of race have not been fundamentally addressed in our society. Meanwhile, the multiracial civil rights movement has fallen apart.

For a generation or more, a group of professional policy "entrepreneurs"—who in any other profession would be in jail—have sold the country simple solutions to complex problems. As they come and go, they are rewarded because they facilitate avoidance of the cancer of segregation that is killing our society. In effect, we are rubbing oil on our heads because we are afraid of the chemotherapy that could save us only at the expense of pain and effort. Any insomniac knows that nighttime television is filled with ways to lose weight without trying, to get a good education without trying, to find love without trying, to find God or heaven without faith. Nothing anyone has ever bought based on an ad at three o'clock in the morning has worked. Yet, these professionals seem to make a go of it, and keep running ads, and keep making money off us, because we want the harm to stop but we don't want to change our own behavior and our own lives in a way that, while difficult, would actually make a difference.

Segregation Holds Us All Down

The idea of new private charter schools has been put forward as a solution to the problem of poor, segregated schools. But these schools are in fact more deeply segregated and more dysfunctional than failing public schools in segregated neighborhoods (Finnegan et al. 2004). Indeed, they are part of the armature of segregation.

We know that most of the most severe problems in our society have to do with racial discrimination and segregation. We know that these problems depress the opportunities of individuals and prevent children from graduating from high school, prevent children from going to college, and prevent them from growing up to have good jobs and a decent living (Orfield 2006a). We know that average black students live in neighborhoods that are overwhelmingly poor and go to schools that are overwhelmingly poor, while their poor, white counterparts live in an integrated society and enjoy the networks and access to opportunity that they do not (Rumberger and Palardy 2005). This is not to say that it is easy for poor whites; it is just that it is a different situation. Poor whites have much more of the fluidity of the American economy at their fingertips.

We know that children in overwhelmingly poor schools—cut off from role models, from connections to the broader economy, from opportunity, from the networks of success, from guidance counselors, from colleges that will accept their high schools—are severely disadvantaged (Kahlenberg 2001). We also know that blacks and whites who live apart from one another have restricted opportunities to get to know one another. They do not go to school with one another; they do not live in the same neighborhoods; they do not communicate across racial lines.

The suspicions of whites who have power, and the fears of blacks who do not have power, warp our society so that we cannot talk to one another or deal with one another. This makes it difficult for us to understand one another's aspirations and hearts and minds in the way that fellow citizens need to understand one another.

We cannot avoid the cancer of segregation any longer. We have to face our medicine as a society. If we do, we will become well. The wealth and dynamism of our melting pot society can bring us all to a new level of human understanding and progress.

We cannot continue to allow ourselves to exist as virtually separate societies. In separate societies where one side is held down, profound harm is done to those who are isolated and held apart. We know that it weakens people to be isolated from the mainstream of ideas, from the flow of commerce, from connections and the interpersonal relationships; in contrast, people are strengthened by connections and knowledge. To be cut off from connection with others is harder than having weights applied to one's ankles. To be pressured to succeed in a society in which you cannot fully participate is extremely difficult and painful. *Segregation has to end.*

What to do about segregation is a hard question, one which for many years has been unaddressable in our country. We have had many periods of moratorium on the discussion on segregation in the course of our nation's history. We had such a moratorium during our country's beginnings, soon after it became an independent nation. It was impossible in Congress even to talk about slavery, but religious leaders like William Lloyd Garrison and Frederick Douglass did raise these moral issues. They learned from religious and moral leaders such as William Clarkson, William Wilberforce, and the great black leaders in England, subsequently raising these issues as moral issues. They stood witness in a time when there was a moratorium on discussing race that was far more serious and violent than the moratorium of today.

When Elijah Lovejoy raised the issue of slavery, he was murdered and his press was thrown into the ocean. William Lloyd Garrison narrowly escaped death many times, as did Frederick Douglass. By standing as moral witnesses, they forced the discussion forward. Their truth was undeniable and transformative. During the more recent civil rights movement, Martin Luther King, Jr., and leaders throughout the country again stood witness and spoke with their hearts to the evils of segregation and disparity. The experience was transformative and led to more rapid change than people had believed possible. At the time that people spoke out, many of these issues were not new, but the concerns became real to people in the process of being articulated and named.

The Fight Ahead

Both during slavery and in the 1950s to 1970s civil rights period, bearing witness, raising a moral voice, bringing the facts to bear on a situation, and changing the moral conscience of America made a huge difference. It is again time—long past time—to do the same. It is time again to reach across the bounds of race and build a multiracial society and a multiracial movement for justice. Just as slavery was wrong, just as a poll tax was wrong, so is segregation, so is housing discrimination that gives a black or a Latino man or woman or child an unequal chance at housing when they move to a middle-class neighborhood, so is the fact that this leaves their children in segregated schools when they are steered to a neighborhood that limits their opportunity. It tells them lies about the

American housing market and the possible opportunities for them. When they are discriminated against by buyers and sellers, it is morally wrong and illegal. This hurts them and their ability to achieve wealth and education for their children, and it hurts the community. When a community becomes racially identified, it is discriminated against just as individuals are discriminated against.

Such acts of discrimination are illegal and immoral, yet they happen every day, and we see them happen. When banks do not lend the same amount of money to black and Latino people who have the same basic educational background as white people, this is illegal and immoral. This restricts their opportunity, it segregates their children, it hurts communities, it leads to sprawl and inequality, and it needs to be changed. It is illegal, it is immoral, it needs to be fought against, and it needs to be enforced. Attorneys general and prosecutors across the nation stand silent at this sweeping injustice that transforms our society and our lives. We cannot allow that any longer.

Often in older U.S. suburbs, school boundary lines have been changed so that poor black and Latino children are kept in schools that are poorer and emptier, with fewer resources, less hope, less energy, and less idealism. Every time that such actions are taken, it is an act of segregation and immorality which is illegal, yet unaddressed. No one stands and bears witness to this, and the poor are not protected in that decision. This must change.

We build freeways, massive public works of infrastructure, into the wealthiest, most powerful and whitest communities that have all the jobs—excluding poor and black and Latino people through lack of affordable housing and public transit. This is a state action of immorality, of segregation, of isolation. Those dollars belong in equal part to the poor people who are being left behind and constrained by the sieve of housing discrimination. Segregation holds them back. It takes their money to build a community that they cannot be part of, a community that they can serve but cannot be served by, a community whose wealth they can help to build but whose benefits of good schools and choice opportunities and medical care and public services they cannot enjoy. They are held down; they are robbed and they are left behind, and this cannot continue. We have to bear witness; we have to stand to change these patterns.

As the world changes, the wealthiest and whitest enclaves at the edges of metropolitan areas become more wealthy and more spectacularly insulated than ever—morally, socially, and physically—from the rest of their society, such that in their mind they feel no responsibility or compassion for the realities that underlie their wealth and power. This has to change. Regionalism is how we can change these things and shape a new metropolitan order that is racially just, that forbids segregation, that makes resources for opportunity available in an equitable manner, and that provides representation for all affected parties.

Soft Path vs. Hard Path

There is a long tradition of moral witness before the evils of an unjust society. There is the witness of Thomas Clarkson and the Quaker movement in England. An impressive book, *Bury the Chains*, talks about this improbable and hopeless movement against a British empire that was deeply dependent on the wealth of slavery (Hochshild 2005). A huge percentage of the colonies' crops and money fueled the English economy and kept workers busy in English factory towns such as Leeds. Although it seemed improbable and impossible, a well-educated and principled man and a self-educated former slave, Olaudah Equiano, wrote books and traveled the country, agitating people so that the working class rose up and the churches rose up, forcing the English government to abolish slavery. It was not in its economic self-interest, but it was in its moral self-interest. This was won through moral witness, clear facts, and clever arguments by a broad, multiracial, multiclass, multireligious constituency. It did not have the power of money and special privilege, but it had the transformative power of truth standing against a wicked evil of society.

The same thing happened in the United States with William Lloyd Garrison, Frederick Douglass, and a multiracial, multifaith religious movement. Abolition was initially viewed as being crazy and impossible. People believed that it could never happen; it was not even possible to talk about slavery in the Congress. Yet, a multiracial, fact-based, morally based transformative speak-truth-to-power coalition changed the country and forced the election of Abraham Lincoln, who changed the aims of the war to abolish slavery. In the end, the very things that they thought would never happen, or would only happen in a smaller way, happened.

In every case, the milder, more palliative alternative did not generate the necessary change. The alternative of colonizing blacks was seen as being more moderate, but it did not transform the country. The idea of doing this in England did not transform the country either. In the 1960s, the multiracial, fact-based civil rights movement led by Martin Luther King, Jr., began in the South bearing witness, using creative conflict, and ultimately moving these issues of race forward to transform the nation. It is a long and important history, teaching us that bold action is the only approach that is likely to achieve transformation in our society. David Rusk notes that in dealing with issues of race in our society, there have always been two approaches—the soft path and the hard path (Rusk 1999). The soft path generates less controversy, is race-neutral, addresses issues less directly, and gains early support from philanthropies and the elite. The hard path takes direct aim at the core of issues and ultimately succeeds in converting many soft-path advocates to more substantive reform (Orfield 2003a).

Today, our approach to racial segregation and concentrated poverty is largely the soft path. Most large national civil rights organizations dedicate their time to preserving the gains of the 1960s and 1970s, particularly in terms of affirmative action, racial profiling, and health care—important issues—but the larger, substantive urban forces of racial resegregation remain largely unchallenged and unaddressed.

We need a hard path approach to complement the soft path. We need to take on—head on—the issues of concentrated poverty and segregation. When we do this, we will face challenges—an uproar will probably arise. Addressing race head on, the hard path, is controversial—in fact, it is one of the most controversial positions one can take within American politics. Many political strategists believe that raising racial issues is too costly for a political party interested in maintaining a political majority coalition (Edsall and Edsall 1992; Greenberg 1995; Judis and Texeira 2002). And addressing race has severe consequences: political strategists point to the losses suffered by the Democratic party in older, blue-collar suburbs when Democrats voted for Ronald Reagan because of racial issues (Edsall and Edsall 1992). The hard path is difficult—real social change often faces powerful opposition. However, we need to walk the hard path to see real, lasting desegregation and racial unity.

Strategies for Change

Given the complexity of all these metropolitan systems and all these structures, some may feel that the possibility of an end to segregation is so far away that we can only dream about it. I do not believe this. I do not feel it is that far from being possible, and I do not believe that the forces that have created this are as strong and as scary as people think they are. It is not that I think it is going to be easy. I do not think that there is anything harder in society, but still, I do not think it is as hard as we are making it out to be.

The intersection of three seemingly unrelated topics—the history of the U.S. civil rights movement, the history of land use planning, and the history of fiscal equity—gives rise to two broad-based strategies for change:

One is to organize and build a coalition, a coalition with a multifaceted strength. We have many potential partners who are hurt by segregation and a coalition of these partners could wield tremendous power: The central cities have huge numbers of older suburbs with as many congressmen and as many legislators and as much power as the central cities. Suburbanites who live in the older, rapidly integrating suburbs have the same interest inner-city residents have in preventing segregation and concentrated poverty from devastating their neighborhoods. We can create a long-lasting coalition that includes African-Americans and Latinos, white people who live in these older suburbs, environmentalists, and the legislators that form a majority. We can build a coherent coalition that raises fundamental issues of justice, lasting for as long as we continue to be forward-thinking.

A second strategy is based in the courts. Every amount of fiscal equity that has been achieved in this country has come about as a result of these issues being raised in the courts (Orfield 2006a, 2006b). In all, twenty-eight supreme courts have declared their school finance systems unconstitutional and have moved money around throughout the system to reduce the disparities (Hunter 2006). This never would have happened if people had not pushed on this effort. That strategy of resorting to the courts—together with an organizing strategy and a legislative strategy, and raising the moral issues of what we are talking about for the poorest people—always has the possibility of achieving reform. Every time a significant broad-based challenge is raised, things have gotten better. They have not gotten perfect, but they have gotten better.

Conclusion

We do not need to create regional governments; we already have them. What we need is to take control of the governments through the courts, through organizing, through legislative strategy, through joining together across false distances, and through keeping our eyes on the ball.

We stand in a new century, within a multiracial society. We see that the rest of the world does not have a panacea to these questions; we see that the world at large struggles with race. We, a more diverse society in many ways than most, have been living with inequalities for centuries. We now have to step up and lead the way to a new type of government, a racially conscious process of reform.

Groups of white European immigrants have disappeared into America. When my children look at their heritage, they are twenty different nationalities, a veritable melting pot of Europe. However, blacks and Latinos have never been allowed to enter America in the same way that whites have. They have been held aside, defined as the "other," existing in a virtually separate society. Blacks and Latinos are held apart in separate neighborhoods, held apart in separate schools, viewed differently walking up and down the streets, and discriminated against in every subtle and unsubtle way that man and the mind can catalog.

Every person has as much of a right to education as his or her ability and discipline can gather. Education becomes more important to participating in a democracy than ever before. We know that in societies when people are not given the benefits of a real education, democracy and the economy falter. One-third of our society is prevented from having a decent education.

A growing part of our population is held apart and held aside. We need to let them in. When we let them in, they will experience the kinds of success that we all expect and long for our children to experience. They will become participants in the American economy. They will not be among those left behind; rather, they will be among those pulling ahead.

Each time and each moment in our society when a group that was previously excluded gains access, when their members become full participants in this society, energy builds and creates beautiful things. New ideas, new culture, new music, new art, and new forms of civilization are created. Imagine when all the energies of all people are set forth— not to hold one another down, but to lift one another up. Imagine when all the energies of our society are geared toward improving our common lot. Imagine a society where all the cities are strong, where the older suburbs renew themselves, where the beautiful open spaces are not torn apart because they are not needed to build a new community where a portion of one community can move away from another portion. Imagine all the time we used to spend in hatred of one another, rededicated to building a new society. Imagine how great a civilization that can be.

This is within our reach. We are closer to this than we have been ever before. We know the harms; we can see the harms. The harms that have occurred, the harms that we thought we could contain in central cities, are not contained. They destroy ring after ring of suburbs. We know that we cannot do this anymore; we cannot afford the costs anymore. The present form is untenable in every respect. The only way that we can solve these problems is to talk about the divisions among ourselves; act against them; build a newer, better county; and end the pattern of hundreds of parochial local governments that exclude and prey upon one another. We can create a new open and representative metropolitan society where everyone participates, where local governments work together under a new regional constitution, where everyone's heart and abilities are harnessed, and where we move the whole society forward into a new and transformative culture.

IV

Resources

Bibliography

*Core readings in the study of sustainable metropolitan communities appear in boldface.

Abt Associates Inc. October 1999. *Out of Breath: Adverse Health Effects Associated with Ozone in the Eastern United States* (Washington, DC: Clear the Air Task Force). Downloadable at http://www.abtassoc.com/reports/ES-ozone.pdf or http://www.abtassoc.net under Publications/Reports/Health.

Agyeman, J. 2005. Sustainable Communities and the Challenge of Environmental Justice. New York: New York University Press.

Agyeman, J., R. D. Bullard, and B. Evans. 2003. Just Sustainabilities: Development in an Unequal World. Cambridge, MA: MIT Press.

American Bar Association. 2005. Rhode Island Agency Violated Environmental Equity Requirements. In *The Law of Environmental Justice*, M. Gerrard, 1999: Update Service. Available at www.abanet.org/environ/committees/envtab/ejweb.html.

American Farmland Trust. 2002a. *Farming on the Edge: Sprawling Development Threatens America's Best Farmland.* Available at http://www.farmlandinfo.org/documents/29393/Farming_on_the_Edge_2002.pdf.

American Farmland Trust. 2002b. The Farmland Protection Toolbox. Available at http://www.farmlandinfo.org/documents/27761/FS_Toolbox_10-02.pdf.

American Lung Association. 2000. Children & Ozone Air Population Fact Sheet (September). Available at http://www.lungusa.org.

American Public Transportation Association. 2003. *The Benefits of Public Transportation: The Route to Better Personal Health*, vol. 2 (n.d.). Washington, DC. Available at http://www.apta.com/research/info/online/documents/better_health.pdf.

Anthony, C. 2006a. Race, Poverty and the Human Metropolis. In *The Human Metropolis: People and Nature in the Twenty-first Century*. R. H. Platt, ed. Amherst: University of Massachusetts Press.

Anthony, C. 2006b. Reflections on the Purpose and Meaning of African American Environmental History. In *To Love the Wind and Rain: African Americans and Environmental History.* D. Glave and M. Stoll, eds. Pittsburg: University of Pittsburg Press.

Anthony, C., and B. Starrett. 2005. *Signs of Promise: Stories of Philanthropic Leadership in Advancing Regional and Neighborhood Equity.* Miami: The Funders' Network for Smart Growth and Livable Communities.

Anthony, C., and M. P. Pavel. 2008. *The Green Economy: Green Collar Jobs and Workforce Development.* San Francisco: Tides Foundation.

Associated Press. 2005. Camden, N.J. Ranked Most-Dangerous City. *USA Today* (November 21).

Baland, J., P. Bardhan, and S. Bowles. 2007. *Inequality, Cooperation and Sustainability.* New York: Russell Sage Foundation.

Barnes, W., and L. C. Ledebur. 1998. *The New Regional Economies: The U.S. Common Market and the Global Economy.* Thousand Oaks, CA: Sage Publications.

Beauregard, R. A. 2003. *Voices of Decline: The Postwar Fate of U.S. Cities.* Florence, KY: Routledge Press.

Benjamin, S. 2000. Governance, Economic Settings and Poverty in Bangalore. *Environment and Urbanization,* vol. 12, no. 1: 35–36.

Benner, C. 1998. Growing Together or Drifting Apart: Working Families and Business in the New Economy. San Jose, CA: Working Partnerships USA, p. 34.

Berube, A., and B. Forman. 2001. *A Local Ladder for the Working Poor: The Impact of the Earned Income Tax Credit in U.S. Metropolitan Areas.* Washington, DC: Brookings Institution.

Berube, A., B. Katz, and R. E. Lang, eds. 2003–2006. *Redefining Urban and Suburban America Evidence from Census 2000,* vols. 1–3. Washington, DC: The Brookings Institution.

Blackwell, A. G. 2001. Promoting Equitable Development. *Indiana Law Review,* vol. 34: 1273.

Blackwell, A. G., and R. K. Fox. 2005. Regional Equity and Smart Growth: Opportunities for Advancing Social and Economic Justice in America. Translation Paper #1, 2nd ed. Miami: The Funders' Network for Smart Growth and Livable Communities.

Blackwell, A. G., S. Kwoh, and M. Pastor. 2002. *Searching for the Uncommon Common Ground: New Dimensions on Race in America.* New York: W. W. Norton.

Blezard, R. 2004. Chicago Hope. *Ford Foundation Report,* Summer: 14–19.

Bogren, S. 1990. A Tale of Two Transit Networks: Separate But Not Equal. *CT Magazine* (September/October). Available at http://www.ctaa.org/webmodules/webarticles/articlefiles/ct/sepoct96/equity.pdf.

Bollens, S. 2003. In Through the Back Door: Social Equity and Regional Governance. *Housing Policy Debate*, vol. 13, no. 4: 631–657.

Bollier, D. 1998. *How Smart Growth Can Stop Sprawl: A Briefing Guide for Funders.* Washington, DC: Essential Books.

Bonnen, J. T. 1992. Why There Is No Coherent U.S. Rural Policy. Washington, DC: *Policy Studies Journal*, vol. 20: 190–201.

Bonorris, S., J. Isaacs, and K. Brown, eds. 2004. *Environmental Justice for All: A Fifty-State Survey of Legislation, Policies, and Initiatives.* Section of Individual Rights and Responsibilities, American Bar Association, Section of Environment, Energy and Resources, Public Law Research Institute Hastings College of the Law, University of California, San Francisco.

Boonyabancha, S. 2005. Baan Mankong: Going to Scale with "Slum" and Squatter Upgrading in Thailand. *Environment and Urbanization*, vol. 17, no. 1: 21–46.

Brecher, J. 1990. *Building Bridges.* New York: Monthly Review Press, pp. 93–105.

Brecher, J., J. Lombardi, and J. Stackhouse. 1982. *Brass Valley.* Philadelphia: Temple University Press.

Briggs, X., ed. 2005. *The Geography of Opportunity: Race and Housing Choice in Metropolitan America.* Washington, DC: Brookings Institution Press.

Brophy, P. C., and A. Shabecoff. 2001. *A Guide to Careers in Community Development.* Washington, DC: Island Press.

Brown, D. L., and L. W. Swanson. 2003. *Challenges for Rural America in the 21st Century.* University Park: Penn State University Press.

Brundtland Report. 1987. Report of the World Commission on Environment and Development: Our Common Future. Transmitted to the General Assembly [of the United Nations] as an Annex to document A/42/427—Development and International Cooperation: Environment.

Bryant, B. ed. 1995. *Environmental Justice: Issues, Policies and Solutions.* Washington, DC: Island Press.

Bryant, B. 2003. *Environmental Advocacy: Working for Economic and Environmental Justice.* Garden City, NY: Morgan James Press.

Bryant, B., and P. Mohai. 1991. *Environmental Racism: Issues and Dilemmas.* Ann Arbor: University of Michigan Office of Minority Affairs.

Bryant, B., and P. Mohai, eds. 1992. *Race and the Incidence of Environmental Hazards: A Time for Discourse.* San Francisco: Westview Press.

Buechler, S. M. 2000. *Social Movements in Advanced Capitalism: The Political Economy and Cultural Construction of Social Activism.* Oxford: Oxford University Press.

Bullard, R. D. 1983. Solid Waste Sites and the Houston Black Community. *Sociological Inquiry*, vol. 53: 273–288.

Bullard, R. D. ed. 1993. *Confronting Environmental Racism: Voices from the Grassroots.* Cambridge, MA: South End Press.

Bullard, R. D. 1994a. Grassroots Flowering: The Environmental Justice Movement Comes of Age. *Amicus Journal* (Spring): 32–37.

Bullard, R. D. 1994b. Overcoming Racism in Environmental Decision-making. *Environment*, 36: 10–20, 39–44.

Bullard, R. D., ed. 2005. *The Quest For Environmental Justice: Human Rights and the Politics of Pollution.* San Francisco: Sierra Club Books.

Bullard, R. D. 2007a. *The Black Metropolis in the Twenty-First Century: Race, Power, and Politics of Place.* New York: Rowman & Littlefield.

Bullard, R. D., ed. 2007b. *Growing Smarter: Achieving Livable Communities, Environmental Justice, and Regional Equity.* Cambridge, MA: MIT Press.

Bullard, R. D., and G. S. Johnson. 1997. *Just Transportation: Dismantling Race and Class Barriers to Mobility.* Gabriola Island, BC: New Society Publishers.

Bullard, R. D., G. S. Johnson, and A. O. Torres. 2000. *Sprawl City: Race, Politics, & Planning in Atlanta.* Washington, DC: Island Press.

Bullard, R. D., G. S. Johnson, and A. O. Torres. 2001. The Costs and Consequences of Suburban Sprawl. Symposium, Urban Sprawl: Local and Comparative Perspectives on Managing Atlanta's Growth. *Georgia State University Law Review*, vol. 935, no. 17.

Bullard, R. D., G. S. Johnson, and A. O. Torres. 2004. *Highway Robbery: Transportation Racism and New Routes to Equity.* Cambridge, MA: South End Press.

Burke, L. 1993a. Race and Environmental Equity: A Geographic Analysis of Los Angeles. *Geographic Information Systems* 4: 44.

Burke, L. 1993b. Environmental Equity in Los Angeles. *Technical Report.* Santa Barbara: National Center for Geographic Information and Analysis, University of California. July: 93–96.

Buzbee, W. W. 1999. Urban Sprawl, Federalism, and the Problem of Institutional Complexity, *Fordham Law Review* vol. 68, no. 112: 57–136.

Byrd, B., and N. Rhee. 2004. Building Power in the New Economy: The South Bay Labor Council. *Working USA*, vol. 8, no. 2 (December).

California Department of Conservation. 1992–2002. Division of Land Resource Protection, Farmland Mapping and Monitoring Program, Land Use Conversion Tables for Riverside County for 1992–1994, 1994–1996, 1996–1998, 1998–2000, and 2000–2002.

Calthorpe, P. 1993. *The Next American Metropolis: Ecology, Community and the American Dream.* New York: Princeton Architectural Press.

Calthorpe, P., and W. Fulton. 2001. *The Regional City: Planning for the End of Sprawl.* Washington, DC: Island Press.

Camia, C. 1993. Poor, Minorities Want Voice in Environmental Choices. *Congressional Quarterly Weekly Report*, vol. 51.

Campbell, S. 1996. Green Cities, Growing Cities, Just Cities? Urban Planning and the Contradictions of Sustainable Development. *Journal of the American Planning Association*, vol. 62(3): 296–312.

Cantave, C., and R. Harrison. 2001. Residence and Region. Washington, DC: Joint Center for Political & Economic Studies. Available at http://www.jointcenter.org/DB/printer/resident.htm.

Capra, F. 1997. *The Web of Life*. New York: Anchor Books.

Cashin, S. 2004. *The Failures of Integration: How Race and Class Are Undermining the American Dream*. New York: Public Affairs.

Castells, M. 1997. *The Power of Identity*. Oxford and Malden, MA: Blackwell Publishers.

Castells, M. 2003. *The Power of Identity*. 2nd ed. Oxford and Malden, MA: Blackwell Publishers.

Cavadini, C., A. Siega-Riz, and B. M. Popkin. 2000. U.S. Adolescent Food Intake Trends from 1965 to 1996. *Archives of Disease in Childhood*, vol. 83: 18–24.

Centers for Disease Control. 2000. Death Rates from 72 Selected Causes by Year, Age Groups, Race, & Sex: United States 1979–1998. *National Center for Health Care Statistics*.

Center for Ecoliteracy. 2004. Waste Management: Rethinking School Lunch Guide. Available at http://www.ecoliteracy.org/programs/pdf/rethinking_waste.pdf.

Center for Food and Justice. 2003. CSA in the Classroom: An Evaluation of the First Year of the Program. Los Angeles: The Urban and Environmental Policy Institute, Occidental College, July.

Center for Justice, Tolerance and Community. 2006. Edging Toward Equity: Creating Shared Opportunity in America's Regions (Report from the Conversation on Regional Equity [CORE]).

Cheeseman Day, J., and E. C. Newberger. 2002. *The Big Payoff: Educational Attainment and Synthetic Estimates of Work-Life Earnings*. Washington, DC: U.S. Census Bureau, 2002.

Chen, D., R. Ewing, and R. Pendall. 2002. *Measuring Sprawl and Its Impact: The Character and Consequences of Metropolitan Expansion*. Washington, DC: Smart Growth America.

Christensen, H. 2003. Juanamaria Healthy Schools Project Final Evaluation Report. Ventura County Superintendent's Office.

Clinton, Pres. W. 1994. Federal Actions to Address Environmental Justice in Minority Populations and Low-Income Populations. U.S. Presidential Executive Order No. 12898. The White House, Washington, DC: February 11. In *Environment*, vol. 36: 4 (May).

Cloward, R. A., and L. E. Ohlin. 1960. *Delinquency and Opportunity: A Theory of Delinquent Gangs*. Glencoe, IL: Free Press.

Cohen, N. 2000. *Charter of the New Urbanism*. New York: McGraw Hill.

Cole, L. 1992. Empowerment as the Means to Environmental Protection: The Need for Environmental Poverty Law. *Ecology Law Quarterly* 19: 619.

Cole, L., and S. Bowyer. 1991. Pesticides and the Poor in California. *Race, Poverty and the Environment*, 2: 1.

Cole, L. W., and S. R. Foster. 2001. *From the Ground Up: Environmental Racism and the Rise of the Environmental Justice Movement.* New York: New York University Press.

Collins, K. S., D. L. Hughes, M. M. Doty, B. L. Ives, J. N. Edwards, and K. Tenney. 2002. Diverse Communities, Common Concerns: Assessing Health Care Quality for Minority Americans. *Commonwealth Fund*, vol. vi, no. 5. Available at http://www.commonwealthfund.org/publications/publications_show.htm?doc _id=221257.

Commission for Racial Justice. 1987. Toxic Wastes and Race in the United States: A National Report on the Racial and Socioeconomic Characteristics of Communities with Hazardous Waste Sites. New York: United Church of Christ.

Committee on Environmental Justice. 1999. Toward Environmental Justice: Research, Education and Health Policy Needs. Washington, DC: Institute of Medicine.

Contento, I., G. I. Balch, Y. L. Bronner, L. A. Lytie, S. K. Maloney, C. M. Olson, S. S. Swadener. 1995. The Effectiveness of Nutrition Education and Implications for Nutrition Education Policy, Programs and Research: A Review of Research. *Journal Nutritional Education*, vol. 27, no. 6: 284–418.

Corcoran, M. 1995. Rags to Rags: Poverty and Mobility in the United States. *Annual Review of Sociology*, vol. 21: 237–267.

Cullen, S. 2004. Little Sprouts: Homegrown Lunch Program Goes Beyond Pilot Schools Today. *Wisconsin State Journal*, Nov. 24.

Davis, M. 2006. *Planet of Slums*. New York: Verso Books.

Dedman, B. 1988. The Color of Money. *The Atlanta Journal-Constitution*, May.

Delta Farm Press. 2005. Farmers, Schools Grow Healthy Relationship. *Delta Farm Press*, May 13, 2005.

Diamond, I. H., and P. Noonan, eds. 1996. *Land Use in America*. Washington, DC: Island Press.

Dittmar, H., and D. Chen. 1995. Equity in transportation investments. *Transportation: Environmental Justice and Social Equity Conference Proceedings*. Washington, DC: Surface Transportation Policy Project.

Dittmar, H., and G. Ohland. 2004. *The New Transit Town: Best Practices in Transit-Oriented Development*. Washington, DC: Island Press.

Dittmar, H., and D. Chen, "Equity in Transportation Investments," Paper presented at the Transportation: Environmental Justice and Social Equity Conference, July 1995.

Drabenstott, M. 2003. Top Ten Ways to Reinvent Rural Regions. *The Main Street Economist*. Kansas City: Center for the Study of Rural America, Federal Reserve Bank (Nov.).

Dreier, P., J. Mollenkopf, and T. Swanstrom. 2001 (2004). *Place Matters: Metropolitics for the Twenty-first Century*, 2nd rev. ed. Kansas: University of Kansas Press.

Duany, A., E. Plater-Zyberk, and J. Speck. 2000. *Suburban Nation: The Rise of Sprawl and the Decline of the American Dream.* New York: North Point Press.

DuBois, W. E. B. 1903. Of Our Spiritual Strivings. In *The Souls of Black Folk.* Chicago: A. C. McClurg and Company.

Du Plessis, J. 2005. The Growing Problem of Forced Evictions and the Crucial Importance of Community-based, Locally Appropriate Alternatives. *Environment and Urbanization*, vol. 17, no. 1: 123–134.

Edsall, T. B., and M. D. Edsall. 1992. *Chain Reaction: The Impact of Race, Rights, and Taxes on American Politics.* New York: W. W. Norton & Company.

Edwards, A. R. 2005. *The Sustainability Revolution: Portrait of a Paradigm Shift.* Gabriola Island, BC: New Society Publishers.

Ehrenhalt, A. 1999. New Recruits in the War on Sprawl. *New York Times* (April 13) A23.

El Nasser, H., and P. Overberg. 2001. A comprehensive look at sprawl in America. *USA Today*, February 22. Available at http://www.usatoday.com/news/sprawl/main.htm

Farley, R., C. Steeh, T. Jackson, M. Krysan, and K. Reeves. 1993. Continued Racial Residential Segregation in Detroit: "Chocolate City, Vanilla Suburbs" Revisited. *Journal Housing Research*, vol. 4, no. 1. Available at http://www.knowledgeplex.org/programs/jhr/pdf/jhr_0401_farley.pdf.

Feagin, J. R. 1998. *The New Urban Paradigm: Critical Perspectives on the City.* Lanham, MD: Rowman & Littlefield.

Feenstra, G., and J. Ohmart. 2004. Yolo County Farm to School Evaluation Report for the California Farm to School Program. Los Angeles: The Urban and Environmental Policy Institute, Occidental College. October.

Ferris, D., and D. Hahn-Baker. 1995. Environmentalists and Environmental Justice Policy. In *Environmental Justice: Issues, Policies and Solutions*, ed. B. Bryant. Washington, DC: Island Press.

Finnigan, K., N. Adelman, L. Anderson, L. Cotton, M. B. Donnelly, and T. Price. 2004. Evaluation of the Public Charter School Program: Final Report. Washington, DC: United States Department of Education.

Fischlowitz-Roberts, B. 2002. Air Pollution Fatalities Now Exceed Traffic Fatalities 3 to 1. *Earth Policy Institute* (September 17). Available at http://www.earth-policy.org/Updates/Update17.htm.

Floyd, N. L. 2003. Transportation Costs. Jackson: *Clarion-Ledger* (August 10): 1C.

Foreman, C. H. Jr. 1998. *The Promise and Peril of Environmental Justice.* Washington, DC: Brookings Institution Press.

Fox, R. K., and K. Rose. 2003. Expanding Housing Opportunity in Washington, DC: The Case for Inclusionary Zoning. Oakland, CA: PolicyLink.

Frankenberg, E., C. Lee, and G. Orfield. 2003. *A Multiracial Society with Segregated Schools: Are We Losing the Dream?* Cambridge, MA: Civil Rights Project.

Franklin, J. H., and A. A. Moss, Jr. 1974. *From Slavery to Freedom: A History of Negro Americans,* 4th ed. New York: Alfred A. Knopf.

Freshwater, D., and E. Scorsone. 2002. *The Search for Effective Rural Policy: An Endless Quest or an Achievable Goal.* Agricultural Economics Staff Paper no. 436. Lexington: University of Kentucky.

Frey, B. 2002. Three Americas: The Rising Significance of Regions. *Journal of the American Planning Association,* vol. 68, no. 4: 349–355.

Frey, W. H. 2001. *Melting Pot Suburbs: A Census 2000 Study of Suburban Diversity.* Washington, DC: Brookings Institution.

Frey, W. H. 2004. The New Great Migration: Black Americans' Return to the South, 1965–2000. Washington, DC: Brookings Institution.

Frey, W. H. 2005. *Metropolitan America in the New Century: Metropolitan and Central City Demographic Shifts Since 2000.* Washington, DC: Brookings Institution.

Funders' Network for Smart Growth and Livable Communities. 2005. *Signs of Promise: Stories of Philanthropic Leadership in Advancing Regional and Neighborhood Equity.* **Coral Gables, FL: Funders' Network for Smart Growth and Livable Communities.**

Garrett, M., and B. Taylor. 1999. Reconsidering Social Equity in Public Transit. *Berkeley Planning Journal,* vol. 13. Available at http://www.ced.berkeley.edu/pubs.bpj/backissues13.html.

Gelobter, M., M. Dorsey, L. Fields, T. Goldtooth, A. Mendiratta, R. Moore, R. Morello-Frosch, P. M. Shepard, G. Torres. 2005. *The Soul of Environmentalism: Rediscovering Transformational Politics in the 21ˢᵗ Century.* Oakland, CA: Redefining Progress.

Gerrard, M., ed. 1999. *The Law of Environmental Justice.* Chicago: American Bar Association.

Gibbs, R., L. Kusmin, and J. Cromartie. 2005. Low-Skill Employment and the Changing Economy of Rural America. USDA Economic Research Service Economic Research Report No. 10.

Gillette, H. Jr. 2005. *Camden After the Fall: Decline and Renewal in a Post-Industrial City.* **Philadelphia: University of Pennsylvania Press.**

Glasmeier, A., and P. Salant. 2006. Low Skilled Workers in Rural America Face Permanent Job Loss. Durham, NH: The Carsey Institute, Policy Brief No. 2. University of New Hampshire.

Glave, D. D., and M. Stoll, eds. 2006. *To Love the Wind and the Rain: African Americans and Environmental History.* **Pittsburgh: University of Pittsburgh Press.**

Goldberg, D. 2004. Making the Case for Mixed Income and Mixed Use Communities. Atlanta: Atlanta Neighborhood Development Partnership. Mixed Income Communities Initiative (July).

Goode, J., J. Maskovsky, and I. Susser. 2002. *The New Poverty Studies: The Ethnography of Power, Politics, and Impoverished People in the United States.* New York: New York University Press.

Gore, A. 2006. *An Inconvenient Truth: The Planetary Emergency of Global Warming and What We Can Do About It.* Emmaus, PA: Rodale Press.

Gottdiener, M. 1985. *The Social Production of Urban Space.* Austin: University of Texas Press.

Gottdiener, M., and L. Budd. 2005. *Key Concepts in Urban Studies.* London: Sage.

Gottdiener, M., and R. Hutchison. 1999. *The New Urban Sociology.* Boston: McGraw Hill.

Gottlieb, R., M. Vallianatos, R. Freer, and P. Dreier. 2006. *The Next Los Angeles: The Struggle for a Livable City.* Berkeley: University of California Press.

Greenberg, M. 1993. Proving Environmental Inequity in Siting Locally Undesirable Land Uses. *Risk-Issues Health and Safety* 4: 235.

Greenberg, M., and R. Anderson. 1984. Hazardous Waste Sites: The Credibility Gap. New Brunswick, N.J.: Rutgers University Center for Urban Policy Research.

Greenberg, S. B. 1995. *Middle Class Dreams: The Politics and Power of the New American Mayoralty.* New York: Random House Books.

Greenblatt, A. 2005. The Washington Offensive. Governing (January).

Greenstein, R., and W. Wiewel, eds. 2000. *Urban-Suburban Interdependencies.* Cambridge, MA: Lincoln Institute of Land Policy.

Gross, J., G. LeRoy, and M. Janis-Aparicio. 2005. Community Benefits Agreements: Making Development Projects Accountable. Washington, DC: Good Jobs First, and the California Partnership for Working Families. Available at http://www.goodjobsfirst.org/publications/index.cfm.

Gyorko, J., and A. Summers. 1997. A New Strategy for Helping Cities Pay for the Poor. Washington, DC: Brookings Institution.

Hall, P. 1988. *Cities of Tomorrow: An Intellectual History of Urban Planning and Design in the Twentieth Century.* Oxford: Blackwell.

Hallsmith, G. 2003. *The Key to Sustainable Cities: Meeting Human Needs, Transforming Community Systems.* Gabriola Island, BC: New Society Publishers.

Hardoy, J. E., D. Mitlin, and D. Satterthwaite. 2001. *Environmental Problems in an Urbanizing World: Finding Solutions for Cities in Africa, Asia and Latin America.* London: Earthscan Publications.

Hare, N. 1970. Black Ecology. *The Black Scholar*, vol. 1 (6): 2–8.

Harvey, D. 1983. *The Urban Experience.* Baltimore: John Hopkins University Press.

Hatch, J., N. Moss, A. Saran, L. Presley-Cantrell, and C. Mallory. 1993. Community Research: Partnership in Black Communities. *American Journal of Preventive Medicine*, vol. 9 (Suppl.): 27–31.

Herbert, A., and M. P. Pavel. 1993. *Random Kindness and Senseless Acts of Beauty*. San Francisco: Volcano Press.

Hershberg, T. 2001. Regional Imperatives for Global Competitiveness. *Planning for a New Century: The Regional Agenda*. Washington, DC: Island Press.

Hirschman, A. O. 1972. *Exit, Loyalty, and Voice: Response to Decline in Firms, Organizations, and States*. Cambridge, MA: Harvard University Press.

Hochshild, A. 2005. *Bury the Chains*. Boston: Houghton Mifflin.

Hunter, L. M., J. D. Boardman, and J. M. Saint Onge. 2005. The Association Between Natural Amenities, Rural Population Growth, and Long-Term Residents' Economic Well-Being. *Rural Sociology*, vol. 70, no. 4: 452–469.

Hunter, M. A. 2006. *State-by-State Status of School Finance Litigations*. New York: National Access Network.

Huq, S., R. S. Kovats, H. Reid, and D. Satterthwaite. 2007. Reducing Risks to Cities from Disasters and Climate Change, *Environment And Urbanization Journal*, vol. 19, no. 1 (April): 3–15.

Isaacs, J. 2007a. Economic Mobility of Black and White Families. Economic Mobility Project: An Initiative of The Pew Charitable Trusts. Washington, DC: The Pew Charitable Trusts.

Isaacs, J. 2007b. Economic Mobility of Families Across Generations. Economic Mobility Project: An Initiative of The Pew Charitable Trusts. Washington, DC: The Pew Charitable Trusts.

Israel, B. A., B. Checkoway, A. Schulz, and M. Zimmerman. 1992. Health Education and Community Empowerment: Conceptualizing and Measuring Perceptions of Individual, Organizational, and Community Control. *Health Education Quarterly*, vol. 21, no. 2: 149–170.

Israel, B. A., S. J. Schurman, and J. S. House. 1989. Action Research on Occupational Stress: Involving Workers as Researchers. *International Journal of Health Services*, vol. 19 (no. 1): 135–155.

Israel, B. A., S. J. Schurman, and M. K. Hugentobler. 1992. Conducting Action Research: Relationships Between Organization Members and Researchers. *Journal of Applied Behavioral Science*, vol. 28: 74–101.

Jackson, K. 1985. *Crabgrass Frontier: The Suburbanization of the United States*. New York: Oxford University Press.

Jackson, R. J., and C. Kochtitzky. 2001. Creating a Healthy Environment: The Impact of the Built Environment on Public Health. Available at http://www.sprawlwatch.org/health.pdf.

Jacobs, J. 1961. *The Death And Life of Great American Cities*. New York: Random House.

Jacobs, J. 2004. *Dark Age Ahead*. New York: Random House.

Jakowitsch, N., and M. Ernst. 2003. Transportation Costs and the American Dream: Why a Lack of Transportation Choices Strains the Family Budget and Hinders Home Ownership. Surface Transportation Policy Project (July 22). Available at http://www.transact.org/library/americandreamdecoder.asp.

Jargowsky, P. A. 2003. Stunning Progress, Hidden Problems: The Dramatic Decline of Concentrated Poverty in the 1990s. Washington, DC: Brookings Institution (May).

Jargowsky, P. A., and R. Steiner. 1997. *Poverty and Place, Ghettos, Barrios, and the American City*. New York: Russell Sage Foundation.

Joassart-Marcelli, P., J. Musso, and J. Wolch. 2005. Fiscal Consequences of Concentrated Poverty in a Metropolitan Region. *Annals of the Association of American Geographers*, vol. 95, no. 2: 336–356.

Johansson, F. 2004. *The Medici Effect: Breakthrough Insights at the Intersection of Ideas, Concepts, and Cultures*. Cambridge, MA: Harvard Business School Press.

Johnson, K. M. 2006. Demographic Trends in Rural and Small Town America. Durham, NH: The Carsey Institute.

Johnson, K. M., and J. B. Cromartie. 2005. The Rural Rebound and Its Aftermath: Changing Demographic Dynamics and Regional Contrasts. In *Population Change and Rural Society*, eds. W. A. Kandel and D. L. Brown. Dordrecht, Netherlands: Springer Publishing.

Judis, J. B., and R. Texeira. 2002. *The Emerging Democratic Majority*. New York: Scribner Press.

Kahlenberg, R. 2001. *All Together Now: Creating Middle Class Schools through Public School Choice*. Washington, DC: Brookings Institution Press.

Kaika, M. 2005. *City of Flows: Modernity, Nature, and the City*. Oxford: Routledge.

Katz, B., ed. 2000. *Reflections on Regionalism*. Washington, DC: Brookings Institution Press.

Katz, B., and A. Altman. 2005. An Urban Renaissance in a Suburban Nation. Ford Foundation Report (Spring/Summer).

Katz, B., R. Lang, and A. Berube. 2006. *Redefining Urban and Suburban America: Evidence from Census 2000* (3 volumes). Washington, DC: Brookings Institution Press.

Katz, B., and R. Puentes, eds. 2005. *Taking the High Road: A Metropolitan Agenda for Transportation Reform*. Washington, DC: Brookings Institution Press.

Katznelson, I. 2005. *When Affirmative Action Was White: An Untold History of Racial Inequality in Twentieth-Century America*. New York: W. W. Norton.

Kochhar, R. 2004. The Wealth of Hispanic Households: 1996 to 2002. Washington, DC: Pew Hispanic Center.

Korten, D. 2006. *The Great Turning from Empire to Earth Community*. San Francisco: Berrett-Koehler Publishers.

Kuhn, T. S. 1996. *The Structure of Scientific Revolutions*. Chicago: University of Chicago Press.

Lane, R. 2006. Architectural Practice of Sustainability: A Question of Worldview. PhD diss., University of California, Berkeley.

Lang, R. E. 2000. *Office Sprawl: The Evolving Geography of Business*, vol. 1. Washington, DC: The Brookings Institution.

Lavelle, M., and M. Coyle. 1992. Unequal Protection: The Racial Divide in Environmental Law. *National Law Journal*, vol. 15: S1–S12.

Leccese, M., and K. McCormick, eds. 2000. *The Charter for the New Urbanism*. New York: McGraw Hill.

Lee, C. 1992. Environmental Justice: Building a Unified Vision of Health and the Environment. *Environmental Health Perspectives* 2: 141–144.

Lee, C. 1993. Beyond Toxic Wastes and Race. In *Confronting Environmental Racism: Voices from the Grassroots*, ed. R. D. Bullard. Boston: South End Press.

Lerner, M. 1986. *Surplus Powerlessness: The Psychodynamics of Everyday Life and the Psychology of Individual and Social Transformation*. Reprint New York: Prometheus Books, 1998.

Lewan, T., and D. Barclay. 2001a. *Torn from the Land: Black Americans' Farmland Taken Through Cheating, Intimidation, Even Murder*. Associated Press. Available at http://www.hartford-hwp.com/archives/45a/393.html.

Lewan, T., and D. Barclay. 2001b. *Torn from the Land: Developers and Lawyers Use a Legal Maneuver to Strip Black Families of Land*. Associated Press. Available at http://www.theauthenticvoice.org/Torn_From_The_LandIII.html.

Lewis, J. 1997. Foreword. In *Just Transportation: Dismantling Race and Class Barriers to Mobility*, eds. R. D. Bullard and G. S. Johnson. Gabriola Island, BC: New Society Publishers.

Liberty Hill Foundation. 2005. Building a Regional Voice for Environmental Justice. Santa Monica, CA: Liberty Hill. Available at http://www.libertyhill.org/reports.

Lin, B.-H., J. Guthrie, and E. Frazao. 1999. Quality of Children's Diets At and Away from Home: 1994–1996. *Food Review*, Jan.–April: 1–10.

Lynch, K. 1981. *Good City Form*. Cambridge, MA: MIT Press.

Macy, J. 1998. *Coming Back to Life: Practices to Reconnect Our Lives, Our World*. Gabriola Island, BC: New Society Publishers.

Macy, J. 2007. *World As Lover, World As Self: Courage for Global Justice and Ecological Renewal*. Berkeley: Parallax Press.

Marsh, D. S. 2003a. Leadership for Policy Change. Oakland, CA: PolicyLink.

Marsh, D. S. 2003b. Promise and Challenge: Achieving Regional Equity in Greater Boston. Oakland, CA: PolicyLink.

Mascarenhas, M., and R. Gottlieb. 2000. The Farmers' Market Salad Bar: Assessing the First Three Years of the Santa Monica–Malibu Unified School District Program. Los Angeles: The Urban and Environmental Policy Institute, Occidental College. Available at http://uepiapps.oxy.edu/cfj.php?x=publications.

Massey, D. S., and N. Denton. 1998. *American Apartheid, Segregation and the Making of the Underclass*. Cambridge, MA: Harvard University Press.

McAdam, D. 1999. *Political Process and the Development of Black Insurgency, 1930–1970*. 2d ed. Chicago: University of Chicago Press.

McConnell, R., K. Berhane, F. Gilliland, S. J. London, T. Islam, W. J. Gauderman, E. Avol, H. G. Margolis, and J. M. Peters. 2002. Asthma in Exercising Children Exposed to Ozone: A Cohort Study. *The Lancet*, vol. 359.

McDonough, W., and M. Braungart. 2002. *Cradle to Cradle*. New York: North Point Press: 68–91.

McGurty, E. 2007. *Transforming Environmentalism: Warren County, PCBs, And the Origins of Environmental Justice*. Piscataway, NJ: Rutgers University Press.

Menegat, R. 2002. Participatory Democracy and Sustainable Development: Integrated Urban Environmental Management in Porto Alegre, Brazil. *Environment and Urbanization*, vol. 14, no. 2: 181–206.

Mills, R. J., and S. Bhandari. 2003. Health Insurance Coverage in the United States: 2002. U.S. Census Bureau. Available at http://www.census.gov/prod/2003pubs/p60-223.pdf.

Mittelmark, M. B., M. K. Hunt, G. W. Heath, and T. L. Schmid. 1993. Realistic Outcomes: Lessons from Community-based Research and Demonstration Programs for the Prevention of Cardiovascular Diseases. *Journal of Public Health Policy*, vol. 14: 437–462.

Mohai, P., and B. Bryant. 1991. Race, Class and Environmental Quality in the Detroit Area. In *Environmental Racism: Issues and Dilemmas*, eds. B. Bryant and P. Mohai. Ann Arbor: University of Michigan Office of Minority Affairs.

Mohai, P., and B. Bryant. 1992. Environmental Injustice: Weighing Race and Class as Factors in the Distribution of Environmental Hazards. *University of Colorado Law Review*, vol. 63: 921–923.

Morello-Frosch, R., M. Pastor, J. Sadd, C. Porras, and M. Prichard. 2003. Citizens, Science, and Data Judo: Leveraging Secondary Data Analysis to Build a Community-Academic Collaborative for Environmental Justice in Southern California. In *Methods for Conducting Community-based Participatory Research for Health*, ed. A. J. Schulz, B. A. Israel, and P. Lantz. San Francisco: Jossey-Bass. 371–392.

Morris, D. E. 2005. It's a Sprawl World After All: The Human Costs of Unplanned Growth, and Visions of a Better Future. Gabriola Island, BC: New Society Publishers.

Moss Kanter, R. 2000. *Metropolitan Business Coalitions as a Force for Regionalism*. Cambridge, MA: Goodmeasure Press.

Moynihan, D. P., ed. 1969. *On Understanding Poverty*. New York: Basic Books.

Muro, M., and R. Puentes. 2004. *Investing in a Better Future: A Review of the Fiscal and Competitive Advantages of Smarter Growth Development Patterns*. Washington, DC: Brookings Institution.

Mutz, K., G. Bryner, and D. Kenney, eds. 2002. *Justice and Natural Resources*. Washington, DC: Island Press.

National Farm to School Website. http://www.farmtoschool.org.

National Lawyers Guild and Maurice and Jane Sugar Law Center for Economic and Social Justice. 2003. *Lucero v. Detroit Public Schools*.

O'Connor, A. 2001. *Poverty Knowledge: Social Science, Social Policy and the Poor in Twentieth Century U.S. History*. Princeton: Princeton University Press.

Ogden, C. L., K. M. Flegal, M. D. Carroll, and C. L. Johnson. 2002. Prevalence and Trends in Overweight Among U.S. Children and Adolescents, 1999–2000. *JAMA*, vol. 288: 14, 1728–1732.

Orfield, M. 1997. *Metropolitics: A Regional Agenda for Community and Stability*. Washington, DC: The Brookings Institution Press.

Orfield, M. 2002. *American Metropolitics: The New Suburban Reality*. Washington, DC: The Brookings Institution Press.

Orfield, M. 2003a. Comment on Scott A. Bollen's "In Through the Back Door, Social Equity and Regional Governance." *Housing Policy Debate* 13, no. 4: 659–668.

Orfield, M. 2003b. Minority Suburbanization and Racial Change. Minneapolis: Institute on Race and Poverty.

Orfield, M. 2005. Social Patterns: Slicing through the jargon: one map speaks volumes. *Yes! a journal of positive futures*, issue 34 (summer): 27.

Orfield, M. 2006a. Choice, Equal Protection, and Metropolitan Integration: The Hope of the Minneapolis Desegregation Settlement. *Law and Inequality*, vol. 24, no. 2: 269–352.

Orfield, M. 2006b. Land Use and Housing Policies to Reduce Concentrated Poverty and Racial Segregation. *Fordham Urban Law Journal*, vol. 33, no. 3: 877–936.

Orfield, M., and T. Luce. 2005. Minority Suburbanization and Racial Change: Stable Integration, Neighborhood Transition, and the Need for Regional Approaches. Minneapolis: Institute on Race and Poverty.

Orfield, M., and L. Thomas. October 2003. An Activist's Guide to Metropolitics. Minneapolis: Social Science and Research Network.

Pack, J. R. 1998. Poverty and Urban Public Expenditures. *Urban Studies*, vol. 35, no. 11: 1995–2019.

Pastor, M. Jr., C. Benner, and R. Rosner. 2006. Edging Toward Equity. Santa Cruz, CA: Center for Community, Justice, and Tolerance.

Pastor, M. Jr., C. Benner, and M. Matsuoka. Forthcoming in early 2009. *This Could Be the Start of Something Big: How Social Movements for Regional*

Equity Are Transforming Metropolitan America. Ithaca, NY: Cornell University Press.

Pastor, M. Jr., P. Dreier, E. Grigsby, and M. López-Garza. 2000. *Regions That Work: How Cities and Suburbs Can Grow Together*. Minneapolis: University of Minnesota Press.

Pastor, M. Jr., J. Sadd, and J. Hipp 2001. Which Came First? Toxic Facilities, Minority Move-in, and Environmental Justice. *Journal of Urban Affairs*, vol. 23, no. 1: 1–21.

Patel, S. 2004. Tools and Methods for Empowerment Developed by Slum Dwellers Federations in India. *Participatory Learning and Action*, issue 50, London: IIED.

Pavel, M. P. 2004. Building Bridges: Case Studies in Metropolitan Regional Equity. Detroit: Funders' Network for Smart Growth and Livable Communities.

Pavel, M. P. 2008. Breaking through to Regional Equity. *Race, Poverty, and the Environment*, vol. 15, no. 2: 29–33.

Pavel, M. P., and K. Kirsch, eds. 2005. Journey to South Africa: Metropolitan Community Leaders Reflect on the World Summit on Sustainable Development. NY: Ford Foundation.

Pavel, M. P., and R. Butler. 2000. *Voices from the Community: Smart Growth and Social Equity* (video documentary). Oakland: Earth House Media for Association of Bay Area Governments.

Perfecto, I., and B. Velasquez. 1992. Farm Workers: Among the Least Protected. *EPA Journal* (March/April).

Pierce, G. F. 1984. *Activism that Makes Sense*. Chicago: ACTA Publications.

Pitts, S. C. 2005. Organize … to Improve the Quality of Jobs in the Black Community. Institute of Industrial Relations, University of California, Berkeley, Center for Labor Research.

Plastrik, P., and J. Cleveland. 2006. Strategic Review of Sustainable Metropolitan Communities Initiative, Final Report to the Ford Foundation, Integral Assets Consulting.

Plessy v. Ferguson. 1896. 163 U.S. 537.

PolicyLink. 2005. Shared Prosperity, Stronger Regions: An Agenda for Rebuilding America's Older Core Cities. Available at http://www.policylink.org/Research/ OlderCoreCities/ or http://www.policylink.org/publications.html.

Portney, K. E. 2003. *Taking Sustainable Cities Seriously: Economic Development, the Environment, and Quality of Life in American Cities*. Cambridge, MA: MIT Press.

powell, j. a. 2000. Addressing Regional Dilemmas for Minority Communities. In *Reflections on Regionalism*, ed. B. Katz. Washington, DC: Brookings Institution Press.

powell, j. a. 2002. Sprawl, Fragmentation, and the Persistence of Racial Inequality: Limiting Civil Rights by Fragmenting Space. In *Urban Sprawl: Causes, Con-*

sequences, and Policy Responses, ed. Greg Squires. Washington, DC: Urban Institute Press.

powell, j. a. 2003. Thinking Big: The Segregated North Forty Years After the Civil Rights Movement, Jim Crow Is Alive and Well and Living in Our Suburbs. *Boston Globe* (March 23) H12.

powell, j. a. 2005. Regional Equity: Camden's Civil Rights Cause for the Twenty-first Century. The Inaugural Richard Goodwin Lecture in Honor of Ethel Lawrence. Camden, NJ: Rutgers University.

powell, j. a., M. Spencer, and M. P. Pavel. 2009. An Outpouring of Spirit: Faith-based Environmental Justice Action in the Gulf Coast. *Race, Place, and the Environment: In the Wake of the Storm.* R. D. Bullard and B. Wright, eds. To be published.

Prakash, S. 2004. Power, Privilege and Participation: Meeting the Challenge of Equal Research Alliances. Race, Poverty and the Environment. Urban Habitat and WE ACT (Winter).

Public Health Institute. 2001. A Special Report on Policy Implications from the 1999 California Children's Healthy Eating and Exercise Practices Survey (Cal-CHEEPS). Sacramento, CA: Public Health Institute. May.

Pucher, J., and J. L. Renne. 2003. Socioeconomics of Urban Travel: Evidence from the 2001 NHTS. *Transportation Quarterly*, vol. 57, no. 3 (Summer): 49–77. Available at nhts.ornl.gov/2001/articles/socioeconomicsOfUrbanTravel.pdf.

Puentes, R., and L. Bailey. 2003. Improving Metropolitan Decision Making in Transportation: Greater Funding and Devolution for Greater Accountability, vol. 4, no. 10. Washington, DC: Brookings Institution Press. Available at http://www.brook.edu/es/urban/publications/200310_Puentes.pdf.

Puentes, R., and R. Prince. 2003. Fueling Transportation Finance: A Primer on the Gas Tax 1. Washington, DC: The Brookings Institution. Available at http://www.brook.edu/es/urban/publications/gastax.pdf.

Puentes, R., and D. Warren. 2006 *One Fifth of America: A Comprehensive Guide to America's First Suburbs.* Washington, DC: Brookings Institution.

Redmond, J. 2002. President of the California Alliance with Family Farmers, Presentation at Occidental College, Los Angeles. March 27.

Rees, W. E., M. Wackernagel, and P. Testemale. 1995. *Our Ecological Footprint: Reducing Human Impact on the Earth.* Gabriola Island, BC: New Society Publishers.

Rifkin, J. 2004. *The European Dream.* New York: Tarcher.

Robinson-Avila, K. 2005. Sysco Breathes New Life into Local Agriculture. Albuquerque. April 26. Available at http://spectre.nmsu.edu/media/news2.lasso?i=745.

Rumberger, R. W., and G. J. Palardy. 2005. Does Resegregation Matter: The Impact of Social Composition on Academic Achievement in Southern High Schools. *School Resegregation: Must the South Turn Back?* eds. J. C. Boger and G. Orfield. Chapel Hill: University of North Carolina Press.

Rusk, D. 1995. *Cities Without Suburbs.* 2nd ed. Washington, DC: Woodrow Wilson Center Press.

Rusk, David. 1999. *Inside Game/Outside Game: Winning Strategies for Saving Urban America.* Washington, DC: Century Foundation/Brookings Institute Press.

Rusk, D. 2003. *Cities Without Suburbs: A Census 2000 Update.* Washington, DC: Woodrow Wilson Center Press.

Sanchez, T. W. 2006. *An Inherent Bias? Geographic and Racial-Ethnic Patterns of Metropolitan Planning Organization Boards.* Transportation Reform Series. Washington, DC: Brookings Institution.

Satterthwaite, D. 2005. Human Settlements Discussion Paper (Urban01)—The Scale of Urban Change Worldwide 1950–2000 and Its Underpinnings. London: International Institute for Environment and Development.

Savitch, H. V., and R. K. Vogel. 2004. Suburbs Without a City: Power and City County Consolidation. *Urban Affairs Review,* vol. 39, no. 6.

Schein, R. H. 2006. *Landscape and Race in the United States.* New York: Routledge Press.

School Nutrition Association. 2005. Nutrition Professionals Brief Congress on Farm to School. April 12. Available at http://www.schoolnutrition.org/Blog .aspx?id=6760&blogid=622.

Schrag, P. 1999. High-Tech Haves and the Have-Nots Could Be the Cause of a Meltdown. Reprinted in the *San Jose Mercury News* (June 4): 6B.

Schwartz, P. E. 2005. PRODUCE: Growing Green Garlic, the Natural Way, Is the Path to Success for Doug Powell. *Riverside Press Enterprise,* March 28.

Scott, A., and M. Storper. 2003. Regions, Globalization, Development. *Regional Studies,* vol. 37, nos. 6–7: 579–593.

Self, R. O. 2005. *American Babylon, Race and the Struggle for Postwar Oakland.* Princeton, NJ: Princeton University Press.

Sendich, E. 2006. *Planning and Urban Design Standards* / American Planning Association. Hoboken, NJ: John Wiley & Sons, Inc.

Sensenbrenner, L. 2003. Kids Eat Up Fresh Produce: Locally Grown Food Served At School. Madison: *The Capital Times,* October 18.

Slusser, W. M., W. G. Cumberland, B. L. Browdy, L. Lange, and C. Neumann. 2007. A School Salad Bar Increases Frequency of Fruit and Vegetable Consumption Among children Living in Low-Income Households. *Public Health Nutrition,* vol. 10, issue 12 (December): 1490–1496 Published online by Cambridge University Press (July 5). Available for a fee at http://journals.cambridge.org/ action/login.

Sohmer, R. 2005. Mind the Gap: Reducing Disparities to Improve Regional Competitiveness. Washington, DC: Brookings Institution.

Speth, G. 2005. *Red Sky at Morning: America and the Crisis of the Global Environment.* New Haven, CT: Yale University Press.

Stauber, K. N. 2001. Why Invest in Rural America—and How? A Critical Public Policy Question for the 21st Century. *Economic Review*, Second Quarter. Kansas City: Federal Reserve Bank of Kansas City.

Stolz, R. 2000. Race, Poverty & Transportation. *Poverty & Race* (March/April). Available at http://www.prrac.org/full_text.php?text_id=91&item_id=1811 &newsletter_id=49&header=Poverty/Welfare.

Storper, M. 1997. *The Regional World: Territorial Development in a Global Economy.* New York: Guilford Press.

Suarez-Balcazar, Y., G. W. Harper, and R. Lewis. 2005. An Interactive and Contextual Model of Community—University Collaborations for Research and Action. *Health Education and Behavior*, vol. 32, no. 1.

Sugrue, T. J. 1996. *The Origins of the Urban Crisis: Race and Inequality in Postwar Detroit.* Princeton, NJ: Princeton University Press.

Susser, I. 2002. Preface. In *The New Poverty Studies: The Ethnography of Power, Politics, and Impoverished People in the United States.* G. Judith, J. Maskovsky, and I. Susser, eds. New York: New York University Press.

Tarrow, S. 1998. *Power in Movement: Social Movements and Contentious Politics.* 2nd ed. Cambridge: Cambridge University Press.

Tarrow, S., D. McAdam, C. Tilly. 2001. Cambridge: Cambridge University Press.

Taylor, R. 2005. Director, Food Services at Riverside Unified School District USDA. Personal Communication with Gottlieb, R., April.

Thomas, G. T. 2002. Racial Income Gap Is More Like a Chasm. Buffalo, New York: *Business First* (December 16). Available at http://www.bizjournals.com/ buffalo/stories/2002/12/16/story2.html.

Tolnay, S. E., E. M. Beck, and F. Brundage. 1995. *A Festival of Violence: An Analysis of Southern Lynchings, 1882–1930*, Urbana/Chicago: University of Illinois Press.

Trotter, J. W., E. Lewis, and T. W. Hunter, eds. 2004. *The African American Urban Experience: From the Colonial Era to the Present.* New York: Palgrave Macmillan Publishing Company.

Trout Unlimited, Naugatuck River Chapter. 2000. Press Release.

Underhill, W. 2007. This Way Forward. *Newsweek International*, April 16.

Underwood, J. D. 2001. Less Oil, More World Security. *Earth Times* (Nov. 9).

Urban Environment Conference, Inc. 1985. Taking Back Our Health: An Institute on Surviving the Toxic Threat to Minority Communities. Washington, DC: Urban Envrionment, Inc.

U.S. Conference of Mayors. 2003. City Livability Award. Available at http:// www.usmayors.org/71stAnnualMeeting/citylivability03.pdf.

U.S. Department of Commerce Minority Business Development Agency. 2000. Minority Purchasing Power: 2000 to 2045 (September).

U.S. Department of Transportation. 2004. Journey to Work Profiles for Large Metropolitan Area. Federal Highway Administration. Available at http://www.fhwa.dot.gov/ctpp/jtw/jtw8.htm#det.

USDA. 1998a. Commission on Small Farms. *A Time to Act.*

USDA. 1998b. Economic Research Service, Small Farms in the U.S. Agricultural Outlook, May 1998.

USDA. 2000. Small Farm/School Meals Initiative. Local Food Connections—From Farms to Schools. Midwest Regional Workshop, Nov. 13.

USDA. 2000. Agricultural Management Resource Survey.

USDA. 2002. National Agricultural Statistical Service, Census of Agriculture.

Voith, R. 1998. Do Suburbs Need Cities? *Journal of Regional Science,* vol. 38, no. 3 (August): 445–464.

Von Hoffman, A., E. Belsky, J. DeNormandie, and R. Bratt. 2004. America's Working Communities and the Impact of Multifamily Housing. Cambridge, MA: Neighborhood Reinvestment Corporation and the Joint Center for Housing Studies of Harvard University. 1–138. Available at http://www.jchs.harvard.edu/publications/communitydevelopment/w04-5.pdf.

Wackernagel, M. 1996. *Our Ecological Footprint: Reducing Human Impact on the Earth.* Gabriola Island, BC: New Society Publishers.

Washington, W. N. 2004. Collaborative/Participatory Research. *Journal of Health Care for the Poor and Underserved,* vol. 15, no. 1: 18–29.

Wells, B. 2002. *Smart Growth at the Frontier: Strategies and Resources for Rural Communities.* Washington, DC: Northeast-Midwest Institute.

Wernette, D. R., and L. A. Nieves. 1992. Breathing Polluted Air: Minorities Are Disproportionately Exposed. *EPA Journal,* vol. 18, no. 1 (March): 16–17.

Wheeler, S. M., and T. Beatley. 2004. *The Sustainable Urban Development Reader.* Oxford: Routledge.

Wilson, W. J. 2007. A New Agenda for America's Ghetto Poor. In *Ending Poverty in America: How to Restore the American Dream,* ed. J. Edwards, M. Crain, and A. Kalleberg, 88–98. New York: New Press.

Wolch, J., M. Pastor, Jr., and P. Dreier, eds. 2004. *Up Against the Sprawl: Public Policy and the Making of Southern California.* Minneapolis: University of Minnesota Press.

Contributor Biographies

Carl Anthony is Ford Foundation Senior Fellow and Visiting Scholar at the Department of Geography at University of California, Berkeley. He is founder of the Earth House Leadership Center in Oakland, California. Prior to his present role he was acting director of the Ford Foundation's Community and Resource Development Unit, where he directed the foundation's Sustainable Metropolitan Communities Initiative and the Regional Equity Demonstration Initiative. He founded and, for twelve years, was executive director of the Urban Habitat program in Oakland, California, promoting multicultural urban environmental leadership for sustainable, socially just communities in the San Francisco Bay Area.

Angela Glover Blackwell is president of PolicyLink, a national nonprofit research, communications, capacity-building, and advocacy organization working to advance a new generation of policies to achieve economic and social equity. Prior to founding PolicyLink, Blackwell was senior vice president for the Rockefeller Foundation and directed the foundation's domestic and cultural divisions.

Robert D. Bullard, Ph.D., is Ware Professor of Sociology and Director of the Environmental Justice Resource Center at Clark Atlanta University. He is the author of numerous articles, monographs, and scholarly papers that address environmental justice and public participation concerns. His book *Dumping in Dixie: Race, Class and Environmental Quality* (Westview Press, 1990, 1994, 2000) has become a standard text in the environmental justice field.

Sheryll Cashin, professor of law at Georgetown University, writes about race relations, government, and inequality in America. Cashin has published widely in academic journals, is a frequent radio and television commentator, and is the author of *The Failures of Integration: How Race and Class Are Undermining the American Dream* (Public Affairs, 2004). Previously she worked in the Clinton White House as an advisor on urban and economic policy, particularly community development in inner-city neighborhoods.

Kizzy Charles-Guzmán is environmental policy coordinator at West Harlem Environmental Action, Inc. (WE ACT), New York's first environmental justice organization created to improve environmental health and quality of life in communities of color. At WE ACT, she assisted in campaign development and implementation, among other duties. Raised in Venezuela, Charles-Guzmán earned a

B.A. in geology, with concentrations in environmental science and cross-cultural studies from Carleton College.

Don Chen is program officer at the Ford Foundation. Prior to joining Ford, he was executive director of Smart Growth America where he led its coalition-building, policy development, communications, and research efforts. He is an internationally recognized expert on land use, transportation, and environmental policy. Chen has published a number of articles and reports, including "Measuring Sprawl and Its Impact" (with Reid Ewing and Rolf Pendall), an analysis of how urban form affects quality of life in the United States.

Celine D'Cruz began battling for the rights of pavement and slum dwellers when she was still a teenager. Organizing poor women in her hometown of Mumbai, India, she helped them bargain collectively to bring education, health, and other vital services to their families. Since 1998, D'Cruz has been the coordinator of Slum/Shack Dwellers International. She travels extensively in Asia and Africa, helping slum dweller organizations negotiate with governments for land security, housing finance, and basic amenities.

Amy B. Dean is cofounder of Building Partnerships USA (BPUSA), a peer-based resource organization for transferring best practices and leadership networking and technical assistance to regionally based organizations. From 1993 to 2003, as the leader of Silicon Valley's labor movement, Dean was the youngest person and first woman to lead a major metropolitan labor federation. Recognized by the *New York Times* as one of the most "innovative figures in Silicon Valley," she earned a "national reputation for being an innovator and entrepreneur in the social justice movement." Dean has published numerous articles on labor and politics and has contributed chapters to several volumes on the revitalization of the American labor movement.

Hattie Dorsey was most recently president and chief executive officer of Atlanta Neighborhood Development Partnership, Inc., an organization promoting sustainable mixed-income communities and ongoing revitalization. Dorsey is a leading advocate for housing issues. She has demonstrated how affordable housing, once thought a "problem" only for the poorest of the poor, affects more of the economic population strata, and she has made clear the connection between housing and quality-of-life issues such as air quality and transportation, and the availability of clean water. Her work has helped to shift perceptions about the importance of revitalized neighborhoods and the critical need for an affordable, mixed-income housing approach that includes all segments of the population—such as the police, teachers, and secretaries.

Cynthia M. Duncan is director of the Carsey Institute at the University of New Hampshire and former director of the Community and Resource Development program at the Ford Foundation. The Carsey Institute supports interdisciplinary research on families and communities, including community, environment, and development issues, especially in small cities and rural communities. Duncan is the author of numerous articles, book chapters, and books on poverty and changing rural communities, including *Worlds Apart: Why Poverty Persists in Rural America* (Yale University Press, 1999).

Juliet Ellis is executive director of Urban Habitat, a nonprofit organization in Oakland promoting multicultural urban environmental leadership for sustainable, socially just communities in the San Francisco Bay Area from a regional perspective. Ellis also serves on the boards of the Bay Area Alliance for Sustainable Communities, the Bay Area Transportation and Land Use Coalition, and the Ella Baker Center for Human Rights.

Danny Feingold, communications director of Los Angeles Alliance for a New Economy (LAANE) uses media, Web site development, video production, and other strategic communications tools to advance the mission of LAANE. Prior to his tenure with LAANE, he worked as a journalist and book editor.

Deeohn Ferris is president of Global Environmental Resources Inc. (GERI), a professional services firm that provides management consulting, technical support services, and training to clients on environmental, natural resources, and public health programs and projects. Concentrating on community involvement, environmental justice, stakeholder engagement, and public participation, GERI assists clients in improving environmental performance, achieving smart growth and sustainability.

Kenneth Galdston has worked with InterValley Project, Inc., as project director/organizer since 1997. IVP uses citizen action and local ownership strategies to save, strengthen, and create jobs, affordable housing, and critical public services for low-income people living in New England regions that have experienced industrial and community decline.

Greg Galluzzo has been a community organizer since 1971 and is cofounder and executive director since 1986 of Chicago-based Gamaliel Foundation, a network of nearly sixty community organizations in fourteen states. Gamaliel works with congregation-centered community organizations on social justice issues, particularly related to housing, transportation, and urban sprawl.

Howard Gillette, Jr., is a history professor at Rutgers University–Camden. He specializes in modern U.S. history, with a focus in urban and regional development. Among other works, he is the author of *Camden After the Fall: Decline and Renewal in a Post-Industrial City* (University of Pennsylvania Press, 2005) and *Between Justice and Beauty: Race, Planning, and the Failure of Urban Policy in Washington, D.C.* (Johns Hopkins University Press, 1995).

David Goldberg is Smart Growth America's communications director. Prior to 2002, Goldberg wrote for the *Atlanta Journal-Constitution* for nearly a decade about the transportation, environmental, and land use issues associated with metropolitan growth in the Atlanta region. A three-time nominee for the Pulitzer Prize, Goldberg has written numerous freelance reports on growth-related issues, and he appears frequently as a speaker or panelist on these topics.

Robert Gottlieb is the Henry R. Luce Professor of Urban Environmental Studies and director of the Urban Environmental Policy Institute at Occidental College.

He is the author and coauthor of ten books, and is the editor of two MIT Press book series, Urban and Industrial Environments and Sustainable Metropolitan Communities.

Bart Harvey is former chairman of the board of trustees and chief executive officer of Enterprise Community Partners. He is also chairman of the board of Enterprise Community Investment and has published in numerous journals and periodicals. Enterprise's mission is to provide decent, affordable housing and a path out of poverty for low-income families. During Harvey's tenure, the organization has become a leading provider of development capital and expertise to create decent, affordable homes and rebuild communities.

William A. Johnson, Jr. was elected the sixty-fourth mayor of Rochester, New York's third-largest city, in November 1993. Mayor Johnson became one of the few African-Americans in the nation ever to lead a predominantly white city. Prior to his election, he served twenty-one years as president and CEO of the Urban League of Rochester, developing and overseeing numerous innovative programs in education, youth development, family services, employment training, and housing development.

Chris Jones is the vice president of research for Regional Plan Association (RPA), the nation's oldest metropolitan planning organization. Since 1994, he has directed RPA's economic research and policy analysis. He coordinates the assessment of alternative economic futures for Lower Manhattan for the Civic Alliance to Rebuild Downtown New York, a coalition of nearly eighty civic organizations in the New York region.

Van Jones is cofounder and president of the Ella Baker Center for Human Rights, which promotes positive alternatives to violence and incarceration. Since its inception, the Ella Baker Center has successfully combined solutions to social inequality and environmental destruction through developing measures to build a "green economy." Locally, the City of Oakland adopted a proposal from the Ella Baker Center and the Oakland Apollo Alliance to create a "Green Jobs Corps." Nationally, the Ella Baker Center worked with members of the U.S. House of Representatives to pass the Green Jobs Act of 2007, which will provide $125 million in funding to train 35,000 people a year in "green-collar jobs." Mr. Jones is based in Oakland, California.

Anupama Joshi is director of the National Farm to School Program based at the Center for Food & Justice, a division of the Urban and Environmental Policy Institute at Occidental College in Los Angeles, California. She has worked for over ten years on nutrition, agriculture, and food systems issues, both nationally and internationally. Joshi is leading a national collaboration with more than thirty organizations across the country, exploring partnerships for promoting the farm-to-school movement.

Bruce Katz is a vice president at the Brookings Institution and founding director of the Brookings Metropolitan Policy Program. The Metro Program seeks to redefine the challenges facing cities and metropolitan areas by publishing cutting-edge research on major demographic, market, development, and governance trends. He is a frequent writer and commentator on urban and metropolitan issues and advises national, state, regional, and municipal leaders on policy reforms that advance the competitiveness of metropolitan areas.

Victoria Kovari has led the public transportation campaign of the Metropolitan Organizing Strategy Enabling Strength (MOSES), a coalition of seventy member churches in the Detroit area. By collaborating on regional transportation planning, MOSES is working to bring about an expanded public transportation system to serve metropolitan Detroit. In addition, Kovari has engaged in fair housing issues in her organizing work spanning the past twenty years.

Mike Kruglik, a longtime organizer, has been developing faith-based citizens' power organizations since 1973. In 1986, Kruglik, with Greg Galluzzo and Mary Gonzales, helped to create the Gamaliel Foundation, and has served since 1999 as its director of metropolitan equity. From 2001 to 2004, he was executive director of Metropolitan Congregations United (MCU) in St. Louis, Missouri.

Steve Lerner is the research director of Commonweal, a research institute that focuses on health and environment issues. He is the author of numerous publications, including *Diamond: A Struggle for Environmental Justice in Louisiana's Chemical Corridor* (MIT Press, 2005), and of *Eco-Pioneers: Practical Visionaries Solving Today's Environmental Problems* (MIT Press, 1998).

Greg LeRoy is the director of Smart Growth America, located in Washington, D.C. LeRoy has been called "the leading national watchdog of state and local economic development subsidies" and "God's witness to corporate welfare." He has been researching, writing, speaking, and consulting on economic development issues for more than twenty years for city and state governments, labor-management committees, unions, community groups, and development associations. His most recent publication, *The Great American Jobs Scam: Corporate Tax Dodging and the Myth of Job Creation* (Berrett-Koehler Publishers, 2005), has been welcomed as a resource for leaders around the country.

Amy Liu is the deputy director and cofounder of the Metropolitan Policy Program, overseeing its research, policy, and communications programs. She is also coauthor and editor of select Brookings publications, including *A Region Divided: The State of Growth in Greater Washington, D.C.*, of which she was the principal author. Previously, Liu was special assistant to Secretary Henry Cisneros at the U.S. Department of Housing and Urban Development.

Steven McCullough is president and chief executive officer of Bethel New Life, Inc., a twenty-six-year-old, faith-based community development corporation that serves the Chicago communities of West Garfield Park and Austin. Previously he served as Bethel's chief operating officer. Prior to that he was executive vice president of the Chicago Association of Neighborhood Development Organizations (CANDO). McCullough's commitment to Bethel's work is informed by his having grown up in the shadow of Garfield Park, observing at close range a vibrant community through its decline and subsequent transformation.

Mary Nelson is president emeritus of Bethel New Life, Inc., a twenty-six-year-old faith-based community development corporation with a national reputation for cutting-edge initiatives and significant change in the lives of people and its West Side Chicago community. Bethel's comprehensive array of community-focused efforts includes affordable housing, employment services, elderly services, family support, and economic development initiatives.

Jeremy Nowak is president of The Reinvestment Fund (TRF). He has served as a consultant to the Ford Foundation, the Pew Charitable Trusts, and the Annie E. Casey Foundation. In addition to his work in the Philadelphia area, he has provided assistance to a dozen development finance institutions in the United States, China, Latin America, and Africa.

Myron Orfield, Ameregis president and founder and associate professor at the University of Minnesota School of Law, is a leader in the use of GIS mapping technology to influence public policy. Previously, Orfield was a legislator in the Minnesota house of representatives and state senate. His book *Metropolitics: A Regional Agenda for Community and Stability* (Brookings Institution, 1997) "redefined the field of regional studies." His subsequent work, *American Metropolitics: The New Suburban Reality*, (Brookings Institution, 2002) analyzes the economic, racial, environmental, and political trends of the twenty-five largest U.S. metropolitan regions.

Manuel Pastor is professor of geography, American studies, and ethnicity at the University of Southern California. Previously, he was professor of Latin American and Latino studies and codirector of the Center for Justice, Tolerance, and Community at the University of California, Santa Cruz. He has authored numerous publications on U.S. urban issues, focusing on the labor market and social conditions facing low-income urban communities. Pastor is coauthor of *Searching for the Uncommon Common Ground: New Dimensions on Race in America* (W. W. Norton, 2002) and of *Regions That Work: How Cities and Suburbs Can Grow Together* (University of Minnesota Press, 2000).

M. Paloma Pavel, founder and president of Earth House, served as director of strategic communications for the Sustainable Metropolitan Communities Initiative at the Ford Foundation. An international consultant, educator, researcher, and media activist, her areas of specialization include urban sustainability, living systems, strategic communications, strategic planning, and leadership development. Pavel has produced several multimedia projects on the theme of regional equity and serves as coeditor of Sustainable Metropolitan Communities Books (SMCB) at the MIT Press.

john a. powell is executive director of the Kirwan Institute for the Study of Race and Ethnicity at Ohio State University and an internationally recognized authority in the areas of civil rights, civil liberties, and issues relating to race, ethnicity, poverty, and the law. Previously, powell founded and directed the Institute on Race and Poverty at the University of Minnesota and was national legal director of the American Civil Liberties Union, where he was instrumental in developing educational adequacy theory.

Rev. Cheryl Rivera, an ordained Baptist minister, has worked for more than ten years with the Gamaliel Foundation doing interfaith faith-based organizing. She was founder and national director of Gamaliel's African-American Leadership Commission, which provides nationwide training and leadership development in regional organizing for African-American clergy and laity. Rev. Rivera is also director and lead organizer of the Metropolitan Alliance of Congregations–West, and is past president of Gamaliel Foundation's National Leadership Assembly.

Faith R. Rivers is assistant professor of law at Vermont Law School. She worked previously with the Coastal Community Foundation of South Carolina to launch the Heirs' Property Preservation Project and was executive director of the South Carolina Bar Foundation from 1998 to 2004. Her current scholarship focuses on the legal history, legislative solutions, and administrative policies pertaining to the preservation of heirs' property in coastal South Carolina.

Nicolas Ronderos is a senior planner with the Regional Plan Association (RPA). His work focuses on the interrelations between transportation, community planning, and real estate development in New York City and the surrounding region.

Rachel Rosner is a research associate and project manager at the Center for Justice, Tolerance & Community (CJTC). She has a background in community and regional development with a master's in public policy and administration from the University of Wisconsin–Madison. Ms. Rosner has also been a board member of the Ecological Rights Foundation. In the past, she served as the associate director for the Social Change Across Borders Summer Institute, which brought together Latino and Latin American community leaders for collaboration and training around issues of transnational organizing.

David Rusk is a former federal official and New Mexico legislator, and was mayor of Albuquerque from 1977 to 1981. A consultant on urban and suburban policy, Rusk is the author of *Cities without Suburbs* and *Inside Game/Outside Game*. Since 1993, he has spoken and consulted in more than 120 U.S. communities.

Priscilla Salant is the coordinator of outreach and engagement in the College of Agricultural and Life Sciences at the University of Idaho. She provides leadership for civic engagement within the university and throughout Idaho. Her research focuses on social and economic trends in the nation's rural areas, community engagement at land grant universities, European rural policy, and community information systems.

David Satterthwaite is a Senior Fellow of the Human Settlements Program at the International Institute for Environment and Development (IIED). He has worked at IIED on issues relating to housing and the urban environment since 1974 and has advised an array of agencies on urban environmental issues, including the Brundtland Commission, WHO, UNICEF, UNCHS, the Department for International Development, and the Swedish International Development Cooperation Agency. Satterthwaite teaches at the University of London and the London School of Economics (LSE). He is also editor of the international journal *Environment and Urbanization*, and was one of the lead authors in the chapter on human settlements for the Intergovernmental Panel on Climate Change.

Ellen Schneider is founder and executive director of Active Voice. Schneider was most recently executive director of American Documentary, which produces P.O.V., PBS's longest-running independent nonfiction film series. She was part of the start-up team for the Independent Television Service and has served on juries ranging from the Sundance Film Festival to the RioCine Festival in Brazil.

Peggy M. Shepard is executive director and cofounder of West Harlem Environmental Action, Inc. (WE ACT), New York's first environmental justice organization, created to improve environmental health and quality of life in communities of color. She frequently lectures at universities and conferences on issues of environmental justice and community-based health research.

L. Benjamin Starrett is executive director of Funders' Network for Smart Growth and Livable Communities, the public policy director of the Growth Partnership, and the managing director of Collins Center for Public Policy. Funders' Network includes more than fifty philanthropic foundations, representing more than $50 billion in assets, who support smart growth efforts through their awards and grants.

Jennie Stephens is director of the Center for Heirs' Property Preservation. She served previously as senior program director at Coastal Community Foundation of South Carolina, a public grant-making foundation. Stephens has extensive experience in nonprofit management and program planning and development.

Elizabeth Tan is director of development and planning at Urban Habitat, a nonprofit organization in Oakland, California, which builds power in low-income communities and communities of color by combining education, advocacy, research, and coalition building to advance environmental, economic, and social justice in the Bay Area.

Petra Todorovich is a senior planner at the Regional Planning Association (RPA). She directs RPA's Lower Manhattan work and coordinates the Civic Alliance to Rebuild Downtown New York, a network of civic organizations that came together after 9/11 to promote a participatory and equitable rebuilding process for the World Trade Center site and Lower Manhattan. Todorovich is also project manager for America 2050, a national project based at RPA that focuses on megaregions as the future centers of population and employment growth.

Andrea Torrice is a public television documentary and educational television producer. Her programs have covered topics such as the environment, health, peace studies, and cultural issues. Her documentaries have been featured in numerous festivals worldwide, including the Berlin, Mill Valley, Women in the Director's Chair, and London International festivals. Her documentary *Forsaken Cries: The Story of Rwanda* won an award at the Rosebud International Human Rights Film Festival, and she was the recipient of a Gold Award from the Corporation for Public broadcasting for her environmental documentary *Bad Chemistry*.

Mark Vallianatos is policy director of the Urban and Environmental Policy Institute (UEPI) at Occidental College. He is also director of UEPI's Center for Food and Justice, where he has focused on food access and school food policies. Vallianatos is coauthor of *The Next Los Angeles: The Struggle for a Livable City*. Before moving to the Los Angeles region, he worked on international trade and environmental issues in Washington, D.C.

Robert Yaro is the president of the Regional Plan Association (RPA), America's oldest independent metropolitan policy, research, and advocacy group. He serves on Mayor Bloomberg's Sustainability Advisory Board, which helped prepare PlaNYC 2030, New York City's new long-range sustainability plan. Since 2001, Mr. Yaro has been professor of practice in city and regional planning at the University of Pennsylvania. He has also taught at Harvard University and the University of Massachusetts.

Authors' Organizations

African American Forum on Race and Regionalism, founded in 2002, is a collaborative project of the Sustainable Metropolitan Communities Initiative founded by Carl Anthony at the Ford Foundation. The forum works to help broaden, strengthen, and promote the involvement of African-Americans and African-American organizations in policy development, decisions, and place-based projects. www.aafrr.org

Atlanta Neighborhood Development Partnership (ANDP) develops mixed-income housing accessible to people of moderate to low incomes, working toward a more equitable distribution of affordable housing throughout the metropolitan Atlanta region. www.andpi.org

Bethel New Life, Inc., is a nationally recognized faith-based community development corporation on Chicago's West Side, launched in 1969. Its projects include the adaptive use of a former 434-bed hospital as an elderly housing complex, energy-efficient affordable housing, transit-oriented development, and the rescue of the Garfield Park Conservatory. www.bethelnewlife.org

The Brookings Institution's Metropolitan Policy Program promotes innovative solutions to help metropolitan communities grow in more inclusive, competitive, and sustainable ways. Formerly the Center on Urban and Metropolitan Policy, it was launched in 1996 to provide decision makers with cutting-edge research and policy analysis on the shifting realities of cities and metropolitan areas. www.brook.edu/metro

Building Partnerships USA (BPUSA) is a nonprofit organization that is taking the concept of peer learning to its next logical step within the labor movement, by bringing experienced coalition partners into different communities as resources in support of progressive regional efforts. www.building-partnerships.org

California Public Subsidies Project (CPSP) is a coordinated program designed to reform the economic development practices of governments and government agencies in four major regions of California. The four organizations managing the project include: the Center on Policy Initiatives in San Diego, the East Bay Alliance for Sustainable Economy in Oakland, the Los Angeles Alliance for a New Economy, and Working Partnerships USA in San Jose.

The Carsey Institute supports interdisciplinary research that informs policy makers and practitioners working to improve the well-being of families and build healthy, equitable communities in New England and nationally. www .carseyinstitute.unh.edu

Center for Justice, Tolerance and Community at University of California, Santa Cruz is a progressive research institute tackling issues of social justice, diversity, and tolerance, and the building of collaborative relationships between the university and local community. Its mission is to promote equity—including studies of the roots of prejudice, sources of economic inequality, and obstacles to the building of community. www.cjtc.ucsc.edu

The City of Rochester is creatively engaging some of the most pressing issues of contemporary regional planning. Its Neighbors Building Neighborhoods planning process has become a model of participatory planning for metropolitan communities. www.cityofrochester.gov

Coastal Community Foundation of South Carolina (CCF) is a public grant-making foundation that fosters philanthropy for the lasting good of the community. Founded in 1974, CCF manages more than $100 million for the charitable good of Charleston and the surrounding region. It promotes collaborative efforts to address community needs, including the Heirs' Property Preservation Project and the Low Country Conservation Loan Fund. www.ccfgives.org

Commonweal is a nonprofit health and environmental research institute in Bolinas, California, founded in 1976. It conducts programs that contribute to human and ecosystem health, including support programs for people with cancer and for physicians, medical students, and other health professionals who wish to restore the principles of healing to the practice of medicine and practice with a deeper sense of meaning, satisfaction, and lineage. www.commonweal.org

Earth House Center is dedicated to building multiracial, multiclass, and cross-sector collaborations for meeting the ecological and social justice challenges of our historic time. Based in Oakland, California, Earth House projects build healthy, just, and sustainable communities through education, research, and multimedia tools. www.EarthHouseCenter.org and MetroEquity.net

Enterprise Community Partners, Inc., is a leading provider of the development capital and expertise for creating decent, affordable homes and rebuilding communities. For more than two decades, Enterprise has pioneered neighborhood solutions through public-private partnerships with financial institutions, governments, community organizations, and others. It has raised and invested $6 billion in equity, grants, and loans and is currently investing in communities at a rate of close to $1 billion a year. www.enterprisecommunity.org

Environmental Justice Resource Center at Clark Atlanta University (EJRC), founded in 1994, is a research, policy, and information clearinghouse working on issues of environmental justice, race, the environment, civil rights, brownfields, transportation equity, suburban sprawl, and Smart Growth. www.ejrc.cau.edu

Film Arts Foundation is a nonprofit leader in the media arts field, providing comprehensive training, equipment, information, and exhibition opportunities to independent filmmakers. www.filmarts.org

The Ford Foundation is an independent, nonprofit, nongovernmental organization working to create political, economic, and social systems that promote peace, human welfare, and the sustainability of the environment on which life depends. It promotes collaboration among the nonprofit, government, and business sectors, mainly by making grants or loans that build knowledge and strengthen organizations and networks. www.fordfound.org

Funders' Network for Smart Growth and Livable Communities works to strengthen the capacity of philanthropic funders to support and link organizations working to promote smart growth and create livable communities. It brings together foundations, nonprofit organizations, and other partners to address the range of environmental, social, and economic problems caused by development strategies that fail to consider the "big picture." www.fundersnetwork.org

Gamaliel Foundation, founded in 1968, is a network of grassroots, interfaith, interracial, multi-issue organizations that works to create a more just and more democratic society. It is empowering people to participate effectively in the political, environmental, social, and economic decisions affecting their lives. www.gamaliel.org

Good Jobs First works with a broad spectrum of organizations to ensure that subsidized businesses are held accountable for family-wage jobs and other effective results. It provides information to the public, the media, public officials, and economic development professionals on best practices in state and local job subsidies, also tracking corporate accountability innovations in all states to develop these practices that can then be employed elsewhere. www.goodjobsfirst.org

Innovative Housing Institute is a nonprofit corporation providing technical assistance and professional support to local governments, private developers, housing agencies, and community organizations. www.inhousing.org

Institute on Race & Poverty at the University of Minnesota Law School works to create greater understanding of racialized poverty and investigates the ways that policies and practices disproportionately affect people of color and the disadvantaged. It is assisting the places where people live to develop in ways that both promote access to opportunity and help maintain regional stability. www.irpumn.org

International Institute for Environment and Development (IIED) is an international policy research institute and nongovernmental organization based in London, focusing on sustainable and equitable global development. Founded in 1971, it and its partnerships have generated close-working relations with many key development actors, from smallholder farmers and big city slum dwellers to national governments and regional NGOs, global institutions, and international processes. IIED acts as a catalyst, broker, and facilitator and helps vulnerable groups find their voice and ensure their interests are heard in decision making. www.iied.org

InterValley Project, Inc. (IVP) brings together congregations, labor union locals, and community and tenant groups through its Sustainable and Equitable Development Project. IVP is working to save, strengthen, and create jobs, affordable housing, and key public services for low-income people living in New

England regions that have experienced industrial and community decline. www
.intervalleyproject.org

Kirwan Institute for the Study of Race and Ethnicity at Ohio State University is a
universitywide interdisciplinary research institute working to deepen understanding of the causes of and solutions to racial and ethnic disparities and hierarchies.
It brings together scholars and researchers from various disciplines to focus on
the histories, present conditions, and the future prospects of racially and ethnically marginalized people. www.kirwaninstitute.org

Los Angeles Alliance for a New Economy (LAANE), founded in 1993, is an innovator in the fight against working poverty and a national authority on issues
affecting the working poor. Combining a vision of social justice with a practical
approach to social change, LAANE has helped set in motion a broad movement
based on the principle that hard work deserves fair pay, good benefits, and decent working conditions. www.laane.org

Metropolitan Organizing Strategy Enabling Strength (MOSES), composed of
seventy member churches, is an urban-suburban coalition advocating for an
expanded public transportation system to serve metropolitan Detroit. MOSES
works together with unions, major auto manufacturers, other corporations, environmentalists, mayors, legislators, and community members on regional mass
transit issues. www.mosesmi.org

PolicyLink is a national nonprofit research, communications, capacity-building,
and advocacy organization working to advance a new generation of policies to
achieve economic and social equity from the wisdom, voice, and experience of
local constituencies. With local and national partners, PolicyLink spotlights
promising practices, supports advocacy campaigns, and helps to bridge the traditional divide between local communities and policy making at the local, regional,
state, and national levels. www.policylink.org

Regional Plan Association (RPA) is an independent, not-for-profit regional
planning organization that improves the quality of life and the economic competitiveness of the thirty-one-county New York–New Jersey–Connecticut region
through research, planning, and advocacy. For more than eighty years, RPA has
been shaping transportation systems, protecting open spaces, and promoting better community design for the region's continued growth. www.rpa.org

The Reinvestment Fund (TRF) is a development financial institution that manages $170 million in assets. TRF's mission is to create economic opportunity
and build wealth for low-wealth communities and low- and moderate-income
families. TRF makes loans and equity investments in support of affordable housing projects, small businesses, community facilities, and energy conservation programs. www.trfund.com

Smart Growth America is a coalition of national, state, and local organizations
working to improve the ways that U.S. towns, cities, and metro areas are planned
and built. It seeks to preserve our built and natural heritage, promote fairness for
people of all backgrounds, fight for high-quality neighborhoods, expand choices
in housing and transportation, and improve poorly conceived development projects. www.smartgrowthamerica.com

Urban and Environmental Policy Institute is a community-oriented research and advocacy organization based at Occidental College in Los Angeles. It serves as the umbrella for a variety of affiliated programs addressing work and industry, food and nutrition, housing, transportation, regional and community development, land use, and urban environmental issues. www.departments.oxy.edu/uepi and www.farmtoschool.org

Urban Habitat addresses issues of social and environmental justice from a regional perspective. In partnership with various Bay Area organizations, Urban Habitat has championed many environmental and social justice issues that affect Bay Area's most disadvantaged communities, including health, food security, energy, military base conversion, transportation, redevelopment, education, and open space. www.urbanhabitat.org

West Harlem Environmental Action, Inc. (WE ACT) is a nonprofit, community-based environmental justice organization building community power to fight environmental racism and improve environmental health in communities of color. WE ACT's mission is to inform, educate, train, and mobilize the predominately African-American and Latino residents of northern Manhattan on issues that impact their quality of life. www.weact.org

Index